Transcription and translation

a practical approach

Edited by

B D Hames

Department of Biochemistry,
University of Leeds, Leeds, England

S J Higgins

Department of Biochemistry,
University of Leeds, Leeds, England

D1473854

IRL PRESS
Oxford · Washington DC

IRL Press Limited
P.O. Box 1,
Eynsham,
Oxford OX8 1JJ,
England

First published June 1984
First reprinting May 1985

British Library Cataloguing in Publication Data

Transcription and translation. — (The Practical approach series)
  1. Biological chemistry — Technique
  I. Hames,B.D. II. Higgins,S.J.
  III. Series
ISBN 0-904147-52-5

Cover illustration. Electron micrographs of rRNA genes from *Notophthalmus viridescens* in the process of transcription (on the left) and polysomes from *Bombyx mori* showing nascent polypeptides (on the right). The photographs were kindly supplied by Steven McKnight and Oscar L.Miller Jr., The Department of Biology, University of Virginia, USA.

Printed in England by Information Printing, Oxford.

Preface

Our present and future understanding of the mechanism and regulation of gene expression depends upon both direct investigations of gene transcription and the assay of specific messenger RNAs. In addition, the techniques associated with molecular biology and molecular genetics will be required by increasing numbers of researchers in the biological sciences. The aim of this book is to provide detailed practical protocols for these major areas of study. Eukaryotic, prokaryotic and viral genes are all covered, with the transcription of eukaryotic genes being considered mainly with regard to RNA polymerase II. Considerable revisions of some chapters were necessary in order to prevent undue repetition whilst including all the important practical topics and we thank the authors concerned for their understanding during this exercise. While our aim has been to cross-reference between chapters rather than to duplicate practical protocols, where several important approaches to the same technique exist these have been provided in full.

B.D.Hames and S.J.Higgins

Contributors

Michael J. Clemens
Cancer Research Campaign Mammalian Protein Synthesis and Interferon Research Group, Department of Biochemistry, St. George's Hospital Medical School, Cranmer Terrace, London SW17 ORE, UK

Alan Colman
Department of Biological Sciences, University of Warwick, Coventry CV4 7AL, UK

R. Stewart Gilmour
Beatson Institute for Cancer Research, Garscube Estate, Switchback Road, Bearsden, Glasgow G61 1BD, UK

J.B. Gurdon
Cancer Research Campaign Molecular Embryology Group, Department of Zoology, Downing Street, Cambridge CB2 3EJ, UK

I. Barry Holland
Department of Genetics, Adrian Building, University of Leicester, University Road, Leicester LE1 7RH, UK

Ru Chih C. Huang
Department of Biology, Johns Hopkins University, 34 Charles Street, Baltimore, MD 21218, USA

James L. Manley
Department of Biological Sciences, Columbia University, New York, NY 10027, USA

William F. Marzluff
Department of Chemistry, The Florida State University, Tallahassee, FL 32306, USA

Stephen J. Minter
Department of Molecular Biology, University of Manchester Institute of Science and Technology (UMIST), Sackville Street, P.O. Box 88, Manchester M60 1QD, UK

Julie M. Pratt
Department of Genetics, Adrian Building, University of Leicester, University Road, Leicester LE1 7RH, UK

Paul G. Sealey
MRC Mammalian Genome Unit, Department of Zoology, University of Edinburgh, West Mains Road, Edinburgh EH9 3JT, UK

Demetrios A. Spandidos
Wolfson Laboratory for Molecular Pathology, The Beatson Institute for Cancer Research, Garscube Estate, Switchback Road, Bearsden, Glasgow G61 1BD, UK

Neil G. Stoker
Imperial Cancer Research Fund Laboratories, P.O. Box 123, Lincoln's Inn Fields, London WC2A 3PX, UK

Neil M. Wilkie
Wolfson Laboratory for Molecular Pathology, The Beatson Institute for Cancer Research, Garscube Estate, Switchback Road, Bearsden, Glasgow G61 1BD, UK

Contents

3. TRANSCRIPTION OF EUKARYOTIC GENES IN A WHOLE-CELL EXTRACT   71
James L. Manley

6. IN VIVO GENE EXPRESSION SYSTEMS IN PROKARYOTES 153
Neil G. Stoker, Julie M. Pratt and I. Barry Holland

10. TRANSLATION OF EUKARYOTIC MESSENGER RNA IN XENOPUS OOCYTES

Alan Colman

APPENDICES

INDEX **319**

Abbreviations

APH	aminoglycoside phosphotransferase
bp	base pairs
BPV	bovine papilloma virus
BSA	bovine serum albumin
CAT	chloramphenicol acetyltransferase
cDNA	complementary DNA
Ci	Curie (3.7 x 10^{10} Bq)
c.p.m.	counts per minute
DEAE	diethylaminoethyl
DEPC	diethylpyrocarbonate
DHFR	dihydrofolate reductase
DMSO	dimethyl sulphoxide
d.p.m.	disintegrations per minute
DTT	dithiothreitol
EDTA	ethylenediamine tetraacetic acid
EGTA	ethyleneglycobis(β-aminoethyl)ether tetraacetic acid
EMC	encephalomyocarditis virus
HAT medium	hypoxanthine-aminopterin-thymidine medium
Hepes	N-2-hydroxyethylpiperazine-N′-2-ethanesulphonic acid
Hg-RNA	mercury-substituted RNA
Hg-UTP	5′-mercurated UTP
HGPRT	hypoxanthine-guanine phosphoribosyltransferase
HMBA	hexamethylene bisacetamide
HnRNA	heterogeneous nuclear RNA
HSV-1	*Herpes simplex* virus type 1
kb	kilobases
LTR	long terminal repeat
MMTV	mouse mammary tumour virus
MoMuSV	Moloney murine sarcoma virus
Mops	3-(N-morpholino)propanesulphonic acid
mRNA	messenger RNA
NHP	non-histone protein
NP-40	Nonidet P-40
PAGE	polyacrylamide gel electrophoresis
PBP	penicillin-binding protein
PBS	phosphate-buffered saline
p.f.u.	plaque forming units
Pipes	piperazine-N,N′-bis-2-ethanesulphonic acid
PMSF	phenylmethylsulphonyl fluoride
p.s.i.	pounds per square inch (lb/in^2)
RNP	ribonucleoprotein
rRNA	ribosomal RNA
SDS	sodium dodecyl sulphate
SDS-PAGE	polyacrylamide gel electrophoresis in the presence of SDS

α-S-RNA	RNA synthesised with α-thionucleotides
γ-S-RNA	RNA synthesised with γ-thionucleotides
SV40	simian virus 40
TCA	trichloroacetic acid
TEMED	N,N,N′,N′,-tetramethylethylenediamine
TET	tetracycline-resistance protein
TK	thymidine kinase
t.l.c.	thin-layer chromatography
TMV	tobacco mosaic virus
Tricine	N-[2-hydroxy-1,1-bis(hydroxymethyl)ethyl] glycine
tRNA	transfer RNA
XGPRT	xanthine-guanine phosphoribosyltransferase

Introduction

J.B. GURDON

The value of experimental systems for the analysis of gene expression will be obvious to all who work in the areas of cell biology and molecular biology, but it may be helpful to distinguish two different objectives of work in this area. One is to determine the *mechanism* of gene expression, and the other to analyse the *control* of this process. The former is concerned with identifying molecules required to obtain the expression of a gene. The type of information sought is which of several DNA clones codes for a certain gene product, and which of many fractions of RNA contain the mRNA required. These answers can be readily provided by the use of appropriate cell-free systems. With cell-free systems containing purified components it is also possible to identify factors required for the accurate transcription of DNA and translation of mRNA. The second, much more difficult, objective is to understand the control of gene expression. This requires a knowledge of the rate at which each step in gene expression proceeds, and identification of the components which are limiting in these steps. The reason why a meaningful analysis of gene control is so hard to achieve is that any component involved in a reaction can become limiting under particular experimental conditions even though most of these conditions may never normally exist *in vivo*. There is no simple way of determining whether a component which is limiting *in vitro* is also limiting *in vivo*. The same problem does not apply to an analysis of the *mechanism* of gene expression since even if the components in a cell-free system are present at concentrations different from normal, the coding capacity and requirement for essential factors should not be altered.

The ideal towards which everyone strives is a cell-free system which reflects normal gene expression and which consists entirely of known components. Very few such systems exist. Nearly all commonly-used cell-free systems involve the use of crude extracts to which purified components, such as cloned DNA or mRNA, are added. The great majority of systems described in this volume fall into this class. However, another type of system which has proved more successful than might have been predicted initially consists of a living cell into which purified components are injected. When a cell is disrupted, the lysate usually contains large amounts of DNase, RNase and proteolyic activities, so that these activities must be removed or reduced in the initial steps in the preparation of cell-free systems. However, when a living cell is injected with a solution of DNA or mRNA comprising as much as 10% of its volume, little degradation of the injected molecules takes place. Not surprisingly, therefore, microinjection of DNA and mRNA into living cells is an important and useful technique in the analysis of gene expression. Various methods and systems for microinjection are described in this volume.

Finally, it is important to be aware of the relative merits of cell-free systems and injected living cells for studying gene expression. Cell-free systems, and especially those whose components are mainly defined, have proved especially valuable in the initial recognition as well as the subsequent purification of transcriptional and translational factors. On the other hand, living cells can be used for such an analysis only under exceptional circumstances, for example, when a type of cell is available which is known to lack a factor which can be extracted from another cell type. The disadvantage of cell-free systems is that the range of steps in gene expression which takes place is limited and that the efficiency (or rate) of each step may be $10^2 - 10^5$ times less than in an injected cell. The significance of this greater efficiency is that the control of a particular reaction in gene expression can be studied more validly when that reaction is proceeding more closely to normal than when it is taking place at less than 1% of the rate *in vivo*. In conclusion, it is important to know the rate of gene expression in any experimental system used for analysis of the control of gene expression but this is not necessary for analysis of the mechanism of gene expression.

During recent years, experimental systems have greatly improved both in the range and efficiency of the gene expression steps which they carry out. Furthermore, there has been a great proliferation in the types and sources of systems which can be usefully applied to a particular problem. I therefore believe that the present volume will be very widely welcomed. The chapters have been contributed by those who have extensive experience of the procedures involved, and who, in many cases, have been directly involved in their development.

CHAPTER 1

Expression of Exogenous DNA in Mammalian Cells

DEMETRIOS A. SPANDIDOS and NEIL M. WILKIE

1. INTRODUCTION

The ability to introduce isolated DNA into cultured mammalian cells (DNA-mediated gene transfer) has proved to be one of the most powerful tools for analysing eukaryotic gene expression. The technique is currently being used to identify and analyse transcription control signals in eukaryote genomes, to investigate RNA splicing, to analyse mechanisms of gene modulation by regulators such as hormones and chemicals and during differentiation, and to identify and analyse cellular oncogenes implicated in carcinogenesis. This chapter describes the methods currently available for DNA-mediated gene transfer into cultured mammalian cells and illustrates how this approach may be used in the analysis of gene expression.

There are several methods in current use for the introduction of isolated DNA into cultured mammalian cells; these are outlined in Section 2. The most widely used procedure by far is the calcium phosphate transfection technique, although other methods may be preferred for particular applications (see Section 2.7). Using the calcium phosphate technique, virtually any source of donor DNA can be used (including phage particles or whole chromosomes) and gene expression can be determined in recipient cells either after a short equilibration period (transient or short-term expression) or after the establishment of long-term cell lines (transformed cells). When donor DNA codes for a selectable marker, that is a gene for which expression can be positively selected, long-term transformed cells are readily obtained by growth in suitable selective media. These approaches are illustrated in Section 2.1 using the calcium phosphate technique for a widely-used selectable marker, the thymidine kinase (*tk*) gene of *Herpes simplex* virus type 1 (HSV-1), with TK$^-$ mutant cells as recipients. The use of other selectable markers is outlined in Section 3. When donor DNA does not contain a suitable marker gene, it is mixed and transfected with a second DNA which does contain a suitable marker. After selection for this marker, transformed cells can be scored for co-expression of genes contained in the first DNA (Section 2.1.8).

The protocols outlined above, which use isolated DNA for transfection, can be regarded as *DNA-based vector systems*. Since these are the most widely used systems at present, most of this chapter is devoted to them. An alternative approach which is being increasingly used is to employ a eukaryotic virus as the vector for the DNA of interest. Such *virus-based vectors* are briefly described in Section 4. Finally, Section 5 of this chapter describes the expression of transferred genes in the recipient cells and approaches which have been used in conjunction with gene transfer systems to analyse the regulation of gene expression in eukaryotic cells.

The more specialised techniques for microinjection of exogenous DNA into *Xenopus* oocytes and use of that system for studies of gene expression are described in Chapter 2.

Table 1. The Calcium Phosphate Technique for Gene Transfer[a].

This technique is illustrated by the transformation of mouse LATK⁻ cells with DNA containing the *tk* gene, for example plasmid pTK1 (*Figure 1*).

1. Prepare the following solutions:

    2 x Hepes-buffered saline (2 x HBS)

    1.63 g NaCl
    1.19 g Hepes
    0.023 g $Na_2HPO_4.2H_2O$
    Distilled water to 100 ml

    Adjust the pH to 7.1 with 0.5 M NaOH.
    Filter sterilise the solution and store it at 4°C.
    The final composition of 2 x HBS is 0.28 M NaCl, 1.5 mM Na_2HPO_4, 50 mM Hepes, pH 7.1.

    2.5 M $CaCl_2$

    Dissolve 10.8 g of $CaCl_2.6H_2O$ (Analar) in 20 ml (final volume) of distilled water, filter sterilise the solution and store it at 4°C.

    0.1 mM EDTA, 1.0 mM Tris-HCl, pH 8.0

    Mix 50 μl of 0.2 M EDTA, pH 8.0, and 100 μl of 1.0 M Tris-HCl, pH 8.0, with distilled water to 100 ml final volume. Filter-sterilise the solution and store it at 4°C.

2. Harvest exponentially-growing mouse LATK⁻ cells by trypsinisation.

3. Re-plate the cells at a density of 1 x 10⁶ per flask (25 cm² growth area) containing 5 ml of pre-warmed fresh medium[b] [SF12 medium (Flow Laboratories) containing 15% Hyclone serum (Sterile Systems Inc.)]. Incubate at 37°C for 24 h.

4. Into a plastic bijoux vial, place 0.5 ml of donor DNA (e.g., pTK1; see *Figure 1*) plus carrier DNA (if required) at a concentration of 80 μg/ml in 0.1 mM EDTA, 1.0 mM Tris-HCl, pH 8.0. Add 0.4 ml of 0.1 mM EDTA, 1.0 mM Tris-HCl, pH 8.0, and 0.1 ml of 2.5 M $CaCl_2$. Mix.

5. Add this DNA solution slowly (~30 sec) with continuous mixing to 1.0 ml of 2 x HBS already in a second bijoux vial. Mix immediately by vortexing and leave the solution at room temperature for 30 min. The DNA concentration at this stage is 20 μg/ml.

6. After the incubation, a fine precipitate will have formed. Add 0.5 ml of this DNA-calcium phosphate suspension to each flask containing cells in 5 ml of growth medium.

7. Incubate the flasks at 37°C for 24 h to allow absorption of the DNA-calcium phosphate co-precipitate by the cells.

8. Pre-selection expression stage. Replace the medium with fresh, pre-warmed medium and incubate at 37°C for a further 24 h to allow expression of the transferred gene(s) to occur.

9. Selection stage. Replace the medium with an appropriate selection medium, in this case HAT medium (SF12 medium containing 15% serum, 0.1 mM hypoxanthine, 0.4 μM aminopterin, 16.0 μM thymidine)[b,c].

10. Renew the selection medium every 2 − 3 days for up to 2 − 3 weeks when colonies are routinely counted[d].

11. To pick individual colonies, remove the growth medium and then cut off the top of the flask using a heated scalpel. Place a stainless-steel cloning ring (diameter 0.4 − 0.8 cm) over each colony and detach the cells by incubation for 2 − 3 min at room temperature with 0.1 ml of 0.25% trypsin in sterile buffered diluent (6 g NaCl, 2.96 g trisodium citrate, 1.79 g Tricine, 0.005 g Phenol Red per 700 ml distilled water). After the cells have detached, use a sterile Pasteur pipette to transfer them to flasks (25 cm² growth area) containing fresh, pre-warmed growth medium.

[a]This technique can be used to introduce any DNA into mammalian cells for either transient expression assays or long-term transformation (see Section 5).
[b]The particular growth medium used will depend on the nature of the recipient cells.
[c]An alternative procedure is given in *Table 2*; see also Section 2.1 (vi).
[d]If a permanent record is desired, the cells in some of the flasks can be fixed in ice-cold methanol for 15 min, then stained with 10% Giemsa stain (prepared in water) for 15 min and finally rinsed in tap water.

2. METHODS OF INTRODUCTION OF DNA INTO MAMMALIAN CELLS

Unless special techniques are used, the uptake and expression of exogenous DNA in mammalian cells is very inefficient. Efforts to improve the biological activity of isolated genes has led to the development of a number of different methods for increasing the efficiency of uptake and subsequent expression. These are described in the following sections.

2.1 The Calcium Phosphate Transfection Technique

The calcium phosphate technique for introducing genes into mammalian cells was first described by Graham and van der Eb (1) and is the most widely used method in current practice. It can be used to introduce any DNA into mammalian cells for transient expression assays or long-term transformation (see Section 5). Basically, exogenous DNA is mixed with calcium chloride and is then added to a solution containing phosphate ions. A calcium phosphate-DNA co-precipitate is formed which is readily taken up by mammalian cells in culture, resulting in high levels of exogenous gene expression.

Several modifications have been made in the original technique since its introduction. In the authors' experience, the best procedure for most purposes is that illustrated in *Table 1* which describes the transformation of mouse L cells lacking thymidine kinase (LATK⁻ cells) with DNA containing the *tk* gene, for example, plasmid pTK1 (*Figure 1*). Several points require further discussion.

(i) The donor DNA, at a concentration of 40 μg/ml in 1 mM Tris-HCl, 0.1 mM EDTA, 0.25 M $CaCl_2$, pH 8.0, is added slowly with constant mixing to an equal volume of 2-fold concentrated Hepes-buffered saline (step 5). Typically, 1 ml of DNA is added to 1 ml of saline over a period of 30 sec. When low amounts of viral genomes or cloned genes are used as

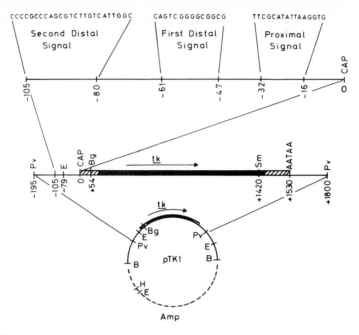

Figure 1. Restriction map of the recombinant plasmid pTK1 showing nucleotide sequences involved in the control of expression of the *tk* gene. This plasmid carries the HSV-1 *tk* gene in plasmid pAT153 (ref. 2). The broken lines represent plasmid pAT153 sequences; the solid lines HSV-1 sequences; the heavy line *tk* coding structural sequences; the hatched line *tk* non-coding structural sequences. The three regulatory sequences located upstream of the cap site (CAP) are designated proximal signal, first distal signal and second distal signal (3). All nucleotides are numbered using the cap site as a reference point. The maps are not drawn to scale. E = *Eco*RI, H = *Hind*III, Pv = *Pvu*II, Sm = *Sma*I, Bg = *Bgl*II, B = *Bam*HI. Amp = bacterial β-lactamase gene which confers resistance to the antibiotic ampicillin. AATAA = consensus sequence for the polyadenylation signal in eukaryotic cells.

the donor DNA, carrier DNA must be added to raise the total DNA concentration to the required value, that is, 20 μg/ml.

(ii) The molecular weight of the carrier DNA affects the final efficiency of uptake and of expression of the donor DNA (see Section 2.1.2).

(iii) After incubation for 30 min at room temperature, a fine precipitate of DNA-calcium phosphate is formed which remains in suspension. Occasionally, with very high molecular weight DNA, a coarse uneven precipitate is formed which results in low efficiency gene transfer. This problem can be overcome by slower mixing of the DNA-CaCl$_2$ with the Hepes-buffered saline, combined with continuous agitation (for example, by bubbling a stream of air through the mixture, or vortexing while mixing). Alternatively the molecular weight of donor DNA can be reduced by passage five times through a 19-gauge syringe needle.

(iv) The DNA-calcium phosphate suspension is then added to cultures of cells without prior removal of the culture medium. Usually, 1 ml of the DNA-calcium phosphate suspension is added to 10 ml of culture. The inoculum should be one-tenth the volume of the medium in order to obtain the desired final concentration of extracellular calcium ions.

(v) After inoculation with the DNA-calcium phosphate suspension, the cultures are incubated at 37°C for 24 h. The medium is then replaced with fresh medium and incubation is continued for another 24 h at 37°C. This medium is then removed and replaced with a suitable selection medium which is changed every 2 − 3 days for up to 2 − 3 weeks. The selection medium used depends on the nature of the experiment, the recipient cells and whether selectable markers are being used (see Section 3). After counting the colonies which grow, the cells are trypsinised and colonies are picked for further analysis.

(vi) An alternative procedure to this selection procedure (*Table 1*, steps 9 − 11) is to trypsinise the cells after pre-selection expression (*Table 1*, step 8) and then plate at appropriate dilutions under selective conditions in liquid or semi-solid media (4) (e.g., methocel-containing medium). This method is outlined in *Table 2*. Plating in methocel has the advantage of being a more quantitative technique since the medium is not changed, thus avoiding the problem of secondary colonies being derived from primary transformants. Since the method avoids repetitive media changes, it is also less time-consuming and considerably less expensive. The methocel protocol can be used for any selectable marker in any cell line which grows in methocel (i.e., cells with an anchorage-dependent phenotype), or for the selection of dominant transforming genes (see Section 3.2). In these cases the growth medium must be tailored to suit the experiment (see Sections 3.1 and 3.2).

Several parameters affect the expression of donor DNA in the recipient cells. These are discussed below.

2.1.1 *Donor DNA*

DNA from any source can be used, provided regulatory sequences appropriate for gene expression in mammalian cells are present (see Sections 3 and 5).

(i) If the donor DNA is genomic DNA isolated directly from eukaryotic cells, it is often important that it has a high molecular weight, since most eukaryote genes are interrupted by introns and therefore may occupy up to 60 kb of genomic DNA. A method for the isolation of high molecular weight chromosomal DNA is described in *Table 3*.

(ii) If recombinant plasmid DNA is to be used as donor, it is best purified from the bacterial host using the alkaline extraction procedure and then banded in a CsCl gradient (*Table 4*). The DNA is sufficiently pure at this stage and needs no further treatment for efficient transformation.

(iii) Lambda phage particles can be used directly in the calcium phosphate method without prior isolation of phage DNA. This method is potentially quite efficient. Recombinant phage containing eukaryote genes have been used in biochemical transformation experiments (see Section 3.1 and ref. 6) using the HSV-1 coded *tk* gene (HSV-1 *tk*).

DNA sequences cloned into recombinant plasmids have been most used as

Table 2. Methocel Selection Procedure[a].

1. Prepare medium containing methocel (methocel medium) as follows:

   (i) Mix 3 g of methocel MC 4000 CP (from Fluka) with 200 ml of distilled water and autoclave. The methocel dissolves to yield a clear solution which can be stored at 4°C for at least 6 months.

   (ii) Just before use, warm the medium to 37°C and add:

   > 22.0 ml 10 x Ham's SF2 medium[b]
   > 4.0 ml 50 x essential amino acids[b]
   > 4.0 ml 0.1 M sodium pyruvate[b]
   > 2.5 ml 0.2 M glutamine[b]
   > 5.0 ml 7.5% sodium bicarbonate[b]

   (iii) Next add 100 ml of serum and the appropriate concentration of the drugs to be used for selection. Thus, for methocel medium containing HAT (0.1 mM hypoxanthine, 0.4 μM aminopterin and 16.0 μM thymidine) add:

   > 3.40 ml 10.0 mM hypoxanthine
   > 0.34 ml 0.4 mM aminopterin
   > 0.34 ml 16.0 mM thymidine

   The final concentration of methocel and serum are ~0.9% and 30%, respectively.

   Note that the composition of the methocel medium in step (ii) depends upon the cell line being used while the nature of the components in step (iii) depends upon the selection marker being used (see Section 3).

2. Start transformation as described in *Table 1*, steps 1 – 8. After allowing for pre-selection expression (see *Table 1*, step 8) trypsinise and then count the cells.

3. Mix 0.2 ml of cells with 20 ml of methocel medium in a plastic Universal vial and plate on bacteriological plates (9 cm diameter) containing methocel medium. Up to 2 x 10⁶ cells per plate can be plated, the choice of plating density depending on the transformation frequency expected.

4. Incubate the plates at 37°C for 7 – 10 days depending on the doubling time of the recipient cell line.

5. Count the colonies using an inverted microscope. If required, pick individual colonies using a Pasteur pipette and grow these in a suitable growth medium (5 ml/flask of 25 cm² growth area).

---

[a]The methocel protocol can be used to select for any biochemical marker in any cell line which grows in methocel or for the selection of dominant transforming oncogenes (see Section 3.2).
[b]These stock solutions can be obtained from Flow Laboratories Inc.

the donor DNA for gene transfer experiments and hence these are chosen as examples to illustrate the techniques described in this chapter.

2.1.2 *Carrier DNA*

Commercially available calf-thymus or salmon-sperm DNA can be used as carrier DNA in cell transformation experiments provided it is first purified further as described in *Table 5*. Alternatively, carrier DNA can be isolated from eukaryotic cells by the method given in *Table 3,* or by the method described by Gross-Bellard *et al.*(7). Eukaryotic carrier DNA prepared in the laboratory usually gives 2- to 3-fold higher efficiencies than commercially-available DNA. Prokaryotic genomic DNA is not recommended as carrier since it may inhibit mammalian cell transformation (8).

Table 3. Isolation of High Molecular Weight Chromosomal DNA.

1. Prepare the following solutions:

   Guanidium chloride buffer

Guanidium chloride	76.40 g
Sodium acetate.3H$_2$O	0.272 g
EDTA.Na$_2$	1.86 g
Distilled water to 90 ml	

   Heat at 65°C to dissolve these components then cool to room temperature. Adjust the pH to 7.0 with 5 M NaOH. Add 5 ml of 2-mercaptoethanol. Make the volume to 100 ml with distilled water. The final composition of this buffer is 8 M guanidine-HCl, 20 mM sodium acetate, 50 mM EDTA, 0.7 M 2-mercaptoethanol.

   Buffer A

   This buffer consists of 10 mM NaCl, 10 mM EDTA, 0.5% SDS, 10 mM Tris-HCl, pH 8.0 and is prepared fresh from concentrated stock solutions.

   Phosphate-buffered saline (PBS)

NaCl	8.0 g
KCl	0.2 g
Na$_2$HPO$_4$	1.15 g
KH$_2$PO$_4$	0.2 g
Distilled water to 1 litre	

2. Harvest the cells, count them and then wash them with PBS by centrifugation.

3. Resuspend 1 − 2 x 10^8 cells in 5 ml of guanidium chloride buffer. Mix by gentle vortexing.

4. Homogenise the cells gently by hand with 8 − 10 strokes of a loose-fitting pestle in a Dounce homogeniser. When DNA is isolated from tissue fragments, disruption by homogenisation in an Omnimixer may be required instead.

5. Add 0.5 ml of 20% SDS (specially pure, BDH Chemical Co.). Heat at 65°C for 1 − 3 min and mix thoroughly but gently.

6. Precipitate the DNA by adding an equal volume of isopropanol. If salt precipitates to give a cloudy solution, add 50% isopropanol in distilled H$_2$O until an almost clear solution is obtained.

7. Spool the DNA onto a glass rod and wash it sequentially with 70% ethanol and 100% ethanol.

8. Dry the DNA in a vacuum lyophiliser for 5 min.

9. Dissolve the DNA in 5 ml of the DNA buffer A containing 50 μg/ml of Proteinase K (Boehringer). Incubate at 37°C overnight.

10. Extract the DNA with an equal volume of chloroform three times, separating the phases by centrifugation (10 000 g) each time.

11. Add 5 M NaCl to the aqueous DNA solution to a final concentration of 0.2 M and then add an equal volume of isopropanol.

12. Spool the DNA onto a glass rod, wash it and dry it as in steps 7 and 8. Alternatively, the DNA can be recovered by centrifugation (10 000 g) and dried as in step 8.

13. Dissolve the DNA in 0.1 mM EDTA, 1.0 mM Tris-HCl, pH 8.0.

If the carrier DNA gives coarse precipitates of DNA-calcium phosphate (Section 2.1), resulting in a low efficiency of gene transfer, reduction of the molecular weight of the carrier DNA by mechanical breakage (e.g., five passages through a 19-gauge syringe needle) usually cures the problem. However, when the average molecular weight of carrier DNA is less than 10 kb there is a detectable reduction in the efficiency of uptake and expression of cloned DNA.

2.1.3 *Recipient Cells*

Different cell lines have markedly different efficiencies of uptake and expression of exogenous genes. The best results have been obtained with mouse L cells (LA cells), mouse NIH3T3 fibroblasts and baby hamster kidney (BHK) fibroblasts. Other cultured cells, especially primary cultures, may give efficiencies

Table 4. Isolation of Recombinant Plasmid DNA[a].

1. Prepare the following solutions:

   L-Broth

NaCl	5 g
Tryptone (Difco)	5 g
Yeast extract (Difco)	2.5 g

   Distilled water to 500 ml final volume.
   Sterilise this medium by autoclaving.

   Lysis buffer

1 M glucose	2.5 ml
1 M Tris-HCl, pH 8.0	1.25 ml
0.2 M EDTA, pH 8.0	2.5 ml
lysozyme (Sigma)	250 mg

   Distilled water to 50 ml final volume.

   Alkaline solution

5 M NaOH	3.2 ml
10% SDS	8.0 ml

   Distilled water to 80 ml final volume.

   Neutralisation solution

   Dissolve 11.8 g potassium acetate in ~30 ml of water and adjust the pH to 5.0 with ~5 ml of glacial acetic acid. Add distilled water to 40 ml final volume.

2. Inoculate 500 ml of L-broth in a 2 litre flask with the bacterial host. The medium should be supplemented with the antibiotic appropriate to the plasmid under study, for example 100 μg/ml (final concentration) of ampicillin. Incubate the culture overnight at 37°C with vigorous shaking.

3. Harvest the bacterial cells by centrifugation at 5000 *g* for 10 min at 4°C. Discard the supernatant.

4. Resuspend the cell pellet in 100 ml of ice-cold 50 mM Tris-HCl, pH 8.0. Centrifuge the bacterial cells as in step 3. This cell pellet can be stored at −20°C.

5. Resuspend the cell pellet in 50 ml of lysis buffer and leave on ice for 30 min.

6. Add 80 ml of alkaline solution (prepared as in step 1). Mix well. The cell suspension should become clear as the cells lyse. Leave on ice for 5 min.

7. Add 40 ml of neutralisation solution. Mix well and leave on ice for 15 min.

8. Centrifuge at 5000 *g* for 10 min and filter the supernatant through cotton gauze.

9. Add 100 ml of cold (4°C) isopropanol. Mix and then centrifuge this solution immediately at 5000 *g* for 10 min. Drain off the supernatant and dry the pellet (*Table 3*, step 8).

10. Dissolve the pellet in 10.8 ml of 1 mM EDTA, 10 mM Tris-HCl, pH 8.0. Add 11.7 g CsCl and heat at 37°C for 10 min to dissolve this. Next add 1.2 ml of 3 mg/ml ethidium bromide. Mix well and then centrifuge the solution in two tubes (10 ml capacity) in a suitable rotor at 140 000 *g* at 15°C for 40 h.

11. After centrifugation, view the DNA bands using u.v. light (305 nm). Collect the DNA band nearest the bottom of each tube using a syringe through the tube wall. This is closed circular plasmid DNA.

12. Remove ethidium bromide from the DNA preparation by three extractions with an equal volume of isopropanol each time. After each extraction, re-adjust the volume of the DNA solution with distilled water.

13. Dialyse the DNA against 1 mM EDTA, 10 mM Tris-HCl, pH 8.0, to remove CsCl.

[a]This procedure (adapted from ref. 5) yields >1 mg of most recombinant plasmid DNAs per 500 ml of bacterial culture. An alternative procedure is given in Chapter 7, *Table 8*, whilst a method for preparation of plasmid on a smaller scale is given in Chapter 6, *Table 10*.

10-fold or even 100-fold lower than these cells (see *Figure 4,* Section 2.1.6).

Invariably, the best results in transformation experiments are obtained when the recipient cells are growing exponentially. Therefore, exponentially-growing cells are trypsinised and replated at the appropriate cell densities in fresh medium 24 h before addition of the donor DNA. There is no need to change the medium again just before the addition of DNA. We have found when using LATK⁻ cells or BHKTK⁻ cells that the optimum plating densities are 1 x 10^6 and 5 x 10^5 cells per flask (25 cm² growth area), respectively.

2.1.4 *Absorption of the DNA-Calcium Phosphate Co-precipitate*

In the past, relatively brief absorption periods of 0.5 – 4 h have been used (1,9). However, detailed examination of the optimum absorption period has revealed that the best transformation efficiencies are obtained upon longer (up to 24 h) exposure of the recipient cells to the DNA-calcium phosphate suspension (10). *Figure 2* shows results for the expression of the HSV-1 *tk* gene in LATK⁻ and BHKTK⁻ cell lines using various absorption times. The donor DNA used in the transformation was the recombinant plasmid pTK1 (*Figure 1*) carrying the HSV-1 *tk* gene in plasmid pAT153.

2.1.5. *Pre-selection Expression Time*

Many gene transfer experiments are monitored by the transformation of cells from one phenotype to another. This often involves the use of a selective medium for a specific biochemical marker (see Section 3). It is thought that, in many cases, a period of expression in non-selective conditions prior to selection is required (pre-selection expression; see *Table 1*, step 8). *Figure 3* shows the effect of varying the pre-selection expression time on the rate of transformation of LATK⁻ cells to an LATK⁺ phenotype, induced by the HSV-1 *tk* gene. The frequency of transformation is maximal with pre-selection expression periods of 24 – 48 h for LATK⁻ and 12 – 24 h for BHKTK⁻ cells. After 48 h the efficiency of transformation is reduced, perhaps reflecting instability

Table 5. Further Purification of Commercially-available DNA.

1. Dissolve the DNA at 1.0 mg/ml in 10 mM Tris-HCl, 1 mM EDTA, pH 8.0.

2. Add 5 M NaCl to a final concentration of 0.2 M and then layer 2.5 volumes of ice-cold ethanol over this. Spool the DNA onto a glass rod by slowly rotating the rod whilst moving it across the interface.

3. Wash the DNA with ice-cold 70% ethanol by swirling the glass rod (with spooled DNA) in this for 1 min. Dry the DNA (*Table 3*, step 8).

4. Dissolve the DNA in 1 mM Tris-HCl, 0.1 mM EDTA, pH 8.0, overnight.

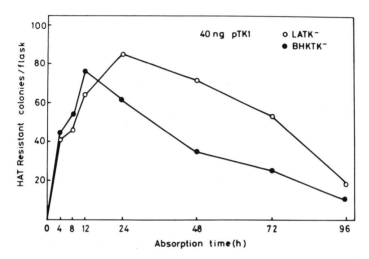

Figure 2. The effect of exposure (absorption) time of recipient cells to donor DNA on transformation efficiency. Mouse LATK⁻ (○) or Syrian hamster BHKTK⁻ (●) cells in flasks (25 cm² growth area) were exposed to 40 ng of donor pTK1 DNA and 10 μg of carrier salmon sperm DNA. Plasmid pTK1 is a recombinant containing the 3.5-kb *Bam*HI fragment of HSV-1 DNA carrying the *tk* gene inserted into the *Bam*HI site of plasmid pAT153 (see ref. 2 and *Figure 1*). A preselection expression time of 24 h was employed (see *Figure 3*). Each point represents the average number of HAT-resistant colonies per flask based on the data from five flasks.

of the donor DNA. Analogous tests should always be carried out to define the optimum conditions for the selectable markers and recipient cells chosen for study.

2.1.6. *Dose-response Curves*

The amount of donor DNA used to transfect mammalian cells affects the amount of gene expression obtained. For example, in long-term cell transformation experiments using biochemically selectable markers such as the HSV *tk* gene, the number of transformed colonies increases with increasing gene dosage. A typical dose-response curve for the *tk* gene using an absorption time of 24 h and a pre-selection expression time of 24 h is shown in *Figure 4*. Dose-response curves vary from gene to gene and depend on factors such as the efficiency of the transcription control signals associated with each gene. *Figure 4* also illustrates that different recipient cells vary in their ability to take up and

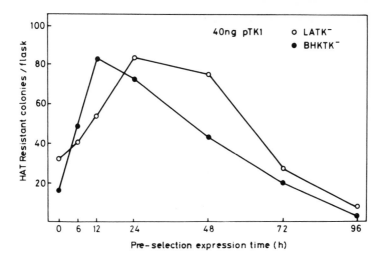

Figure 3. The effect of pre-selection expression time on the rate of transformation. LATK$^-$ (O) or BHKTK$^-$ (●) cells were exposed to 40 ng pTK1 and 10 μg carrier salmon sperm DNA/flask for 24 h and 12 h, respectively. Each point represents the average number of HAT-resistant colonies per flask based on the data from five flasks.

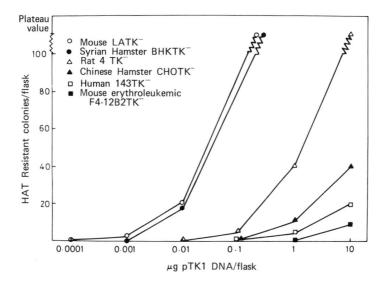

Figure 4. Dose-response curves for transformation by pTK1 DNA. The recipient cells were mouse LATK$^-$ (O), Syrian hamster BHKTK$^-$ (●), rat 4TK$^-$ ($\triangle$), Chinese hamster CHOTK$^-$ (▲), human 143TK$^-$ ($\square$) or mouse erythroleukaemic F4-12B2TK$^-$ (■) cells. The final DNA concentration for each determination was 20 μg/ml during the calcium phosphate co-precipitation step. This was achieved by the addition of salmon sperm DNA as carrier when 1 μg or less of recombinant DNAs was used. Each point represents the average number of HAT-resistant colonies per flask based on the data from 4 – 8 flasks.

express exogenous DNA and exhibit markedly different transformation efficiencies. The authors have even observed these effects in different clones isolated from the same cell line.

11

Table 6. Facilitators of Transformation and Their Effectiveness on Different Cell Lines.

Facilitator	Donor DNA	Recipient cells	Fold increase in transfection efficiency over control	Reference
DMSO	HSV-1	BHK	10 – 100	11
	HSV-*tk*	Mouse erythroleukaemic	20	2
Glycerol	HSV-*tk*	Mouse erythroleukaemic	25	2
	SV40	Monkey kidney	10 – 200[a]	12
Colchicine	HSV-2	Rabbit kidney	6	13
Colcemid	HSV-2	Rabbit kidney	4	13
Cytochalasin D	HSV-2	Rabbit kidney	4	13
Colchicine + Colcemid + Cytochalasin D	HSV-2	Rabbit kidney	27	13

[a]The liposome technique (Section 2.4) was used in these studies. In all the other studies listed here, the calcium phosphate technique (Section 2.1) was used.

2.1.7 *Facilitators of Transformation*

Several chemical 'facilitators' have been described which increase the expression of transferred donor DNA in recipient cells (2,11 – 13). A list of facilitators and their effectiveness in increasing the transformation efficiency in different cell lines is given in *Table 6*. The facilitator is usually added 4 – 8 h after absorption of the DNA-calcium phosphate co-precipitate. Some facilitators, such as DMSO, may significantly increase the expression of exogenous DNA in some cells but exhibit toxic effects in others. The concentration of facilitator which is most effective, the optimum time of exposure after transfection and the optimum duration of the exposure after transfection must be empirically determined for each facilitator and each new cell line. Typical results for the effect of DMSO on the infectivity of intact HSV DNA is shown in *Figure 5*.

2.1.8 *Co-transformation*

The technique of co-transformation was developed for the introduction and expression in mammalian cells of DNA sequences that do not code for a selectable marker (14). To obtain co-transformants, the recipient cells are exposed to the donor DNA along with another DNA which does code for a selectable marker, for example the HSV *tk* gene. Transformation of the recipient cells and selection is carried out as usual; in this case, the TK$^+$ transformants are selected (*Tables 1 and 2*). These are then scored for co-transformation with the donor DNA using a hybridisation probe or some other assay, depending on what is available.

Co-transformation experiments with the HSV-1 *tk* gene and bacteriophage ϕX174 or plasmid pBR322 or a cloned chromosomal rabbit β-globin gene (14)

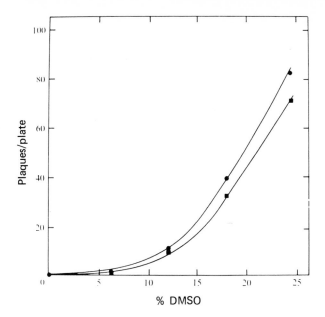

Figure 5. The effect of DMSO on the infectivity of HSV-1 DNA. BHK cells were infected with 0.04 μg of virus DNA per plate and exposed for 4 min to varying concentrations of DMSO dissolved in Hepes-buffered saline (●) or Eagle's medium (■) (see ref. 11).

result in the isolation of mouse cell lines which contain multiple copies of the co-transferred genes. The frequency of co-transformation is very high; over 90% of the transformants contain the co-transforming DNA. This co-transformation system allows the introduction and stable integration of virtually any defined gene into cultured cells without the need for ligation of the gene to either viral vectors or genes coding for selectable biochemical markers.

2.2 The DEAE-dextran Technique

The DEAE-dextran technique was first developed for assaying the infectivity of poliovirus RNA (15) and subsequently was extended to assay infectivity of SV40 (16) and polyoma (17) DNA. For these viruses, this is still the method of choice. Infectivity of other viral DNAs (BK virus, AAV-1, adenovirus 5, HSV-1) has also been demonstrated using the DEAE-dextran technique.

There are two main variations of the technique (18). In the first, viral DNA is mixed with DEAE-dextran and the mixture is added to the recipient cells. In the second, recipient cells are pre-treated with DEAE-dextran solutions and then exposed to the viral DNA. In both cases, the recipient cells are usually washed with isotonic saline before treatment and again after exposure to the DNA or DNA plus DEAE-dextran. The DEAE-dextran is used at a concentration in the range of 100 – 1000 μg/ml depending on the nature of the recipient cells and the exact protocol chosen. With these standard protocols for DNA transfections, only a small fraction (∼ 4%) of monkey cells exposed to SV40 DNA can be transfected. However, recently a protocol has been devised in

Table 7. The DEAE-Dextran Technique for Efficient Infection of Monkey Cells with SV40 DNA[a].

---

1. Plate 3 x 10^5 log-phase BSC-1 monkey cells per Petri dish (3.5 cm diameter) in 3 ml of Dulbecco's modified Eagle's medium (DME medium)[b] supplemented with 7% calf serum. Incubate at 37°C for 24 h in a CO_2 incubator.

2. When the cells are slightly subconfluent, wash them twice with serum-free DME medium[c].

3. Incubate the cells with 0.7 ml of 100 μg/ml SV40 superhelical DNA in DME medium, 50 mM Tris-HCl (pH 7.3), 200 μg/ml DEAE-dextran (mol. wt. 2 x 10^6) at 37°C in a CO_2 incubator.

4. Eight hours after infection, wash the cells once with serum-free DME medium and incubate them at 37°C with DME medium supplemented with 7% calf serum.

5. *Either* count plaques *or* examine the cells for transient expression of SV40 T antigen.

   Standard Plaque Assay

   (i) One day after infection, trypsinise the cells, dilute them and replate onto subconfluent monolayers of BSC-1 cells such that each Petri dish (6.0 cm diameter) receives sufficient cells to produce ~50 plaques.

   (ii) Incubate at 37°C for 2 h to allow the cells to attach.

   (iii) Remove the medium and replace it with 5 ml of DME medium containing 2−5% serum and 1% Noble agar. Incubate the plates at 37°C for 6−8 days.

   (iv) Stain the cells by adding an overlay of 3 ml of 0.1% neutral red in DME medium containing 0.5% Noble agar. Plaques are visible the next day for counting.

   Transient Expression of SV40 T Antigen

   (i) 8−24 h after infection, fix the cells with acetone:methanol (9:1, v/v) at −70°C for 10 min.

   (ii) Incubate the fixed cells with hamster anti-SV40 tumour serum in a humid incubator at 37°C for 30 min.

   (iii) Wash the cells with PBS[d] for 30 min at 37°C.

   (iv) Incubate the cells for 30 min at 37°C with fluorescein-conjugated goat anti-hamster IgG together with ethidium bromide at a final concentration of 2 μg/ml.

   (v) Wash the cells twice with PBS[d] and then once with distilled water.

   (vi) Immerse the cells in glycerol:PBS[d] (2:1, v/v) and examine them microscopically under u.v. light (mercury lamp). Cells expressing SV40 T antigen are fluorescent.

---

[a]From ref. 19.
[b]From Gibco.
[c]This step is important for good efficiency of transfection.
[d]The composition of this buffer is given in *Table 3*, step 1.

which the time period of exposure of BSC-1 monkey cells to DNA in the presence of a low concentration of DEAE-dextran was extended over that normally used (19). With an 8 h exposure to 200 μg/ml of DEAE-dextran, 25% of the cells were reproducibly infected. This more efficient protocol for the infection of BSC-1 monkey cells with SV40 DNA is described in *Table 7*.

The mechanism of action of DEAE-dextran is not known. Several possibilities have been considered, for example that DEAE-dextran binds to the DNA and protects it from nucleases, or interacts with cell membranes and thus brings the DNA close to the cell surface ready for uptake, or stimulates pinocytosis. Whatever the mechanism, up to 25% of the cells can be

Table 8. Microinjection of Mammalian Cells[a].

1. Grow the recipient cells on small glass slides (1.0 cm square).

2. Prepare glass micropipettes from glass capillary tubing (Omega Dot Tubing, 1.2 mm OD, W.P. Instruments) on a micropipette puller (e.g., Model p77 Brown-Flanning apparatus from Sutter Instruments). The glass micropipettes must have tip diameters in the range of 0.1 – 0.5 microns.

3. Prepare the apparatus for microinjection as follows. Set up the micropipette so that the microinjection can be carried out under direct visual control on a fixed stage of an inverted phase-contrast microscope (e.g., Leitz Diavert, 400 x objective). Arrange for movement of the micropipette to be controlled in micromanipulators (e.g., Narishige MO-15, modified to have an hydraulic microdrive along the pipette axis). The apparatus should permit the DNA solution in the micropipette to be forced into the cells under constant pressure from a Hamilton threaded-plunger syringe (model 8700).

4. Suck up the solution of DNA (0.1 mg/ml in PBS[b]) into the micropipette.

5. Using the Hamilton threaded-plunger syringe, force the DNA solution into the cell at constant pressure. The amount of fluid injected into each cell is controlled with moderate precision (within a factor of two) by visually monitoring changes in the cellular refractive index as the fluid enters the cell and regulating the time that the micropipette remains in the cell.

6. After the cells have been microinjected with the donor DNA, allow pre-selection expression of the transferred gene(s) and then select transformants as described in *Table 1*.

[a]From ref. 24.
[b]The composition of this buffer is given in *Table 3*, step 1.

transfected with SV40 DNA (19) whereas with the calcium phosphate technique only 15% of the cells can be transfected at best (20).

2.3 Microinjection

2.3.1 *True Microinjection*

Techniques employing microinjection of nucleic acids using glass micropipettes have been developed to introduce recombinant DNAs into mammalian cells (21 – 25). When *tk* DNA is injected into mouse LATK⁻ cells, 50 – 100% of the cells transiently express thymidine kinase enzyme activity (24). The number of *stable* transformants obtained depends on the nature of the recombinant DNA injected. For example, using pBR322/HSV-1 *tk* recombinant DNA, one in 500 – 1000 cells which have received the DNA become stable transformants whereas when specific SV40 sequences are linked to the pBR322/HSV-1 *tk* recombinant, the transformation frequency increases to one in five injected cells (see also Section 5.3).

Microinjection of mammalian cells can be accomplished as described in *Table 8*. The average volume (10 – 20 femtoliters) injected into each cell is determined by injecting [³H]dTTP (10 μCi/μl) into 5000 cells, washing the cells with PBS and then measuring the radioactive content by liquid scintillation counting. For L cells, this volume corresponds to 1 – 2% of the cell volume. With practice, one can inject 500 – 1000 cells per hour with a successful transfer of material apparently being assured in 50 – 100% of these (24). Similar procedures for injecting macromolecules into cultured mammalian

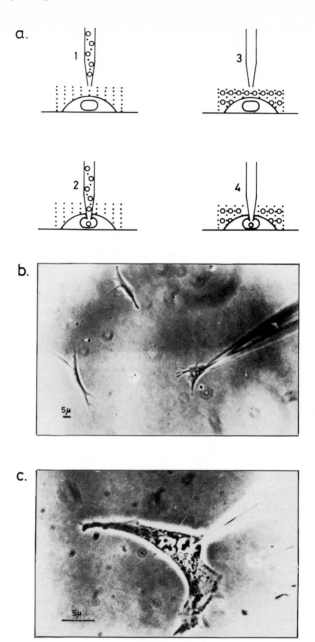

Figure 6. (a) Schematic representation of microinjection (**1** and **2**) and 'pricking' (**3** and **4**). DNA molecules are shown by the circles. In microinjection the micropipette is filled with the DNA solution whereas in 'pricking' the DNA is present in the external medium. **(b)** and **(c)**; photographs showing BHKTK$^-$ cells in the process of being 'pricked'.

Table 9. The 'Pricking' Procedure for Introduction of DNA into Mammalian Cells[a].

1. Prepare microneedles from glass capillaries using a microelectrode puller (e.g., Narishige PN-3 used under the following settings: heater 8, magnet 10, main target 10). The diameter of the microneedle tip should be ~0.1 μm.

2. Trypsinise cells in the logarithmic phase of growth with 0.25% trypsin, 0.02% EDTA for 15 min at 37°C.

3. Wash the trypsinised cells with fresh medium.

4. Suspend the cells in fresh medium and determine the cell density.

5. Seed 100 – 200 cells over a marked area (~0.5 cm²) of each Petri dish.

6. Incubate the cells for 5 h at 37°C to allow them to adhere to the substratum.

7. Just before 'pricking' the cells, discard the medium and wash the cells twice with 100 μl of PBS[b].

8. Next, cover the cells with 50 μl of PBS containing 5 – 1000 μg DNA/ml.

9. Use an injectoscope (e.g., Olympus IMT-YF type 1) to 'prick' the cells. Fix the microneedle to the holder which is located in the perpendicular hole through the optical axis of the condenser. 'Prick' the cells by the up-and-down movement of the condenser lens.

10. After the cells have been 'pricked', replace the medium with normal growth medium and allow pre-selection expression of the transferred genes to occur (see *Table 1*, step 8).

11. Finally, carry out selection for the transferred gene, for example as in *Table 1*, steps 9 – 11 for the *tk* gene.

[a]From ref. 25.
[b]The composition of this buffer is given in *Table 3*, step 1.

cells have been described by others (21 – 23). A schematic representation of microinjection is given in *Figure 6*.

2.3.2 *'Pricking'*

A variation of the standard microinjection technique is the 'pricking' method which mechanically introduces DNA into the nuclei of cultured cells (25). This procedure is described in *Table 9* and illustrated in *Figure 6*. Briefly, cells in the logarithmic stage of growth are treated with trypsin and EDTA, washed with medium and then $1 - 5 \times 10^3$ cells are seeded into Petri dishes (6.0 cm diameter) and incubated at 37°C overnight. Just before 'pricking' the cells, the medium is discarded and the cells are washed twice with PBS, pH 7.2, and then covered with PBS containing the donor DNA at a concentration in the range 5 – 1000 μg/ml. All of the cells in a certain (marked) region of the dish are 'pricked' once in the nuclear domain using a glass microneedle until the tip of the microneedle comes in contact with the substratum. The cells in the other parts of the dish are the 'unpricked' controls. Immediately after 'pricking', the medium is replaced with the normal growth medium for pre-selection to occur and then this in turn is replaced with a suitable selective medium.

That DNA suspended in the external medium actually is introduced into the nuclei of recipient cells by 'pricking' has been demonstrated using HSV-1 *tk*

Table 10. Preparation and Use of DNA-loaded Liposomes[a].

Preparation of DNA-loaded liposomes

1. Mix phosphatidylserine and cholesterol (Sigma) in a 1:1 molar ratio. Dissolve 10 μmol of each in 1 ml of diethyl ether.

2. Add 0.33 ml of PBS[b] containing $20-200$ μg of recombinant plasmid DNA.

3. Sonicate the two-phase system briefly (10 sec) using a B-15 Branson Sonifier fitted with a microtip and set at 500 msec frequency. This forms a one-phase dispersion.

4. Remove the ether by placing the dispersion in a rotating flask connected to a lyophiliser. This leaves a homogeneous, opalescent suspension of liposomes.

5. Separate free DNA (i.e., not trapped in the lipsomes) by centrifugation of the preparation at 100 000 g for 30 min. About $10-20\%$ of the DNA is usually encapsulated into liposomes.

6. Resuspend the pellet (DNA-loaded lipsomes) in 1 ml of PBS.

Transfection using DNA-loaded liposomes

1. Prepare serial dilutions of the liposome suspension in growth medium.

2. Add 100 μl aliquots of liposomes containing $10-500$ ng of DNA directly to recipient cells in flasks (25 cm² growth area) prepared exactly as described in *Table 1* for the calcium phosphate transfection technique.

3. Leave the liposomes in contact with the cells overnight in an incubator at 37°C.

4. Select for stable transformants. The exact protocol required will depend on the selectable marker gene involved (Sections 3 and 5, *Tables 1, 2, 15* and *16*). Alternatively, transient expression may be studied (Section 5.1).

---

[a]Adapted from ref. 27.
[b]The composition of this buffer is given in *Table 3*, step 1.

donor DNA and mouse LATK$^-$ cells. Approximately 25% of the cells 'pricked' in the presence of the recombinant DNA showed *tk* gene expression and 2% of the cells became TK$^+$ transformants (25).

2.4 Liposomes as Gene Carriers

The use of liposomes as vehicles for gene transfer and subsequent expression studies in mammalian cells is well documented (12,26,27). Wong *et al.* (26) have shown that the β-lactamase gene from plasmid pBR322 can be introduced into negatively-charged liposomes by sonication. Incubation of these vesicles with a variety of cultured cells results in the appearance of β-lactamase activity in cell extracts. More recently, liposome-mediated gene transfer was used to stably transform mouse LATK$^-$ cells with a recombinant plasmid carrying the HSV-1 *tk* gene (27).

The preferred method for the preparation of DNA-loaded liposomes is reverse-phase evaporation (28) since this yields large unilamellar vesicles with a high efficiency of DNA entrapment. Details of this procedure are given in *Table 10* together with the use of DNA-loaded liposomes for transfection of mammalian cells.

The potential of liposomes for introducing DNA molecules into mammalian cells has been thoroughly studied using SV40 DNA and monkey kidney cells (12); DNA uptake, resulting in virus production, can be monitored by sensitive plaque and fluorescence assays. The SV40 DNA entrapped in liposomes is at least 100-fold more infectious (1.8×10^3 p.f.u./μg DNA) than free naked DNA. However, this efficiency is much less than that which can be obtained (5×10^6 p.f.u./μg DNA) with the DEAE-dextran protocol (Section 2.2). The infectivity of the DNA in DNA-loaded liposomes can be increased 10- to 200-fold by exposing the cells for 90 sec to high concentrations of polyethylene glycol or glycerol 30 min after addition of the liposomes (12). Although this is still at least 10-fold less efficient than the DEAE-dextran technique, an efficient liposome method is worth pursuing since it could offer some advantages over other conventional techniques such as low toxicity and possible use *in vivo*. The uptake of liposomes by cells seems to occur by endocytosis rather than liposome fusion with the cell membrane (29).

2.5 Protoplast Fusion

A promising procedure to overcome the problems of uptake and expression of DNA by mammalian cells is protoplast fusion, first described by Schaffner (30). According to this method, protoplasts derived from *Escherichia coli* carrying recombinant DNA molecules are fused with cultured eukaryotic cells using polyethylene glycol as a facilitator of fusion. Since then, the protoplast fusion technique has been further refined so that all of the cells in the culture take up and transiently express the exogenous DNA (31). The efficiency of focus formation after transfer of recombinant DNA containing polyoma virus or SV40 early genes is at least equal to that observed after infection with virions. In optimal conditions, transfer by fusion seems to be 10- to 20-fold more efficient than DNA transfection by the calcium phosphate precipitation technique for the introduction of DNA into mammalian cells. A suitable protoplast fusion procedure is described in *Table 11*.

2.6. Other Methods of Gene Transfer

2.6.1 *Chromosome-mediated Gene Transfer*

Chromosome-mediated gene transfer involves the introduction of metaphase chromosomes into recipient cells. Isolated metaphase chromosomes are usually transferred using either the calcium phosphate technique (32,33) or packaging into liposomes (34). The system can be highly efficient but suffers from the present inability to store chromosomes stably or to reconstitute cloned DNA into them. If cloned DNA could be reconstituted into structures containing signals which control the stable segregation and inheritance of genetic information, the system would be potentially so powerful that much future work would be dedicated to this area.

2.6.2 *In Vitro Packaging of Eukaryote Virus DNA*

Transfer systems based on *in vitro* packaging of eukaryotic virus DNA sequences, analogous to lambda phage vectors in prokaryotes, are also highly attractive. Recently, gene transfer by polyoma virus-like particles has been described (35).

Table 11. Procedure for Protoplast Fusion[a].

---

1. Grow the *E. coli* host carrying the recombinant plasmids to a density of 2 x 10⁸ cells/ml.

2. Amplify the plasmids by incubating the culture overnight with chloramphenicol (200 μg/ml final concentration)[b].

3. Centrifuge 10 ml aliquots of the culture at 5000 g for 10 min and resuspend each pellet of bacteria in 100 μl of ice-cold 20% sucrose, 50 mM Tris-HCl, pH 8.0.

4. Add 20 μl of freshly-prepared lysozyme (10 mg/ml; Boehringer) and leave the suspension on ice for 10 min.

5. Add 50 μl of 0.25 M EDTA and dilute each suspension by the progressive addition of 1.3 ml of 9% sucrose, 50 mM Tris-HCl, pH 8.0.

6. Meanwhile the recipient cells should be grown in Dulbecco's modified Eagle's medium (DME medium; from Gibco) in 3.5 cm diameter Petri dishes to a density of 2 x 10⁵ cells/dish.

7. Remove the medium and wash the cells once with 2 ml of Tris-buffered saline (0.15 M NaCl, 20 mM Tris-HCl; pH 7.2).

8. Add 2 ml of the *E. coli* protoplast preparation (from step 5) to each dish and centrifuge the protoplasts onto the layer of cells (1500 r.p.m. for 3 min in a large bucket rotor of a bench centrifuge).

9. Remove the supernatant and add 0.5 ml of 48% (w/w) of polyethylene glycol (PEG 1000) in DME medium. Incubate at room temperature for 1 min.

10. Carefully wash the cells five times with Tris-buffered saline.

11. Add fresh growth medium (DME medium supplemented with 10% newborn calf serum). Incubate at 37°C.

12. Renew the medium on alternate days. Check the cultures after 10 days for the appearance of transformed colonies.

---

[a]From ref. 31.
[b]The antibiotic used for amplification will depend on the recombinant plasmid involved.

2.6.3 *Use of Electrical Stimulation*

The introduction of exogenous DNA into cells using electrical pulses has also been described (36) but requires further development before it can be considered seriously as a practical technique.

2.7 **Choice of Gene Transfer Method**

The choice of method obviously depends on the application. For most purposes, the calcium phosphate technique using cloned DNA, or DNA isolated directly from eukaryotic cells, is the method of choice. The technique is easy, rapid, highly reproducible and reasonably efficient. The use of DEAE-dextran can be recommended only for particular applications such as transfection of poliovirus RNA or papovavirus DNA. Original claims that microinjection results in much higher efficiencies of transformation than the calcium phosphate technique may have to be modified in the light of more recent studies in which the efficiencies were more comparable. However, when the number of recipient cells is limiting, or the donor DNA is very scarce, microinjection offers real advantages. The use of liposomes is obviously more complicated than the calcium phosphate technique and is no more efficient. Its usefulness may lie in certain *in vivo* applications, but this remains to be proven. Protoplast fusion is probably more efficient than calcium phosphate precipitation and may form the basis for applications in which recombinant plasmid libraries can be screened directly for gene expression in eukaryotic cells.

3. SELECTABLE MARKERS

Selectable markers are genes whose expression in cells can be positively selected. In the narrow sense they code for gene products (biochemical markers) whose presence can be selected for using simple chemical additions to the growth medium and are therefore of great use in the selection of transformants in gene transfer experiments (Section 2). However, many other phenotypes such as growth in agar or low serum requirements (i.e., dominant transforming genes which alter growth regulation) can also potentially be utilised. In addition, studies of selectable genes have yielded insight into the mechanisms underlying transfection procedures and provided assays for the analysis of transcriptional control. Several selectable genes of practical importance are described in Section 3.1 whilst dominant transforming genes which alter growth regulation are considered in Section 3.2.

3.1 Biochemical Markers

3.1.1 *Thymidine Kinase*

Thymidine kinase (*tk*) genes are expressed in most eukaryotic cells. The enzyme is part of the salvage pathway for synthesis of thymidine nucleotides and converts thymidine to thymidine monophosphate. TK$^-$ mutants of a number of cultured cells have been isolated (refs. 37 – 39; see also *Figure 4*). Szybalska and Szybalski (40) and Littlefield (41) were first to describe a medium (HAT medium) which selects for TK$^+$ cells because it contains hypoxanthine, aminopterin and thymidine. The aminopterin inhibits one-carbon metabolism and thus inhibits *de novo* synthesis of TTP from dUMP as well as the *de novo* synthesis of dATP and dGTP. Hypoxanthine is a substrate for the salvage pathway for dATP and dGTP and therefore allows these nucleotides to be synthesised. However, with HAT medium the synthesis of TTP is totally dependent on an exogenous source of thymidine and an active *tk* gene. HAT medium therefore selects for TK$^+$ cells. The usual composition of HAT medium suitable for most mammalian cells, and the selection procedure, are given in *Table 1*. Thymidine kinase genes can be thus introduced into TK$^-$ cells and HAT medium used to select TK$^+$ transformants. The best known TK$^-$ recipient is mouse LATK$^-$ cells which have a negligible rate of spontaneous reversion and a high efficiency of transfection. The best studied *tk* gene is that encoded by HSV-1 which has been cloned and sequenced. *Figure 1* describes the gene and its transcription control sequences.

The active HSV *tk* gene can be detected in transfected cells by various assays for active HSV-coded thymidine kinase enzyme (refs. 10,42). The enzyme

Table 12. Assay for HSV-1 Thymidine Kinase Activity[a].

1. Trypsinise and then wash the cells with PBS (see *Table 3*, step 1).

2. Resuspend the cells at a concentration of $0.2-2.0$ x 10^7 cells/ml in 50 mM Tris-HCl (pH 7.5), 5 mM 2-mercaptoethanol, 5 μM thymidine.

3. Disrupt the cells by sonication using a bath sonicator and store at $-70°C$ if necessary. Assay for thymidine kinase (TK) activity within a week.

4. Assay TK activity in 100 μl reaction mixtures containing $5-20$ μl of lysed cells, 50 mM Tris-HCl (pH 7.5), 10 mM $MgCl_2$, 10 mM ATP, 4.4 μM unlabelled thymidine, 5 μCi of 30 Ci/mmol [^{3}H]thymidine plus 0.2 mM TTP. Cellular TK is inhibited at this concentration of TTP whereas viral TK is not. Incubate the reaction mixture at 37°C.

5. Remove 25 μl aliquots at 1, 2 and 4 h for determination of the extent of conversion of thymidine to TMP as follows.

6. *Either*

    (i) Mix each aliquot with 1 ml of ice-cold 4 mM ammonium formate (pH 6.0), 5 μM unlabelled thymidine to stop the reaction.

    (ii) Apply each sample to a DE81 filter paper disc (Whatman; 2.3 cm diameter) under suction.

    (iii) Wash each disc twice with 2 ml of the ice-cold ammonium phosphate-thymidine solution.

    (iv) Further wash each disc twice each with 2 ml of distilled water and then ethanol. Phosphorylated thymidine residues remain bound to the paper while thymidine is washed off.

    (v) Dry each filter and determine the bound radioactivity by scintillation spectrometry.

  or

    (i) Spot the reaction aliquots directly onto DE81 discs, previously numbered in pencil.

    (ii) Immediately drop the discs into ice-cold water and wash well with large volumes of water.

    (iii) Wash the discs with acetone, dry them and determine the bound radioactivity by scintillation spectrometry.

[a]Adapted from ref. 42.

assay described in *Table 12* is based on measuring the conversion of radio-labelled thymidine to thymidine phosphates. Since the virus-coded enzyme has different substrate specificities and feedback control from the cellular enzyme, residual cell-coded activity present in TK$^-$ cells can be selectively inhibited by 0.2 mM TTP allowing the assay of only HSV-coded thymidine kinase (*Figure 7*). Anti-HSV TK antisera can also be used as a further test of enzyme specificity. An alternative strategy for detecting the active HSV *tk* gene is to assay for HSV *tk* mRNA (10,43). *Table 13* describes the isolation of total cellular RNA and an assay for *tk* mRNA by hybridisation is given in *Table 14*.

3.1.2 *Hypoxanthine-guanine Phosphoribosyl Transferase (HGPRT)*

Hypoxanthine-guanine phosphoribosyl transferase (HGPRT) is an enzyme in

Table 13. Isolation of Total Cellular RNA[a].

1. Prepare the following solutions:

   Guanidium chloride

   The preparation of this solution is described in *Table 3*.

   CsCl solution (5.7 M CsCl, 50 mM EDTA).

   CsCl (BRL, Ultrapure reagent) 47.98 g
   EDTA.Na$_2$ 0.93 g
   Distilled water to 50 ml final volume.

   If necessary, adjust the refractive index of the solution to a value of 1.3995 by the addition of solid CsCl or distilled water.

   10 x Mops (0.2 M Mops, 50 mM sodium acetate, 10 mM EDTA, pH 7.0)

   Mops (sodium salt)   20.93 g
   Sodium acetate.3H$_2$O   3.40 g
   EDTA.Na$_2$ 1.86 g
   Distilled water to 500 ml final volume.

   Adjust the pH to 7.0 with acetic acid.

   LiCl-urea solution (4.0 M LiCl, 8.0 M urea)

   LiCl (BDH)   8.48 g
   Urea (Sequanal grade, Pierce Chemical Co.) 24.00 g
   Distilled water to 50 ml final volume.

2. Resuspend $1-2$ x 10^8 cells in 5 ml of guanidinium chloride solution and disrupt them by homogenisation as described in *Table 3*, step 4.

3. Add 0.5 ml of 20% SDS (specially pure, BDH) and heat at 65°C for $1-3$ min.

4. Layer the mixture over a 3 ml cushion of CsCl solution and centrifuge for 48 h at 140 000 *g* at 15°C (e.g., in a 10 x 10 ml fixed-angle titanium rotor at 40 000 r.p.m.).

5. After centrifugation, remove the protein at the top of the tube and discard it. Collect the DNA banded in the middle of the tube, since this can be precipitated with isopropanol and further purified for use as described in *Table 3*. The RNA is pelleted at the bottom of the tube.

6. Drain the tube of its liquid contents and redissolve the RNA in 0.7 ml of distilled water.

7. Transfer the RNA into a microcentrifuge tube and add 0.7 ml of LiCl-urea solution immediately.

8. Mix slightly by vortexing and leave overnight at 4°C to precipitate the RNA.

9. Centrifuge for 10 min in a microcentrifuge to pellet the RNA.

10. Dissolve the RNA in 1.0 ml of 0.1 x Mops and then dialyse this solution for 2 h against 0.1 x Mops. This yield of RNA should be ~ 1 mg RNA per 1 x 10^8 cells[b].

[a]Care should be taken to use glassware or plasticware which has been rinsed with 0.05% diethylpyrocarbonate in water followed by baking at 120°C for 2 h to remove RNase contamination. All solutions which come into contact with the RNA should, if possible, be autoclaved before use and the investigator should wear disposable plastic gloves to avoid contamination with RNase (see Chapter 8).
[b]RNA can be stably stored for long periods at -20°C in the form of an ethanol precipitate or freeze-dried.

the salvage pathway for adenine and guanine. HGPRT$^-$ cells have been isolated, and HGPRT$^+$ revertants and transformants can be selected in HAT medium in exactly the same way as TK$^+$ cells (ref. 40, 41; *Table 1*).

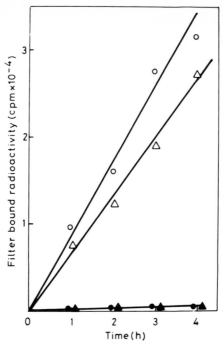

Figure 7. HSV TK activity in mouse LATK⁺ and hamster BHKTK⁺ cell lines 3 – 4 months after transformation with recombinant plasmid pTK1. TK activity in the presence of 0.2 mM TTP was determined in extracts of 1 x 10⁵ cells as described in *Table 12*. ○, LAT-1 and △, BT-1 mouse and hamster TK⁺ transformants of ●, LATK⁻ and ▲, BHKTK⁻ non-transformed cell lines, respectively.

Table 14. Spot Hybridisation Assay for Thymidine Kinase Messenger RNA[a].

1. Prepare the following solutions:

    Pre-hybridisation buffer

    50% deionised formamide[b]
    5 x SSC[c]
    5 x Denhardt's solution[d]
    250 μg/ml denatured salmon sperm DNA[e]
    50 mM sodium phosphate buffer, pH 6.5.

    Hybridisation buffer

    50% deionised formamide
    5 x SSC
    1 x Denhardt's solution
    100 μg/ml denatured salmon sperm DNA
    10% sodium dextran sulphate 500 (Sigma)
    20 mM sodium phosphate buffer, pH 6.5.

2. Dissolve total cellular RNA (prepared as in *Table 13*) in distilled water at a concentration up to 4 mg/ml.

3. Pre-treat a nitrocellulose filter (10 cm x 10 cm) by wetting it with water and then soaking it in 20 x SSC for a few minutes or until needed. Blot the filter free of excess liquid, air dry and then apply 5 μl aliquots of the RNA to marked areas of the filter.

4. Bake the filter at 80°C for 2 – 3 h.

5. Transfer the filter to a heat-sealable plastic bag and add 10 ml of pre-hybridisation buffer. Squeeze as much air as possible from the bag and heat-seal the open end of the bag with a commercial bag sealer. Incubate the bag submerged in a water bath at 42°C for 3 – 5 h.

6. Remove the bag from the water bath and open it by cutting off one corner using scissors. Squeeze out as much of the pre-hybridiation solution as possible.

7. Add 5 – 10 ml of hybridisation solution containing 5 – 10 ng/ml of denatured DNA probe previously labelled *in vitro* with ^{32}P by nick-translation. Squeeze as much air from the bag as possible and re-seal the cut edge. Incubate the bag submerged in a water bath at 42°C for 12 – 15 h.

8. Remove the bag from the water bath and cut along the length of three sides. Wearing disposable plastic gloves, remove the filter and immediately submerge it in 250 ml of 0.5% SDS, 2 x SSC in a plastic box at room temperature.

9. Shake the box gently at room temperature for 15 min.

10. Pour off the liquid and wash the filter again at room temperature in fresh 0.5% SDS, 2 x SSC as in steps 8 and 9.

11. Add 250 ml of fresh 0.5% SDS, 2 x SSC and incubate at 60°C for 30 min in a shaking water bath.

12. Replace the washing solution with 250 ml of 0.5% SDS, 0.1 x SSC and incubate at 60°C for 30 min in the shaking water bath.

13. Repeat step 11.

14. Dry the filter at room temperature on a sheet of Whatman 3MM paper.

15. Transfer the filter into a plastic bag and expose it to X-ray film for autoradiography.

---

[a]Adapted from ref. 43.
[b]Deionise formamide by stirring 5 g of a mixed bed ion-exchange resin (e.g. Amberlite MB-3 or Dowex MR-3 from Sigma) with 100 ml formamide for 30 min at room temperature in a fume hood. Filter the formamide through 3MM Whatman chromatography paper and store at −20°C until use.
[c]1 x SSC = 0.15 M NaCl, 15 mM trisodium citrate.
[d]A stock of 50 x Denhardt's solution is prepared by adding 1 g Ficoll (mol. wt. 400 000), 1 g polyvinylpyrrolidone, 1 g BSA (Fraction V) to 100 ml H_2O.
[e]To prepare denatured salmon sperm DNA, dissolve the DNA (Sigma Type-III, sodium salt) in water at a concentration of 10 mg/ml. Shear the DNA by passing it 10 times through a 19-gauge hypodermic needle. Boil the DNA for 10 min, cool rapidly and then store at −20°C.

3.1.3 *Aminoglycoside Phosphotransferase (APH)*

An attractive dominant selection system for mammalian cells employs the pro-karyotic aminoglycoside 3′-phosphotransferase gene (*aph*). Transposon Tn601 (903) contains the *aph*-I gene and transposon Tn5 carries the *aph*-II gene. These genes confer resistance to the aminoglycoside antibiotics kanamy-cin, neomycin and 3′-deoxystreptamine (also known as geneticin or G418) which inhibit protein synthesis in prokaryotes. They are inactivated by APH which phosphorylates the aminoglycoside residue. These antibiotics also in-hibit protein synthesis in eukaryotic cells and thus provide a dominant selec-tion system for which no special cell mutants are required.

Mammalian cells have no endogenous aminoglycoside phosphotransferases, but the bacterial enzymes can be expressed in mammalian cells by fusion to eukaryote-derived transcription-control sequences (44). The Tn5 *aph*-I gene has been fused to the control sequences of the HSV *tk* gene; the structure of this hybrid gene is shown in *Figure 8a*. The *aph*-II gene has been inserted into

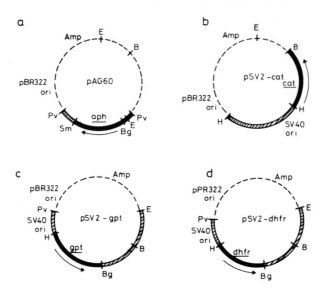

Figure 8. Structures of plasmid vectors containing biochemically selectable markers. (**a**) Plasmid pAG60 contains the aminoglycoside 3′-phosphotransferase (*aph*) gene (44); (**b**) plasmid pSV2cat contains the chloramphenicol acetyltransferase (*cat*) gene (47); (**c**) plasmid pSV2-*gpt* contains the xanthine-guanine phosphoribosyltransferase (*gpt*) gene (49); (**d**) plasmid pSV2-*dhfr* contains the dihydrofolate reductase (*dhfr*) gene (51). The broken lines represent plasmid sequences; the solid lines the *aph, cat, gpt* and *dhfr* sequences; the hatched lines the control sequences of the HSV-1 *tk* gene in pAG60, the SV40 sequences in pSV2CAT, pSV2-*gpt* and pSV2-*dhfr*; the arrows indicate the direction of transcription. The maps are not drawn to scale. E = *Eco*RI, B = *Bam*HI, Pv = *Pvu*II; Sm = *Sma*I, Bg = *Bgl*II, H = *Hind*III, ori = origin of replication, Amp = β-lactamase gene.

SV40-based vectors and used to study interferon gene expression (45).

Geneticin (G418) can be used to select cells expressing either bacterial APH-I or APH-II and is now commercially available (Gibco). Cells transformed for APH expression after gene transfer can be selected by the addition of 100 – 500 μg/ml of geneticin in the growth medium. The normal APH⁻ cells are killed. Cultured cell lines differ in their sensitivity to geneticin and the most suitable concentration of geneticin to use must be empirically determined (D.A. Spandidos and S. Campo, unpublished observations).

3.1.4 *Chloramphenicol Acetyltransferase (CAT)*

Transposon Tn9 encodes chloramphenicol acetyltransferase (CAT), the enzyme which inactivates chloramphenicol by acetylating it. Eukaryotic cells never express this enzyme. However, the *cat* gene can be expressed from eukaryotic promoters and there is a sensitive enzymatic assay for it (46,47). Furthermore, SV40 recombinant vectors carrying the *cat* gene have been constructed (*Figure 8b*) and used to assess the strength of a variety of eukaryotic promoters (46,47). The CAT system has also recently been used in short-term transfection assays to study the host specificity of enhancer sequences of SV40 and Moloney murine sarcoma virus (48).

Table 15. Selection for *gpt* Transformants[a].

1. Seed ~1 x 10^6 recipient cells per Petri dish (10 cm diameter). Incubate at 37°C for 24 h.

2. Transfect the cultures with *E. coli gpt* recombinant plasmid DNA using the calcium phosphate technique (see *Table 1*, steps 1 – 7).

3. Replace the medium with fresh, pre-warmed medium and incubate at 37°C for 3 days to allow pre-selection expression to occur.

4. Trypsinise the cells and disperse 5 x 10^5 cells per Petri dish (10 cm diameter) in Eagle's medium containing 10% dialysed foetal calf serum, 250 μg/ml xanthine, 15 μg/ml hypoxanthine, 10 μg/ml thymidine, 2 μg/ml aminopterin and 25 μg/ml mycophenolic acid (from Eli-Lilly Co.). Incubate at 37°C for 24 h.

5. Replace the culture medium with fresh medium containing the same supplements and renew the medium thereafter every 3 days.

6. Colonies can be isolated 2 weeks later using cloning rings as described in *Table 1*, step 11. The number of *gpt* transformants can be determined after fixation in methanol and staining with 10% Giemsa (see *Table 1*, footnote d).

[a]Adapted from ref. 50.

3.1.5 *Xanthine-guanine Phosphoribosyltransferase (XGPRT)*

The *E. coli gpt* gene encodes the enzyme xanthine-guanine phosphoribosyltransferase (XGPRT) which is the bacterial analogue of the mammalian enzyme HGPRT. XGPRT will efficiently convert xanthine into XMP, which is a precursor of GMP, whereas HGPRT will utilize only hypoxanthine and guanine. The *E. coli gpt* gene was cloned by Mulligan and Berg (49) into an SV40 late replacement vector (*Figure 8c*). Monkey cells infected with the recombinant virus stock expressed the *E. coli gpt* gene whose product can be distinguished from the endogenous mammalian HGPRT enzyme by its insensitivity to inhibition by hypoxanthine and its electrophoretic mobility (49).

The ability to express the *E. coli gpt* gene in mammalian cells allowed Mulligan and Berg (50) to develop a positive selection system. Mycophenolic acid is an inhibitor of IMP dehydrogenase which blocks the conversion of IMP into XMP and thus the *de novo* synthesis of GMP. When aminopterin, which also blocks the synthesis of IMP, is added to the medium this inhibition can be made more effective. Mammalian cells can grow in the presence of these two inhibitors if the medium is further supplemented with guanine and either hypoxanthine or adenine; the salvage pathway phosphoribosyltransferases can convert these bases to the corresponding mononucleotides. The inhibition is not overcome if the medium is supplemented with adenine and xanthine because the mammalian HGPRT cannot convert xanthine to XMP and thus GMP cannot be produced. If, however, the mammalian cells express the *E. coli gpt* gene, growth on adenine plus xanthine is possible. Like resistance to geneticin (Section 3.1.3), *gpt* selection can be applied to any type of cell. A protocol of *gpt* selection in non-mutant cells is given in *Table 15*. However, when HGPRT⁻ cells are used as recipients (Section 3.1.2), HAT selection (see *Table 1*) is sufficient for the isolation of *gpt*-transformed cells.

Table 16. Selection for *dhfr* Transformants[a].

1.	Seed ~1 x 10⁶ DHFR⁻ CHO cells per Petri dish (10 cm diameter) 12 – 18 h prior to transfection.
2.	Transfect the cells with *dhfr* recombinant plasmid DNA using the calcium phosphate technique (see *Table 1*, steps 1 – 7) using an absorption period of 8 h.
3.	Next, incubate the cells in (non-selective) Ham's F12 medium at 37°C for 3 days to allow for pre-selection expression.
4.	Trypsinise the cells and disperse one-half from each confluent plate in selective medium (Ham's F12 medium containing 0.2 µg/ml methotrexate) on five fresh dishes (10 cm diameter). Incubate at 37°C for 24 h.
5.	Replace the medium with fresh selective medium containing the same supplement and renew the selective medium thereafter every 3 days.
6.	After 2 weeks[b], individual colonies can be isolated using cloning rings (*Table 1*, step 11).

[a]Adapted from ref. 51.
[b]At this stage a permanent record, if desired, can be obtained by fixing and staining the cells in some of the dishes as described in *Table 1*, footnote d.

3.1.6 *Dihydrofolate Reductase (DHFR)*

The mouse *dhfr* gene has also been used as a selectable marker based on the fact that cellular dihydrofolate reductase (DHFR) is inhibited by methotrexate. Introduction into the cell of an altered *dhfr* gene or an increased copy number of the normal gene results in increased resistance to methotrexate. Mouse *dhfr* cDNA has been cloned both in SV40 late replacement vectors and in the pSV series of plasmid vectors (51) (*Figure 8d*). Transfection of Chinese hamster ovary (CHO) cells with these recombinants and selection of DHFR⁺ cells by growth in methotrexate is described in *Table 16*.

Use of the *dhfr* gene as a selectable marker has the advantage that the transformed cells produce very large amounts of wild-type DHFR because of gene amplification (51) and the segment of chromosomal DNA amplified is large and includes a considerable amount of DNA sequences flanking the *dhfr* gene (52). Kaufman and Sharp (53) have constructed hybrid *dhfr* genes linked to the SV40 genome and transferred the resulting recombinants into DHFR⁻ CHO cells and then selected for growth in the presence of methotrexate. Under these conditions, cell lines were obtained expressing greater than 10% of the total soluble protein as a polypeptide related to the SV40 small t-antigen. When an altered *dhfr* gene is used as donor DNA there is no need for DHFR⁻ recipient cells since the mutant gene is directly selectable and behaves dominantly (32).

A prokaryotic DHFR has been also cloned and used to induce resistance to methotrexate in mouse cells (54).

3.2 Dominant Transforming Genes Which Alter Growth Regulation

DNA-mediated gene transfer has been one of the most important techniques used in the identification and analysis of a number of virus-coded and cellular genes which cause changes in the growth pattern and morphology of cells.

Table 17. DNA-mediated Transfer of Cellular Oncogenes to NIH3T3 Cells.

1. Plate 1.5 x 10^5 NIH3T3 cells in 10 ml of Ham's SF12 medium containing 15% 'Hyclone' foetal calf serum per flask (25 cm² growth area) and incubate at 37°C.

2. After 24 h incubation, prepare a DNA-calcium phosphate co-precipitate using the donor DNA (see *Table 1*, steps 4 – 6) and add 1.0 ml of this to each flask of cells.

3. After 24 h exposure of the cells to the DNA, replace the medium with fresh medium.

4. After an additional 24 h, change the medium according to one of the following protocols:

    (i) Low serum medium (Ham's SF12 containing 5% 'Hyclone' serum) changed ever 3 – 4 days for 2 – 3 weeks. Foci formation should be observed at 10 days post-transfection.

    (ii) Trypsinise the cells and plate them in Ham's SF12 medium containing 30% 'Hyclone' serum and supplemented with 0.9% methocel or 0.6% agar. Colonies of transformed cells should be detected 7 – 10 days post-transfection.

    (iii) Trypsinise the cells and inject 1 – 2 x 10^6 cells into each of several experimental animals (e.g., 'nude' mice). Tumours should appear at the site of injection 2 – 10 weeks later depending on the donor DNA.

Theoretically, transformation by these genes could be detected by two kinds of assay based upon the rescue from senescence or the recognition of 'second stage' transformation markers.

3.2.1 *Rescue from Senescence*

This assay depends on the observation that most primary cultures of mammalian cells stop dividing after a limited number of passages, that is they 'senesce'. In some circumstances, such cells can be rescued from senescence by treatment with various agents, usually viruses or virus-coded genes (55,56). To date, cells have not been transformed in this fashion using chromosomal DNA, although an early report (57) suggested that this was possible using whole chromosomes.

3.2.2. *'Second Stage' Transformation Markers*

This type of assay involves the recognition in the recipient cells of such diverse phenotypes as morphological change, the ability to grow in semi-solid agar and tumour induction in animals (58 – 62). The recipient cells in such experiments are usually permanent lines of rodent fibroblasts which efficiently take up and express exogenous DNA transfected by the calcium phosphate method. They include NIH3T3 fibroblasts, Fisher rat embryo fibroblasts, CHO cells and the C13 clone of BHK cells (58 – 62).

NIH3T3 cells have recently been used in the identification and analysis of activated transforming genes (oncogenes) present in different kinds of cancer cells (63 – 67). They possess, to an extreme degree, properties of ease of transfection combined with a high susceptibility to 'second stage' transformation. An outline of the procedures to assay transforming genes present in total tumour cell DNA is given in *Table 17*. It should be noted that NIH3T3 cell populations can be highly heterogeneous. Different clones obtained from different laboratories vary in such phenotypic properties as growth medium requirements and serum requirements, ease of transformation and background of spontaneous transformants. Therefore the potential user should carefully check these different parameters prior to use.

4. VIRUS-BASED VECTORS

As described in previous sections of this Chapter, the most common procedure for the introduction of cloned DNA sequences of interest into a recipient mammalian cell is by transfection with naked recombinant plasmid or phage DNA, i.e., the use of DNA-based vectors. Alternatively, in some cases, it is possible to use a eukaryotic virus to carry the cloned DNA into the cell during normal viral infection. Recent reviews on the use of these virus-based vectors have been published elsewhere (68,69). Two classes of viral vectors are available and are described here; the lytic viruses such as polyoma and SV40 and the so-called 'shuttle vectors' based on retroviruses and papillomaviruses. The latter represent systems in which eukaryotic cells contain circular episomal copies of recombinant vector which can be extracted and re-isolated directly by bacterial transformation.

4.1 Lytic Virus Vector Systems

The extensive structural and functional analyses available for SV40 (55) have made this virus an attractive source of genetic material (e.g., promoters and 'enhancers') for DNA-based vectors as well as virus vectors. Recombinant SV40 genomes can be packaged into infectious virus provided helper functions are supplied. After infection of the host cells with the recombinant SV40, the hybrid mRNAs are correctly processed and translated (70,71). An important improvement in SV40 vectors came with the development of COS cell lines (72). COS cells carry integrated origin-defective SV40 genomes and permit helper virus-independent propagation of early-region-defective SV40 vector viruses (73). However, actively replicating SV40 is cytotoxic and has the additional limitation of accepting only relatively small (up to 2.5 kb) fragments of foreign DNA. More recently, other lytic virus vector systems based on adenoviruses (74,75), herpes viruses (76) and vaccinia viruses (77) have been investigated which have the potential to carry much larger fragments of foreign DNA.

4.2 Retrovirus-based Vector Systems

Most eukaryotic vector systems which employ virions to propagate the recombinant genome use lytic viruses (Section 4.1) which have the serious disadvantage of killing the cell. Ideally, one would like to produce a stable cell line which continuously replicates and expresses the donor DNA. Retroviruses offer many attractions as virus vectors for recombinant DNA. They replicate via a circular DNA provirus intermediate which integrates efficiently into the host cell chromosome by a transposon-like mechanism, and are not necessarily cytotoxic. Several laboratories are active in developing retrovirus-based vectors and successful propagation of recombinant retroviruses containing the

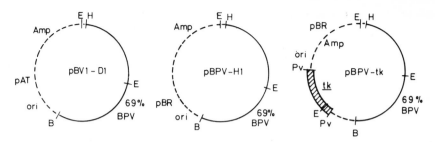

Figure 9. Papillomavirus-based vectors. Construction of pBVI-DI has been described by Campo and Spandidos (89), pBPV-HI by DiMaio *et al.* (91) and pBPV-*tk* by Sekiguchi *et al.* (92). The broken lines represent plasmid sequences; the solid lines BPV sequences; the hatched lines the HSV *tk* gene sequences. The maps are not drawn to scale. E = *Eco*RI, B = *Bam*HI, H = *Hind*III, Pv = *Pvu*II, ori = origin of replication of pAT153, pBR = plasmid pBR322 sequences, Amp = β-lactamase gene, pAT = plasmid pAT153 sequences.

HSV *tk* gene (Section 3.1.1) or the *E. coli gpt* gene (Section 3.1.5) has been described (78 – 81). Replication signals and selectable markers for eukaryotic cells could be included in such vectors which would permit their re-isolation directly into bacterial hosts. Retrovirus-packaging mutants are now available which can be used to produce helper-virus-free recombinant viruses (81,82). An interesting possibility is the use of retrovirus vectors to introduce genes into the germ line of animals (83).

4.3 Papillomavirus-based Vectors

Apart from the use of retroviruses (Section 4.2), a second approach to the production of a stable cell line which continually replicates and expresses the donor DNA is the development of systems in which the donor DNA replicates as an episome. Episomal systems have another advantage in that they are not subject to packaging constraints. Papillomaviruses behave differently from other papovaviruses in that papillomavirus-transformed cells do not contain integrated viral DNA (84 – 86). Instead they contain 50 – 300 copies of circular unintegrated viral DNA. As an example, consider bovine papilloma virus (BPV). Recombinant plasmids containing the transforming region of BPV DNA can efficiently induce morphological transformation of mouse C127 or NIH3T3 fibroblastic cell lines. Only 69% of the BPV-1 genome is required for transformation. The transformed cells carry the recombinant genomes as episomes and correctly express eukaryotic genes which have been ligated into the recombinant plasmid, for example human γ-interferon (90) or human β-globin (91). Furthermore, DiMaio *et al.* (91) and Sekiguchi *et al.* (92) have shown that BPV recombinant plasmids may be recovered from low molecular weight Hirt (93) supernatant DNA for subsequent transformation of *E. coli*. Thus BPV vectors combined with the appropriate host cells constitute an established mammalian shuttle vector system. The structures of some BPV-based shuttle vectors are illustrated in *Figure 9* and the use of BPV-based shuttle vectors, is described in *Table 18*. It should be noted that this system applies mainly to mammalian cells susceptible to BPV-induced morphological transformation, such as NIH3T3 cells or C127 cells.

Expression of Exogenous DNA in Mammalian Cells

Table 18. Use of BPV-based Shuttle Vectors.

After the transformation of mammalian cells with a BPV-based shuttle vector, BPV DNA-linked sequences are found as multiple episomal copies in the nucleus. This DNA can be isolated in a supernatant fraction after extraction by the method of Hirt (93) and can then be used directly to transform a bacterial host.

Transformation of Mammalian Cells

1. Transfect mouse NIH3T3 or C127 cells with the recombinant plasmids containing the transforming region of BPV-1 (see *Figure 9*) using the calcium phosphate technique (*Table 1*).

2. Isolate transformed colonies and pick these using cloning rings (*Table 1*), transferring the trypsinised cells to flasks (25 cm² growth area) each containing 5 ml of Ham's SF12 medium with 5% 'Hyclone' serum.

3. Culture the cells in this growth medium until a sufficient number of cells are obtained for isolation of DNA. For bulk culture, the cells are most easily grown in larger flasks (75 cm² growth area).

Isolation of DNA from Hirt Supernatant[a]

1. Remove the growth medium from the cells and wash the cells twice with 0.15 M NaCl, 10 mM Tris-HCl, pH 7.4, being careful not to disturb the cell monolayer.

2. Carefully add 20 ml of 0.6% SDS, 10 mM EDTA, 10 mM Tris-HCl, pH 7.5 per flask (75 cm² growth area containing $1-2 \times 10^7$ cells) so that it spreads evenly. Incubate at room temperature for 20 min without disturbance.

3. Gently pour off the solution into a Corex glass tube (DuPont). Use a rubber policeman if necessary.

4. Close the tube with Parafilm and mix the contents by inverting the tube gently 10 times. Leave overnight, or preferably up to 72 h, at 4°C.

5. Centrifuge at 15 000 g for 30 min. Remove the supernatant and set it aside, while the residual contents of the tube are re-centrifuged to recover additional supernatant if necessary.

6. Extract the pooled supernatants with an equal volume of phenol saturated with 10 mM NaCl, 10 mM EDTA, 50 mM Tris-HCl, pH 8.0.

7. Recover the aqueous layer by centrifugation and then extract it with an equal volume of chloroform-isoamyl alcohol (24:1, v/v).

8. Recover the aqueous layer by centrifugation. Precipitate the DNA by addition of 2.5 volumes of ethanol and store overnight at $-20°C$.

9. Collect the DNA by centrifugation at 12 000 g for 10 min.

10. Dissolve the DNA pellet in 1 mM EDTA, 10 mM Tris-HCl, pH 8.0 (1 ml per original 75 cm² flask of cells) and dialyse against this buffer.

   The procedure usually yields ~10 µg of DNA of which the BPV-related DNA comprises only a small proportion.

Transformation of Bacteria by Plasmid DNA[b]

1. Prepare the following:

   L-Broth

   This is prepared as described in *Table 4*

   Transformation buffer

   0.10 M $CaCl_2$
   0.25 M KCl
   5 mM $MgCl_2$
   10 mM RbCl

5 mM   Tris-HCl, pH 7.6
Filter sterilise this solution.

Storage buffer

0.10 M $CaCl_2$
0.25 M Cl
5 mM $MgCl_2$
10 mM RbCl
5 mM   Tris-HCl, pH 7.6
15% Glycerol
Filter sterilise this solution.

Agar plates

Prepare bacteriological agar:

5.0 g   NaCl
5.0 g   Tryptone (Difco)
2.5 g   Yeast extract (Difco)
6.0 g   Bacto-agar (Difco)
Distilled water to 500 ml

Autoclave and allow to cool to 50°C. Then add the necessary antibiotic(s). The antibiotic(s) to be used depend(s) on the selectable bacterial gene present in the plasmid DNA; for β-lactamase, add ampicillin to 100 μg/ml final concentration. Mix well and pour ~20 ml per Petri dish (10 cm diameter).

2. Inoculate 200 ml of L-broth in a 1 litre flask with 1 ml of an overnight bacterial culture. Grow the cells with shaking until the A_{650} reaches 0.6.

3. Harvest the bacteria by centrifugation at 5000 g for 10 min at 4°C.

4. Gently resuspend the bacteria in 100 ml of transformation buffer. Then leave on ice for 20 min.

5. Harvest the cells by centrifugation as in step 3.

6. Gently resuspend the cells in 20 ml of storage buffer. Place aliquots (e.g., 1 ml) into sterile tubes and store at −70°C[c].

7. Dilute the plasmid DNA to be used for transformation (up to 40 ng) to 0.2 ml final volume with transformation buffer. Add 0.2 ml of freshly-prepared or thawed bacteria from step 6 and leave on ice for 25 min.

8. Transfer to a water bath pre-heated to 37°C. Incubate for 5 min at this temperature.

9. Add 1 ml of L-broth to each tube and incubate at 37°C for 1 h without shaking. This period allows the bacteria to recover and to begin to express antibiotic resistance.

10. Spread 0.2 ml aliquots of cells onto the prepared bacteriological agar plates. Leave the plates at room temperature for 10−15 min until the liquid has been absorbed.

11. Invert the plates and incubate them at 37°C. Colonies should appear in 12−15 h. As an indication of the expected transformation frequency, this protocol should yield $5 \times 10^5 - 5 \times 10^6$ *E. coli* HB101 transformants per μg of intact pBR322 DNA.

---

[a]Adapted from ref. 93.
[b]Adapted from refs. 94, 95.
[c]Competent HB101 cells can be stored at −70°C without serious loss in transformation efficiency. With other bacterial strains, freshly-prepared cells may have to be used in step 7.

5. EXPRESSION OF TRANSFERRED GENES

The mechanisms involved in the uptake and expression of donor DNA in recipient cells are poorly understood. Shortly after the introduction of DNA there

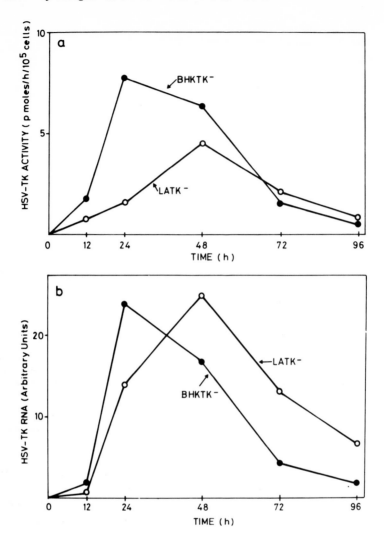

Figure 10. Transient expression of the HSV *tk* gene in mouse and hamster cells. **Top panel.** HSV-TK enzyme activity in lysates of LATK⁻ (○) or BHKTK⁻ (●) cells transfected with a recombinant plasmid containing the HSV *tk* gene. The plasmid used was pTKMOLTR1 (see *Figure 12*) and transfection of cells was carried out as described in *Table 1*. The DNA-calcium phosphate co-precipitate was removed from the cells after 6 h exposure and the medium was replaced with fresh non-selective medium. At various time intervals the cells were harvested, resuspended at a concentration of 2 x 10⁷ cells/ml in 50 mM Tris-HCl (pH 7.5), 5 mM β-mercaptoethanol and 5 μM deoxythymidine and sonicated. The lysates were then stored at −70°C until assayed for HSV-TK activity as described in *Table 12*. **Bottom panel.** HSV-*tk* mRNA in LATK⁻ (○) or BHKTK⁻ (●) cells transfected with recombinant plasmid pTKMOLTR1. Further aliquots of cells from the experiment described above were used to isolate RNA as described in *Table 13*. This was analysed for HSV-*tk* mRNA by spot hybridisation (ref. 43 and *Table 14*); 20 μg of each RNA were spotted onto duplicate nitrocellulose filters and the filters hybridised with a 0.6-kb *Bgl*II-*Sst*I *tk*-specific DNA fragment ³²P-labelled by nick-translation. The filters were autoradiographed and the autoradiographs were analysed by quantitative densitometry. The average value of duplicate dots is given in arbitrary units for each time point.

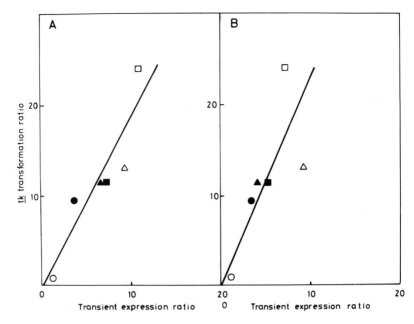

Figure 11. Correlation between short-term expression and transformation efficiency for recombinant plasmids containing transcription-control sequences from different eukaryote genes. **Panel A** shows short-term levels of *tk* mRNA compared with transformation efficiency. **Panel B** shows short-term TK enzyme levels compared with transformation efficiency. Plasmids pTKε1 (○), pTK1 (●), pTKεSV1 (△), pTKεSV2 (▲), pTKεMOE1 (□) and pTKεMOE2 (■) contain the HSV *tk* gene under the control of different transcription control sequences including the SV40 enhancer (pTKεSV1 and pTKεSV2) and the MoMuSV enhancer (pTKεMOE1 and pTKεMOE2). The results are expressed as ratios of values relative to a standard plasmid, pTKε1, which lacks enhancer signals (*Figure 12*). These data are derived from the results of Lang *et al.* (98).

is a transient phase of gene expression. During this phase it seems likely that the newly-introduced genes are converted to minichromosomes (96). Under selective conditions, transformed colonies then appear after 1 − 3 weeks at a frequency which appears to reflect the initial level of gene expression during the transient phase.

5.1 Transient Expression

Using the calcium phosphate technique for gene transfer (Section 2.1), the transient phase of expression lasts between 2 and 3 days. The kinetics of expression, as studied by using an assay for TK enzyme activity (*Table 12*) and a spot hybridization assay for TK mRNA (ref. 43 and *Table 14*) are illustrated in *Figure 10* for BHKTK⁻ and LATK⁻ cells. As shown, maximum expression of the *tk* gene in BHK and LA cells is obtained 24 h and 48 h after transfection, respectively.

The available evidence suggests that *tk* gene transcripts are correctly initiated, terminated and processed (90,91,97). Thus *Figure 11* shows that there is an excellent correlation between the relative amount of *tk* mRNA produced and the amount of TK enzyme activity expressed. Similar results have been reported using the *cat* gene in short-term expression assays (see refs. 46 – 48 and Section 3.1.4).

5.2 Long-term Transformation

Under selective conditions, transformed colonies of cells can be observed 1 – 3 weeks after transfection by the calcium phosphate method. In the case of transfection with the HSV *tk* gene, TK$^+$ transformed colonies can be counted and picked after 7 days when using BHK cells as recipients and 10 – 14 days when using LATK$^-$ cells. The normal HSV *tk* gene gives a transformation frequency of 2000 – 4000 colonies per μg. However, as explained in Section 5.3, the *tk* gene can also be placed under the control of different eukaryotic transcription regulation signals. The different hybrid genes have widely differing activities in both the transient assay and the transformation assay (10,98). Transformation frequencies vary over four orders of magnitude and so the assay is two orders of magnitude more sensitive than the transient assay (Section 5.1). *Figure 11* shows a direct comparison between the level of transient *tk* gene expression and the transformation efficiency observed for a number of different hybrid *tk* genes (10,98). There is a marked correlation between the transient level of expression and the subsequent transformation frequency. The initial rate of expression may, therefore, have a major effect in determining the transformation efficiency. Nevertheless, despite this correlation, very little difference in the amounts of *tk* DNA, *tk* mRNA or TK enzyme can be found in low-passage, stably-transformed cell lines (98). This strongly suggests that additional genetic controls are superimposed in cells grown in selective HAT medium. Similar tests have not yet been thoroughly applied to other selectable genes.

When cells are grown in culture, complex changes in the copy number and state of the donor DNA occur (2,4). To some extent, the state of the donor DNA in long-term transformed cells depends on the DNA sequence controlling *tk* gene expression. The results of Perucho *et al.* (99) for *tk* gene transfection suggest that at some stage during the transfection process, active recombination between input DNA molecules can occur, leading to complexes of extrachromosomal DNA in which the selectable gene is linked to carrier DNA. When hybrid *tk* genes with low transformation frequencies are used, long-term transformed cells normally contain 5 – 10 copies of tandemly-duplicated DNA with little evidence of sequence rearrangement (4,98). In most cases it is difficult to determine whether or not the tandem array is integrated into chromosomes, but some evidence for extrachromosomal copies of the *tk* gene have been obtained (2,4,100). In contrast, when transcription-control DNA sequences giving a high transformation frequency are used, long-term cultures contain only 1 – 5 copies of the *tk* gene, independently integrated into carrier or recipient cell DNA. In some cases limited duplication of some integrated copies, including carrier or host-cell DNA flanking sequences, has been observed (101).

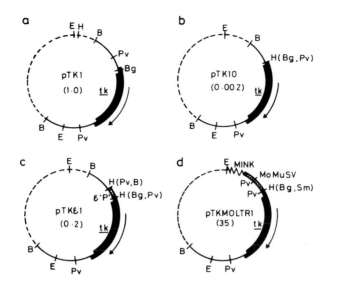

Figure 12. Recombinant plasmids carrying hybrid genes containing the HSV-1 structural thymidine kinase sequences linked to heterologous eukaryotic promoters. The transformation efficiency of each plasmid was tested on mouse LATK⁻ cells as described in *Table 1*. This efficiency, expressed as a ratio relative to the value obtained for the normal gene in pTK1, is shown in parentheses for each recombinant plasmid (**a**) pTK1 recombinant containing the 3.5-kb *Bam*HI fragment of HSV-1 DNA carrying the intact *tk* gene. (**b**) pTK10; the promoter region of the *tk* gene has been deleted. This 249-bp *Pvu*II-*Bgl*II fragment was removed and the *Pvu*II and *Bgl*II sites were converted to a *Hind*III site. (**c**) pTKε1; a 197-bp *Bam*HI-*Pvu*II fragment containing the promoter sequences from the human epsilon (ε) globin gene was inserted as a *Hind*III fragment into the *Hind*III site of pTK10. (**d**) pTKMOLTR1; an *Eco*RI-*Hind*III fragment containing 473 bp from the MoMuSV LTR and adjacent 5'-flanking mink DNA sequences places *tk* transcription under the control of the promoter which normally regulates MoMuSV viral gene expression. The broken lines represent plasmid pAT153 sequences; the solid lines HSV sequences; the outer heavy lines the structural *tk* gene, the inner heavy line the *tk* coding sequences; the hatched lines the human and MoMuSV sequences carrying the promoter elements; the wavy line the mink sequences; the arrows indicate the direction of transcription. The maps are not drawn to scale. E = *Eco*RI, B = *Bam*HI, H = *Hind*III, Pv = *Pvu*II, Bg = *Bgl*II and Sm = *Sma*II. ε'p' = human epsilon globin promoter, MoMuSV = Moloney murine sarcoma virus. A more detailed description of these plasmids is given elsewhere (98).

5.3 Assays of Transcription Control Sequences

Potentially, any gene for which suitable quantitative assays are available could be used to identify and analyse transcription regulatory signals in genomic DNA. In general, the main approach has been to make deletions or substitutions in, or add exogenous regulatory signals to, the normal regulatory sequences and then to measure the effects on gene expression using DNA-mediated gene transfer. A simple example using *tk* gene transformation is shown in *Figure 12*. The normal *tk* gene promoter is located between a *Pvu*II and a *Bgl*II site of plasmid pTK1. Deletion of the promoter and its replacement by a single *Hind*III cloning site, to yield plasmid pTK10, results in a 500-fold inactivation of gene expression. Insertion of regulatory sequences from the human ε-globin gene to yield plasmid pTKε1, or the long terminal

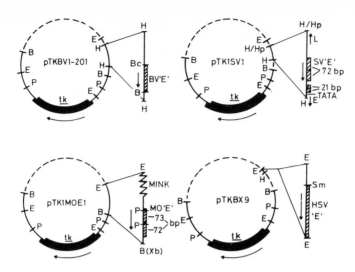

Figure 13. Recombinant plasmids carrying viral enhancer sequences. Plasmids pTKBV1-201, pTK1SV1 and pTK1MOE1 carry the bovine papilloma virus (BPV-1), simian virus 40 (SV40) and Moloney murine sarcoma virus (MoMuSV) enhancers, respectively, and have been described elsewhere (10). Recombinant plasmid pTKBX9 carries a 220-bp *Sma*I-*Eco*RI fragment containing an enhancer sequence located 100 bp upstream of the transcriptional start site of an immediate early mRNA of HSV-1. The *Sma*I site has been converted to an *Eco*RI site using molecular linkers. The broken lines represent plasmid pAT153 sequences; the solid lines viral sequences; the outer heavy lines the structural *tk* gene; the inner heavy line the *tk* coding sequences; the hatched lines the viral sequences carrying the enhancer elements; the wavy line mink DNA sequences; TATA = TATA box (promoter homology sequences); the arrows indicate the direction of transcription; enhancer regions are designated as 'E'. Thus BV'E', SV'E', MO'E' and HSV'E' are the enhancers for BPV-1, SV40, MoMuSV and HSV-1, respectively. The maps are not drawn to scale. E = *Eco*RI, H = *Hind*III, P = *Pvu*II, Sm = *Sma*I, Bc = *Bcl*I, H/Hp = *Hind*III/*Hpa*II, B = *Bam*HI; B(xb) = *Bam*HI/*Xba*I. For definition of MINK, see *Figure 12* legend.

repeat (LTR) sequence from Moloney murine sarcoma provirus DNA to yield plasmid pTKMOLTR1 results in reactivation of gene expression. Plasmid pTK10 can therefore be used as a replacement vector for regulatory DNA sequences. For example, it has been used to analyse transcription-control signals in a papillomavirus genome (102). The high level of gene expression obtained with the Moloney murine sarcoma virus (MoMuSV) LTR sequence compared with the normal *tk* promoter, or the ε-globin promoter, is due to the presence of an additional regulatory sequence of a type variously termed 'enhancers' or 'modulators'. Enhancers are short sequences which stimulate the activity of promoters. They act independently of their orientation in the genomic DNA and at a distance with respect to promoters (10,98,103,104). Most enhancers so far studied have been derived from eukaryotic viruses, but cellular enhancer sequences are probably common. *Figure 13* shows the structure of plasmids in which 'enhancer' regions from several eukaryotic viruses are placed approximately 1.0 kb upstream of the normal *tk* structural gene. The ability of these sequences to increase transcription of the *tk* gene in the transient phase of gene expression (Section 5.1) can be illustrated by Northern blot analysis of total cellular RNA from transfected cells (*Figure 14*). A procedure for isolation of

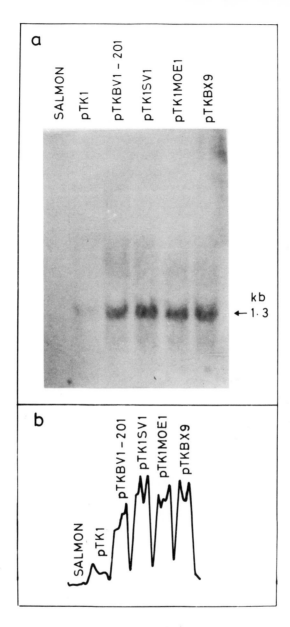

Figure 14. Northern blot analysis of *tk* mRNA in transfected cells. Total cell RNAs (20 µg) from mouse LATK⁻ cells transfected with the recombinant plasmid DNAs indicated were fractionated on a 1% formaldehyde-agarose gel, then blotted onto a nitrocellulose filter and probed with a 0.6-kb *Bgl*II-*Sst*I *tk*-specific cDNA fragment labelled with ^{32}P (10). (**a**) The autoradiograph and (**b**) densitometric scanning of the autoradiograph across the 1.3-kb band (*tk* mRNA). Salmon = DNA from salmon sperm. The structure of pTK1 (a plasmid containing the *tk* gene but lacking an enhancer sequence) is given in *Figure 1*. The structures of pTKBV1-201, pTK1SV1, pTK1MOE1, and pTKBX9 (plasmids containing the enhancers of BPV-1, SV40, MoMuSV and HSV-1, respectively) are given in *Figure 13*.

total cellular RNA from transfected cells for studies of this kind is described in *Table 13*. The technique of Northern blotting is described in Chapter 2, Section 5.2.2. Briefly the isolated RNA transcripts are fractionated by agarose gel electrophoresis and then transferred to a nitrocellulose membrane filter. RNA transcripts of interest are subsequently detected by hybridisation to an appropriate radiolabelled DNA probe followed by autoradiography. The normal *tk* transcript is 1.3 kb long, and RNA of this size is found in higher amounts (5- to 20-fold; see *Figure 14*) when enhancers are present (plasmids pTKBV1-201, pTK1SV1, pTK1M0E1, pTKBX9; see *Figure 13*) than in their absence (plasmid pTK1; see *Figure 1*). Many laboratories have used similar techniques using different gene systems in the analysis of promoters and enhancers but, to date, these techniques have not been used to investigate transcription-termination signals.

5.4 Inducible Genes

A major topic of interest is the mechanism of gene activation. A number of genes are known whose expression can be induced by hormone treatment, treatment with metal ions, stimulation of cells to differentiate or which are subject to regulation by virus-coded proteins. Gene transfer techniques are now being applied to investigate the transcription-control regions of such genes. The following sections describe some of these systems as an illustration of the practical application of gene transfer assays.

5.4.1 *Metallothionein Genes*

Metallothioneins are ubiquitous proteins of vertebrates. They bind heavy metals and are thought to be involved in zinc homeostasis and resistance to heavy metal toxicity. The metallothionein-1 gene (*mt*-1) of mice is controlled at the transcriptional level by heavy metals and by glucocorticoid hormones which are inducers in this system. Because the *mt*-1 gene serves a 'housekeeping' function, its regulation probably does not require tissue-specific factors and it can therefore be studied in a variety of systems. When recombinant vectors containing the mouse *mt*-1 gene and the *E. coli gpt* gene were used to transform HGPRT⁻ HeLa cells, the *mt*-1 gene retained transcriptional regulation by cadmium but not by glucocorticoids (105). To determine whether the lack of response to glucocorticoids was due to the inability of the human glucocorticoid receptors to interact with mouse DNA sequences, mouse LATK⁻ cells were transformed (105) with a vector containing the intact *mt*-1 gene and the HSV-1 *tk* gene, or a hybrid gene with the *mt*-1 promoter regulatory region fused to a viral thymidine kinase structural gene (plasmid pMK1). Again it was found that transcription of the *mt*-1 gene was regulated by cadmium but not by glucocorticoids. Moreover, thymidine kinase activity was regulated by cadmium when the fusion gene was transfected into mouse cells. Deletion-mapping experiments indicated that the DNA sequences necessary for regulation of the mouse *mt*-1 gene by cadmium lie within 148 bp of its transcriptional start site (105).

The 'promoter' regions of the *mt*-1 gene have also been investigated using gene transfer by microinjection. When mouse eggs were microinjected with pMK plasmid and then incubated with cadmium, thymidine kinase activity increased approximately 10-fold compared with control eggs not exposed to cadmium. Analysis of a set of deletion mutants revealed that the minimum sequence required for cadmium regulation lies within 90 bp of the transcription start site (106). The same pMK fusion plasmid was also introduced into mice by microinjection into fertilised eggs followed by re-insertion of the eggs into foster mothers (107). Ten of the sixty nine progeny mice carried pMK sequences and seven of these expressed high levels of viral thymidine kinase in the liver. This enzyme was inducible by heavy metals, as indicated by assay of thymidine kinase activity following sequential partial hepatectomies with or without cadmium treatment. However, glucocorticoid treatment was ineffective in all transgenic mice tested. The generality of introducing genes into animals via this technology has been further tested by using metallothionein-growth hormone fusion genes. A DNA fragment containing the promoter of the mouse *mt*-1 gene fused to the structural gene of rat growth hormone was microinjected into the pronuclei of fertilised mouse eggs (108). Seven out of twenty one mice that developed from these eggs carried the fusion gene and six of these grew significantly larger than their littermates.

5.4.2 *Glucocorticoid-responsive Genes*

Steroid hormones are thought to function through a common molecular mechanism in which the hormone first associates with a tissue-specific, cytoplasmic receptor protein. This interaction causes an alteration in the receptor that increases its affinity for binding sites in the cell nucleus. The final result of steroid action is to alter the pattern of proteins synthesised by the cell, probably by altering the transcription of a few specific genes. The introduction of hormonally-responsive genes into cells provides an experimental system to determine whether inducibility is a property inherent in defined nucleotide sequences. For example, mouse mammary tumor virus (MMTV), a retrovirus, is responsive to glucocorticoid stimulation. Cloned DNA from integrated or unintegrated provirus DNA, introduced into mouse LATK$^-$ cells by co-transformation with the HSV *tk* gene, was found to be transcribed into normal viral RNA transcripts. The intracellular levels of these transcripts were increased after exposure of the cells to dexamethasone, a synthetic glucocorticoid hormone (109,110). When fused to either the gene coding for the transforming protein of Harvey murine sarcoma virus or mouse dihydrofolate reductase, the MMTV LTR sequence conferred dexamethasone inducibility (111,112). This strongly suggests that the sequences necessary for hormonal control of MMTV map within the viral LTR.

Similar experiments have been carried out to analyse the regulatory sequences necessary for the hormonal control of the rat liver α_{2u}-globulin gene (113) and the human growth hormone gene (114).

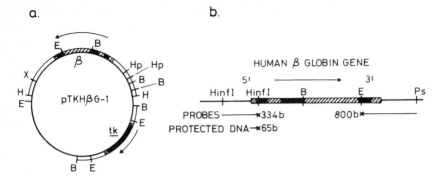

Figure 15. (a) Recombinant plasmid pTKHβG-1 containing the human β-globin and the HSV-1 *tk* genes. This recombinant has been described in detail elsewhere (2). Exons are represented as solid bars and introns and untranslated regions as hatched bars. β is the human β-globin gene. The arrow indicates the direction of transcription. (b) Restriction map of the human β-globin gene [shown with opposite polarity to that in (a)]. Also shown are the 800 nucleotides long DNA probe used for Northern blot hybridisation analysis of human β-globin RNA transcripts (*Figure 16*), the 334 nucleotides long DNA probe used for nuclease S1 mapping of these transcripts (see *Figure 17* legend) and the 65 nucleotides long DNA fragment of this probe protected from nuclease S1 by the human β-globin RNA transcript (*Figure 17*). The maps are not drawn to scale. E = *Eco*RI, H = *Hind*II, B = *Bam*HI, X = *Xba*I, Hp = *Hpa*I, Ps = *Pst*I.

5.4.3 *Globin Genes*

The globin gene family provides an excellent model system for the study of tissue-specific gene expression and the modulation of gene expression during development. In both mice and humans, the β-major globin gene is normally only expressed in differentiated erythroid cells, but when globin genes containing up to 1 kb of DNA sequence 5′ to the mRNA cap site were introduced into mouse fibroblast cells using the calcium phosphate technique, they were expressed (115,116). Therefore, some element of normal control was missing from the exogenous globin genes introduced. Nonetheless, in some systems, induction of the introduced globin genes has been observed during erythroid differentiation (2). For example, Friend erythroleukaemia cells are permanent cell lines derived from Friend virus-induced murine tumours. They are erythroid precursors which, when treated with chemical inducers such as DMSO or hexamethylene bisacetamide (HMBA), differentiate to more mature cells which express large amounts of globin (mainly adult globin). When the human β-globin gene linked to the HSV *tk* gene (*Figure 15* and ref. 2) was introduced into TK⁻ Friend cells, human β-globin mRNA was correctly initiated and terminated (D.A. Spandidos, J. Grindlay and J. Paul, unpublished results). Induction with HMBA resulted in a 20- to 50-fold increase in globin-specific RNA (*Figure 16*). Furthermore, the site of initiation of the globin RNA transcripts was examined by mapping with nuclease S1. This technique is described in detail in Chapter 5, Section 5.2. Briefly it involves hybridisation of the RNA transcripts with a suitable radiolabelled DNA probe, in this case a 334 nucleotide long DNA fragment which is complementary to the 5′ end of the human β-globin gene and some 5′-flanking sequences (see

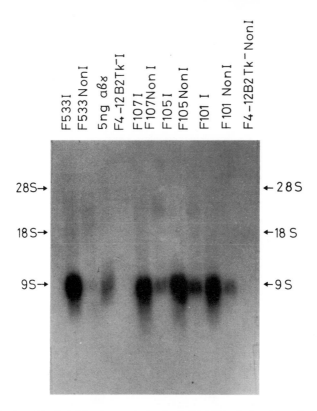

Figure 16. Northern blot hybridisation analysis of poly(A)$^+$ RNA isolated from induced and non-induced Friend erythroleukaemia cells. F4-12B2TK$^-$ cells were transformed with recombinant plasmid pTKHβG-1 and yielded cell lines F101, F105 and F107. Similarly, transformation with pTKHβGHp-3 yielded cell line F533. Poly(A)$^+$ RNA was isolated from 200 μg total RNA of recipient (F4-12B2TK$^-$) and transformed cells (F101, F105, F107 and F533) each either un-induced (denoted as NonI) or induced (I) for 3 days with 3 mM HMBA. The RNAs were fractionated on a 1% formaldehyde-agarose gel then blotted onto a nitrocellulose filter. The *Eco*RI-*Pst*I 800 nucleotides long DNA fragment containing the 3' region of the human β-globin gene and its 3' extragenic flanking region (see *Figure 15b*) was labelled *in vitro* with ^{32}P by nick-translation and used as a hybridisation probe. α, β, γ globin RNA (5 ng) isolated from foetal exchange blood served as a marker for globin 9S RNA. The resulting autoradiogram is shown. The positions of 28S and 18S rRNA markers were determined by staining the gel with ethidium bromide.

Figure 15b). The hybrids are then digested with nuclease S1, which digests only single-stranded DNA and not DNA in hybrids, and resistant DNA fragments are sized by gel electrophoresis. If the RNA transcript has the correct 5' end, it will hybridize to and thereby protect a 65-nucleotide portion of the DNA probe from digestion (see *Figure 15b*), the rest of the DNA remaining single-stranded and hence is destroyed by the nuclease. As shown in *Figure 17*, the DNA fragment protected by the globin-specific RNA transcripts was 65 nucleotides long and hence these transcripts had the correct 5' end and must have been transcribed from the correct initiation site. Similar results have been obtained with the human ϵ-globin gene (2,118, 119).

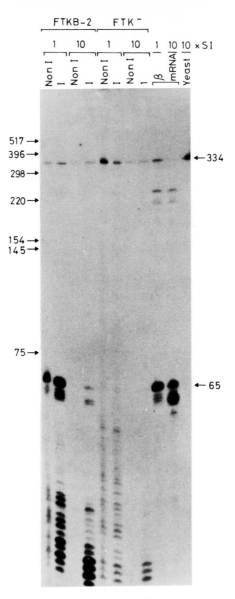

Figure 17. Mapping of the 5′ end of human β-globin mRNA in mouse erythroleukaemic cells using nuclease S1. A single-stranded *Hinf*I fragment (334 nucleotides long) which spans the 5′ end of the human β-globin gene (*Figure 15b*) was ³²P-labelled at the 5′ end and hybridised to total RNA from non-transformed recipient F4-12B2TK⁻ cells (FTK⁻) or these cells transformed with pTKHβG-1 DNA (FTKB-2 cells). Nuclease S1 analysis was performed as previously described (ref. 117; see also Chapter 2, Section 5.2.2 and Chapter 5, Section 5.2). Two different concentrations of nuclease S1 (1000 and 10 000 units) were used and are indicated as 1 or 10 above the lanes of the autoradiogram shown. NonI = non-induced; I = induced cells. Human β-globin mRNA from foetal exchange blood and yeast RNA were used as positive and negative controls, respectively. The lengths of the DNA fragments protected by the RNA from nuclease S1 digestion are given in base pairs at the right-hand side of the autoradiogram whereas the sizes of DNA markers (*Hinf*I fragments of plasmid pBR322) are given on the left-hand side.

44

5.4.4. *Other Inducible Genes*

Many other inducible gene systems are known which are under study using very similar techniques to those described above. They include interferons [inducible by virus infection or treatment with poly(I), poly(C); refs. 45,120,121], heat-shock proteins (122 – 124), immunoglobulin genes (125) and histocompatibility-locus antigens (126,127) as well as virus-coded genes such as the HSV-coded immediate early gene (128) and the HSV *tk* gene itself (129,130).

6. ACKNOWLEDGEMENTS

The authors are indebted to the Cancer Research Campaign of Great Britain for support, to Jas Lang for stimulating discussion and permission to quote unpublished observations, to Joan Grindley for the nuclease S1 mapping experiments quoted and to Ephraim Kam for *Figures 6b* and *c*.

7. REFERENCES

1. Graham,F.L. and van der Eb,A.J. (1973) *Virology,* **52**, 456.
2. Spandidos,D.A. and Paul,J. (1982) *EMBO J.,* **1**, 15.
3. McKnight,S.L. (1982) *Cell,* **31**, 355.
4. Spandidos,D.A., Harrison,P.R. and Paul,J. (1982) *Exp. Cell Res.,* **141**, 149.
5. Birnboim,H.C. and Doly,J. (1979) *Nucleic Acids Res.,* **7**, 1513.
6. Ishiura,M., Hirose,S., Uchida,T., Hamada,Y., Suzuki,Y. and Okada,Y. (1982) *Mol. Cell Biol.,* **2**, 607.
7. Gross-Bellard,M., Oudet,P. and Chambon,P. (1973) *Eur. J. Biochem.,* **36**, 32.
8. Yoder,J.I. and Ganesan,A.T. (1983) *Mol. Cell Biol.,* **3**, 956.
9. Wigler,M,. Pellicer,A., Silverstein,S. and Axel,R. (1978) *Cell,* **14**, 725.
10. Spandidos,D.A. and Wilkie,N.M. (1983) *EMBO J.,* **2**, 1193.
11. Stow,N.D. and Wilkie,N.M. (1976) *J. Gen. Virol.,* **33**, 447.
12. Fraley,R., Subramani,S., Berg,P. and Papahadjopoulos,D. (1980) *J. Biol. Chem.,* **255**, 10431.
13. Farber,F.E. and Eberle,B. (1976) *Exp. Cell Res.,* **103**, 15.
14. Wigler,M., Sweet,R., Sim,G.K., Wold,B., Pellicer,A., Lacy,E., Maniatis,T., Silverstein,S. and Axel,R. (1979) *Cell,* **16**, 777.
15. Vaheri,A. and Pagano,J.S. (1965) *Virology,* **27**, 435.
16. McCutchan,J.H. and Pagano,J.S. (1965) *J. Natl. Cancer Inst.,* **41**, 351.
17. Warden,D. and Thorne,H.V. (1968) *J. Gen. Virol.,* **3**, 371.
18. Pagano,J.S. (1970) *Prog. Med. Virol.,* **12**, 1.
19. Sompayral,L.M. and Danna,K.J. (1981) *Proc. Natl. Acad. Sci. USA,* **78**, 7575.
20. Chu,G. and Sharp,P.A. (1981) *Gene,* **13**, 197.
21. Diacumakos,E.G. (1973) in *Methods in Cell Biology,* Vol. 7, Prescott,D.M. (ed.), Academic Press, NY and London, p. 287.
22. Graessmann,M. and Graessmann,A. (1976) *Proc. Natl. Acad. Sci. USA,* **73**, 366.
23. Stacey,D.W. and Alfrey,V.G. (1976) *Cell,* **9**, 725.
24. Capecchi,M. (1980) *Cell,* **22**, 479.
25. Yamamoto,F., Furusawa,M., Furusawa,I. and Obingta,M. (1982) *Exp. Cell Res.,* **142**, 79.
26. Wong,T.K., Nicolau,C. and Hofschneider,P.H. (1980) *Gene,* **10**, 87.
27. Schaefer-Ridder,M., Wong,Y. and Hofschneider,P.H. (1982) *Science (Wash.),* **215**, 166.
28. Szoka,F.,Jr. and Papahadjopoulos,D. (1978) *Proc. Natl. Acad. Sci. USA,* **75**, 4194.
29. Straubinger,R.M., Hong,K., Friend,D.S. and Papahadjopoulos,D. (1983) *Cell,* **32**, 1069.
30. Schaffner,W. (1980) *Proc. Natl. Acad. Sci. USA,* **77**, 2163.
31. Rassoulzadegan,M., Binetruy,B. and Guzin,F. (1982) *Nature,* **295**, 257.
32. Spandidos,D.A. and Siminovitch,L. (1977) *Proc. Natl. Acad. Sci. USA,* **74**, 3480.
33. Miller,C.L. and Ruddle,F.H. (1978) *Proc. Natl. Acad. Sci. USA,* **75**, 3346.
34. Mukherjee,A.B., Orloff,S., Butler,J.D.B., Triche,T., Lalley,P. and Schulman,J.D. (1978)

Proc. Natl. Acad. Sci. USA, **75**, 1361.
35. Slilaty,S.N. and Aposhian,H.V. (1983) *Science (Wash.),* **220**, 725.
36. Neumann,E., Schaefer-Ridder,M., Wang,Y. and Hofschneider,P.H. (1982) *EMBO J.,* **1**, 841.
37. Kit,S., Dubbs,D.R., Piekarski,I.J. and Hsu,T.C. (1963) *Exp. Cell Res.,* **31**, 297.
38. Bacchetti,S. and Graham,F.L. (1977) *Proc. Natl. Acad. Sci. USA,* **74**, 1590.
39. Topp,W.C. (1981) *Virology,* **113**, 408.
40. Szybalska,E. and Szybalski,W. (1962) *Proc. Natl. Acad. Sci. USA,* **42**, 2026.
41. Littlefield,J.W. (1964) *Science (Wash.),* **145**, 709.
42. Jamieson,A.T. and Subak-Sharpe,J.H. (1974) *J. Gen. Virol.,* **24**, 481.
43. Spandidos,D.A., Harrison,P.R. and Paul,J. (1981) *Biosci. Rep.,* **1**, 911.
44. Colbere-Garapin,F., Horodniceanu,F., Kourilsky,F. and Garapin,A.C. (1981) *J. Mol. Biol.,* **150**, 1.
45. Canaani,D. and Berg,P. (1982) *Proc. Natl. Acad. Sci. USA,* **79**, 5166.
46. Gorman,C.M., Moffat,L.F. and Howard,B.H. (1982) *Mol. Cell Biol.,* **2**, 1044.
47. Gorman,C.M., Merlino,G.T., Willingham,M.C., Pastan,I. and Howard,B.H. (1982) *Proc. Natl. Acad. Sci. USA,* **79**, 6777.
48. Laimins,L.A., Khoury,G., Gorman,C., Howard,B. and Gruss,P. (1982) *Proc. Natl. Acad. Sci. USA,* **79**, 6453.
49. Mulligán,R.C. and Berg,P. (1980) *Science (Wash.),* **209**, 1422.
50. Mulligan,R.C. and Berg,P. (1981) *Proc. Natl. Acad. Sci. USA,* **78**, 2072.
51. Subramani,S., Mulligan,R. and Berg,P. (1981) *Mol. Cell Biol.,* **1**, 854.
52. Shimke,R.T., Brown,P.C., Kaufman,R.J., McGrogan,M. and Slate,D.L. (1981) *Cold Spring Harbor Symp. Quant. Biol.,* **45**, 785.
53. Kaufman,R.J. and Sharp,P.A. (1982) *J. Mol. Biol.,* **159**, 601.
54. O'Hare,K., Benoist,C. and Breathnach,R. (1981) *Proc. Natl. Acad. Sci. USA,* **78**, 1527.
55. Tooze,J., ed. (1980) *DNA Tumor Viruses,* (2nd Edn.), published by Cold Spring Harbor Laboratory Press, NY.
56. Weiss,R.A., Teich,N., Varmus,H.E. and Coffin,J.M., eds. (1982) *RNA Tumor Viruses,* published by Cold Spring Harbor Laboratory Press, NY.
57. Spandidos,D.A. and Siminovitch,L. (1978) *Nature,* **271**, 259.
58. Spandidos,D.A. and Siminovitch,L. (1977) *Cell,* **12**, 675.
59. Shih,C., Shilo,B.-Z., Goldfarb,M.P., Dannenberg,A. and Weinberg,R.A. (1979) *Proc. Natl. Acad. Sci. USA,* **76**, 5714.
60. Krontiris,T.G. and Cooper,G.M. (1981) *Proc. Natl. Acad. Sci. USA,* **78**, 1181.
61. Smith,B.L., Anisowicz,A., Chodosh,L.A. and Sager,R. (1982) *Proc. Natl. Acad. Sci. USA,* **79**, 1964.
62. Padhy,L.C., Shih,C., Cowing,D., Finkelstein,R. and Weinberg,R.A. (1982) *Cell,* **28**, 865.
63. Goldfarb,M., Shimizu,K., Perucho,M. and Wigler,M. (1982) *Nature,* **296**, 404.
64. Parada,L.F., Tabin,C.J., Shih,C. and Weinberg,R.A. (1982) *Nature,* **297**, 474.
65. Santos,E., Tronick,S.R., Aaronson,S.A., Pulciani,S. and Barbacid,M. (1982) *Nature,* **298**, 343.
66. Hall,A., Marshall,C.J., Spurr,N.K. and Weiss,R.A. (1983) *Nature,* **303**, 396.
67. Goubin,G., Goldman,D.S., Luse,J., Neiman,P.E. and Cooper,G.M. (1983) *Nature,* **302**, 114.
68. Gluzman,Y., ed. (1982) *Eukaryotic Viral Vectors,* published by Cold Spring Harbor Laboratory Press, NY.
69. Rigby,P.W.J. (1983) *J. Gen. Virol.,* **64**, 255.
70. Goff,S.P. and Berg,P. (1976) *Cell,* **9**, 695.
71. Mulligan,R.C., Howard,B.H. and Berg,P. (1979) *Nature,* **277**, 105.
72. Gluzman,Y. (1981) *Cell,* **23**, 175.
73. Mellon,P., Parker,V., Gluzman,Y. and Maniatis,T. (1981) *Cell,* **27**, 279.
74. Gluzman,Y., Reichl,H. and Scolnick,D. (1982) in *Eukaryotic Viral Vectors,* Gluzman,Y. (ed.), Cold Spring Harbor Laboratory Press, NY, p.187.
75. Berkner,K.L. and Sharp,P.A. (1982) in *Eukaryotic Viral Vectors,* Gluzman,Y. (ed.), Cold Spring Harbor Laboratory Press, NY, p.193.
76. Spaete,R.R. and Frenkel,N. (1982) *Cell,* **30**, 295.
77. Panicali,D. and Paoletti,E. (1982) *Proc. Natl. Acad. Sci. USA,* **79**, 4927.
78. Wei,C.M., Gibson,M., Spear,P.G. and Scolnick,E.M. (1981) *J. Virol.,* **39**, 935.
79. Shimotohno,K. and Temin,H.M. (1981) *Cell,* **26**, 67.
80. Tabin,C., Hoffman,J., Goff,S. and Weinberg,R. (1982) *Mol. Cell. Biol.,* **2**, 426.

81. Mann,R., Mulligan,R.C. and Baltimore,D. (1983) *Cell,* **33**. 153.
82. Watanabe,S. and Semin,H.M. (1982) *Proc. Natl. Acad. Sci. USA,* **79**, 5986.
83. Jaenisch,R., Jahner,D., Nobis,P., Simon,I., Lahler,J., Harbers,K. and Grotkopp,D. (1981) *Cell,* **24**, 519.
84. Law,M.F., Lowy,D.R., Dvoretzky,I. and Howley,P.M. (1981) *Proc. Natl. Acad. Sci. USA,* **78**, 2727.
85. Moar,M.H., Campo,M.S., Laird,H.M. and Jarrett,W.F.H. (1981) *J. Virol.,* **39**, 945.
86. Moar,M.H., Campo,M.S., Laird,H.M. and Jarrett,W.F.H. (1981), *Nature,* **293**, 749.
87. Lowy,D.R., Dvoretzky,I., Shober,R., Law,M.F., Engel,L. and Howley,P.M. (1980) *Nature,* **287**, 72.
88. Sarver,N., Gruss,P., Law,M.F., Khoury,G. and Howley,P.M. (1981) *Mol. Cell. Biol.,* **1**, 486.
89. Campo,M.S. and Spandidos,D.A. (1983) *J. Gen. Virol.,* **64**, 549.
90. Zinn,K., Mellon,P., Ptashne,M. and Maniatis,T. (1982) *Proc. Natl. Acad. Sci. USA,* **79**, 4897.
91. Di Maio,D., Treisman,R. and Maniatis,T. (1982) *Proc. Natl. Acad. Sci. USA,* **79**, 4030.
92. Sekiguchi,T., Nishimoto,T., Kai,R. and Sekiguchi,M. (1983) *Gene,* **21**, 267.
93. Hirt,B. (1967) *J. Mol. Biol.,* **26**, 365.
94. Mandel,M. and Higa,A. (1970) *J. Mol. Biol.,* **53**, 154.
95. Kishnec,S.R. (1978) in *Genetic Engineering*, Boyer,H.B. and Nicosia,S. (eds.), Elsevier/ North-Holland, Amsterdam, p. 17.
96. Gilmour,R.S., Gow,J.W. and Spandidos,D.A. (1982) *Biosci. Rep.,* **2**, 1031.
97. McKnight,S.L., Gavis,E.R. and Kingsbury,R. (1981) *Cell,* **25**, 385.
98. Lang,J.C., Wilkie,N.M. and Spandidos,D.A. (1983) *J. Gen. Virol.,* **64**, 2697.
99. Perucho,M., Hanahan,D. and Wigler,M. (1980) *Cell,* **22**, 309.
100. Kretchmer,P.J., Browman,A.H., Huberman,M.H., Souders-Haigh,L., Killos,L. and Anderson,W.F. (1981) *Nucleic Acids Res.,* **9**, 6199.
101. Rayes,G.R., McLane,N.W. and Hayward,G.S. (1982) *J. Gen. Virol.,* **60**, 209.
102. Campo,M.S., Spandidos,D.A., Lang,J. and Wilkie,N.M. (1983) *Nature,* **303**, 77.
103. Banerji,J., Rusconi,S. and Schaffner,W. (1981) *Cell,* **27**, 299.
104. Moreau,P., Hen,R., Wasylyk,B., Everett,R., Gaub,M.P. and Chambon,P. (1981) *Nucleic Acids Res.,* **9**, 6047.
105. Mayo,K.E., Warren,R. and Palmiter,R.D. (1982) *Cell,* **29**, 99.
106. Brinster,R.L., Chen,H.Y., Warren,R., Sarthy,A. and Palmiter,R.D. (1982) *Nature,* **296**, 39.
107. Palmiter,R.D., Chess,H.Y. and Brinster,R.L. (1982) *Cell,* **29**, 701.
108. Palmiter,R.D., Brinster,R.L., Hammer,R.E., Trumbauer,M.E., Rosenfield,M.G., Birnberg, N.C. and Evans,R.M. (1982) *Nature,* **300**, 611.
109. Buetti,E. and Diggelmann,H. (1981) *Cell,* **23**, 335.
110. Hynes,N.E., Kennedy,N., Rahmsdorf,V. and Groner,B. (1981) *Proc. Natl. Acd. Sci. USA,* **78**, 2038.
111. Huang,A.L., Ostrowski,M.C., Berard,D. and Hager,G.L. (1981) *Cell,* **27**, 245.
112. Lee,F., Mulligan,R., Berg,P. and Ringold,G. (1981) *Nature,* **294**, 228.
113. Kurtz,D.T. (1981) *Nature,* **291**, 629.
114. Robins,D.M., Pack,I., Seeburg,P.H. and Axel,R. (1982) *Cell,* **29**, 623.
115. Mantei,N., Boll,W. and Weissmann,C. (1979) *Nature,* **290**, 26.
116. Wold,B., Wigler,M., Lacy,E., Maniatis,T., Silverstein,S. and Axel,R. (1979) *Proc. Natl. Acad. Sci. USA,* **76**, 5684.
117. Alan,M., Grindlay,G.J., Stefani,L. and Paul,J. (1982) *Nucleic Acids Res.,* **10**, 5133.
118. Paul,J., Allan,M., Grindlay,J. and Spandidos,D. (1982) in *Biochemistry of Differentiation and Morphogenesis,* Jaenicke,L. (ed.), 33 Coloquium Mosbach, Springer-Verlag, Berlin, Heidelberg, p. 142.
119. Paul,J. and Spandidos,D.A. (1982) in *Stability and Switching in Cellular Differentiation. Advances in Experimental Medicine and Biology,* Vol. **158**, Clayton,R.M. and Truman,D.E.S. (eds.), Plenum Press, NY, p. 89.
120. Mantei,N. and Weissmann,C. (1982) *Nature,* **297**, 128.
121. Hauser,H., Gross,G., Bruns,W., Hochkeppel,H.-K., Mayr,V. and Colins,J. (1982) *Nature,* **297**, 650.
122. Corces,V., Pellicer,A., Axel,R. and Meselson,M. (1981) *Proc. Natl. Acad. Sci. USA,* **78**, 7038.
123. Burke,J.F. and Ish-Horowicz,D. (1982) *Nucleic Acids Res.,* **10**, 3821.
124. Pelham,H.R.B. (1982) *Cell,* **30**, 517.

125. Rice,D. and Baltimore,D. (1982) *Proc. Natl. Acad. Sci. USA,* **79**, 7862.
126. Mellor,A.L., Golden,L., Weiss,E., Bullman,H., Hurst,J., Simpson,E., James,R.F.L., Townsend,A.R.M., Taylor,P.M., Schidt,W., Ferluga,J., Leben,L., Santamaria,M., Atfield, G., Festernstein,H. and Flavell,R.A. (1982) *Nature,* **298**, 529.
127. Rabourdin-Combe,C. and March,B. (1983) *Nature,* **303**, 670.
128. Post,L.E., Mackem,S. and Roizman,B. (1981) *Cell,* **24**, 555.
129. Wilkie,N.M., Clements,J.B., Boll,W., Mantei,N., Lonsdale,D. and Weissmann,C. (1979) *Nucleic Acids Res.,* **7**, 859.
130. Zipser,D., Lipsich,L. and Kwoh,J. (1981) *Proc. Natl. Acad. Sci. USA,* **78**, 6276.

CHAPTER 2

Expression of Exogenous DNA in Xenopus Oocytes

ALAN COLMAN

1. INTRODUCTION

The oocyte nucleus contains enough of the three eukaryotic RNA polymerases to furnish the needs of the developing embryo at least until the 30 000 cell stage. This maternal store can also be used to transcribe exogenous DNAs injected into the oocyte nucleus, an ability which was first demonstrated[1] by Gurdon and Mertz in 1977 (1). These early experiments demonstrated only that the injected DNAs were transcribed. However, subsequent experiments involving injection of DNA coding for 5S rRNA showed that the transcripts were initiated and terminated at the correct DNA residues (3). This led to a spate of DNA injection experiments (see reviews 4 – 6) some of which are listed in *Table 1*. The results of these numerous studies can be summarised as follows.

(i) *RNA polymerase III transcripts.* The striking demonstration of faithful transcription (by RNA polymerase III) following injection of 5S rRNA genes was soon followed by similar success with tRNA genes (16). Moreover, it has been shown that the complicated post-transcriptional modifications (e.g. base modifications, RNA processing and, in some cases, RNA splicing) which normally accompany the maturation of these primary tRNA transcripts also occur to the transcription products of tRNA genes injected into oocytes (17). It has even proved possible to inject purified tRNA precursors into oocytes and obtain correct processing (6).

(ii) *RNA polymerase II transcripts.* When protein-coding genes lacking introns (e.g. histone genes) are injected, some transcription (by RNA polymerase II) begins and terminates at the correct sites, and a stable transcript is exported to the cytoplasm where it is translated (12). Incorrect transcripts, especially those from the non-coding strand, are degraded within the nucleus and never exported. In contrast, the expression of injected protein-coding genes containing introns has been less successful (6). Either faulty initiation of transcription or faulty splicing could be responsible (see below).

The injection of genes normally transcribed by polymerase II into oocyte nuclei is not always followed by transcriptional initiation. The response depends on the particular gene injected. At present, the exact reasons for this

[1]An earlier demonstration of specific transcription of DNA injected into *Xenopus* oocytes (2) did not identify the cellular site of transcription.

Table 1. Transcription of Injected Genes by *Xenopus* Oocytes[a].

Source of DNA	Gene injected	RNA poly-merase[b]	Correct transcripts[c]	Proteins synthesised	Reference
Xenopus	rDNA	I	Some[d]	−	7
SV40	Viral DNA	II	19S and 25S	Viral proteins and T antigens	8
Polyoma virus	Viral DNA	II	19S and 25S	Viral proteins and T antigens	9,10
Adenovirus	Whole viral genome	II	−	Several early proteins	11
Sea urchin	Histones	II	H2A, H2B, H3	H2A, H2B, H3	12
Drosophila	Histones	II	−	H2A	9
Xenopus	Histones	II	H4	H4, H2A, H2B, H3, H1	13
Herpes virus	Thymidine kinase	II	Some[d]	Thymidine kinase	14
Chicken	Ovalbumin	II	−[e]	Ovalbumin	15
Xenopus	5S rRNA	III	5S rRNA	−	3
Xenopus	tRNA	III	tRNA and precursors	−	16
Yeast	tRNA	III	tRNA and precursors	−	17
Nematode	tRNA	III	tRNA and precursors	−	17

[a]This table is modified and extended from ref. 6.
[b]The RNA polymerase indicated is that which normally transcribes the gene *in vivo* and does not necessarily imply that the RNA polymerase used by the *Xenopus* oocyte was determined.
[c]The absence of an entry in this column indicates that no data were available from the referenced source.
[d]Some transcripts were initiated from the correct site as judged by mapping with nuclease S1.
[e]No correct length transcripts were detected.

variability are unclear although it is evident that structural aspects of the 5'-flanking DNA regions of the studied gene are critical (D.A.Melton, personal communication). However, when an injected gene is transcriptionally initiated, it appears that the site of initiation corresponds to the normal site of initiation of that gene *in vivo* (18 – 20). This demonstration has facilitated studies of DNA sequences upstream of the site of transcriptional initiation, which influence initiation. Thus, Birnstiel and his colleagues (18,21) have delineated several defined regions of DNA upstream of sea urchin histone genes where mutation has specific effects on both the exact position of transcriptional initiation (the so-called 'cap' site) and the amount of initiation. McKnight and Kingsbury (20) have looked for more subtle effects, caused by sequence modification, on the transcription of the *Herpes simplex* thymidine kinase gene, using a competition assay where the unmodified gene and modified derivatives were co-injected into oocyte nuclei. This regime ensured that observed differences in the expression of modified genes were not attributable to differences in volumes injected, oocyte variation, etc. The detailed methods involved in the analysis of initiation of transcription are described later in this chapter.

Similar strategies and analytical techniques have been used to identify and investigate regions involved in transcriptional termination and to identify the

3′ end of RNA transcripts (12,19). However, it is a feature of the oocyte system that only steady-state transcript populations are ever examined and therefore it is very difficult to be certain that the 5′ and 3′ ends of transcripts correspond to the sites of transcriptional initiation and termination, respectively.

As described above, studies using injected tRNA genes have established that the oocyte nucleus will splice tRNA genes containing introns very efficiently. It is now clear that splicing will also occur with transcripts of genes transcribed by RNA polymerase II. Although it has been known for some time that *some* correct splicing must occur, since the correct translation products have been obtained from genes whose protein-coding sequences are interrupted by introns (8,15), it is only recently that the fidelity of splicing has been examined in detail (19). The proportion of primary transcripts which are correctly spliced is variable but the fact that it does occur to a detectable extent makes studies on the effect of sequence modification on splicing a realistic possibility.

(iii) *RNA polymerase I transcripts.* Injected 18S and 28S rRNA genes have been successfully transcribed (by RNA polymerase I) in oocytes. The first indication of this was the observation, using electron microscopy, of characteristic transcription complexes on injected DNA (22). Biochemical confirmation of this transcription and proof that RNA polymerase I is involved have been recently provided by Moss (7).

Injection of 5S rRNA genes into the oocyte nucleus was the first stage towards what is undoubtedly the best understood example of eukaryotic gene expression at the transcriptional level (3,23,24). However, much of this elegant work was accomplished using cell-free systems derived from isolated *Xenopus* nuclei. Similarly, it is likely that cell-free systems for *in vitro* transcription will prove more satisfactory in the long term for looking at the control of transcription of genes transcribed by RNA polymerase II since the amount of transcription of this class of injected genes is rather variable in oocytes. Nevertheless, using oocytes, considerable insight has been obtained into the control sequences involved in the transcription of certain eukaryotic genes, for example histone genes (12,18,21) and the thymidine kinase gene (20). Moreover the oocyte still retains a decided advantage for studies of splicing and post-transcriptional processing and transport since these events occur poorly or not at all in currently available cell-free systems. The author therefore anticipates the continued use of the *Xenopus* oocyte system in studies of post-transcriptional processing and where the special advantage of the oocyte as a translational system is exploited for coupled transcription-translation assays (Section 7).

2. OBTAINING AND PREPARING OOCYTES FOR MICROINJECTION

Xenopus laevis frogs are readily available from a number of commercial sources. They can be easily maintained in the laboratory in water in large, covered tanks at 19−21°C and require feeding only twice per week. Oocytes are obtained from a mature *Xenopus* female by either partial or complete removal of the ovary followed by manual dissection or enzymatic 'stripping' to liberate the oocytes from their follicles. The advantage of only partial ovariectomy is that it does not kill the frog which can then be re-used during a series of experiments. Details of the purchase and maintenance of *X. laevis* frogs, removal of ovarian tissue and the preparation of oocytes for injection are given in Chapter 10 (Section 2).

3. MICROINJECTION OF OOCYTE NUCLEI WITH DNA

3.1 Essential Equipment

The essential equipment for microinjection of DNA into *Xenopus* oocyte nuclei consists of a stereomicroscope, micromanipulator, syringe, light source, micropipette-making apparatus and a microforge. Chapter 10 (Section 3) describes criteria for the selection of suitable equipment as well as directions for the construction, calibration and storage of micropipettes.

3.2 Conformation of DNA used for Injection

With the advent of recombinant DNA technology, most DNAs for injection can be obtained in circular form. Circular DNA is the best template for injection (6); the injection of linear DNA molecules smaller than 5000 bp yields 10- to 20-fold less RNA than the same molecules in circular form, and often the transcription is unfaithful. If possible, the eukaryotic DNA sequence should be excised from the recombinant plasmid DNA and re-circularised using DNA ligase. For genes normally transcribed by RNA polymerase II, this manipulation can avoid the problem of read-through by this enzyme from plasmid promoters although this is not always a problem (13). It does not matter whether closed or nicked DNA circles are injected since an efficient repair mechanism exists in the oocyte nucleus. Even single-stranded circular DNA provides an effective template for transcription since it is converted to a double-stranded form within 6 h of injection.

3.3 Stability of Injected DNA

Circular, double-stranded DNA injected into the nucleus of *Xenopus* oocytes is stable for up to several weeks whereas when injected into the cytoplasm, where no transcription occurs, it is degraded within 48 h. Linear DNA is less stable than its circular counterpart in both the cytoplasm and the nucleus. However, the rate of exonucleolytic attack on linear DNA within the nucleus does not account for its poor performance as a template. Further discussion of these interesting observations can be found in reference 6.

3.4 Concentration of DNA used for Injection

For genes normally transcribed by RNA polymerases II or III, maximal synthesis is elicited by an injection of approximately 10 ng of DNA per nucleus (6). In contrast, Moss (7) has reported that this amount of injected ribosomal DNA (transcribed by RNA polymerase I) results in no detectable transcription

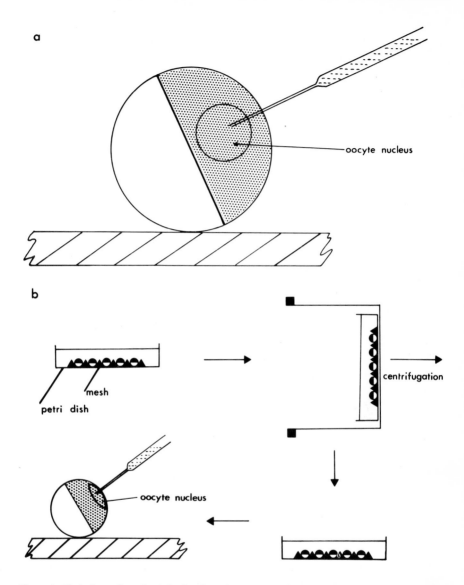

Figure 1. Techniques for microinjection into the oocyte nucleus. **(a)** The relative positions of the micropipette and oocyte when injecting DNA by the 'blind' method (Section 3.6.1). **(b)** Details of the centrifugation procedure for microinjection (Section 3.6.2). The oocytes are supported by nylon netting fixed to the bottom of a Petri dish, with the pigmented half of each oocyte uppermost. The Petri dish is then centrifuged at the bottom of a 1 litre bucket of a swing-out rotor. After centrifugation, the Petri dish is removed and each oocyte is injected by inserting the micropipette into the clearly-visible (flattened) nucleus. Drawings were provided by S.Bhamra.

of the injected DNA. Rather, much lower amounts of ribosomal DNA must be injected; 25 – 50 pg of DNA per nucleus seems optimal. Interestingly, the injection of similarly low amounts of genes transcribed by RNA polymerases II

Table 2. Microinjection of DNA and Labelling of Oocytes.

The DNA to be injected can be dissolved in a variety of buffers, for example 88 mM NaCl, 0.5 mM EDTA, 10 mM Hepes (pH 7.4 with KOH) *or* 10 mM NaCl, 0.5 mM EDTA, pH 7.4. However, dissolution in distilled water yields perfectly satisfactory results.

'Blind' Method

1. Transfer the oocytes to a microscope slide raised up from the microscope stage on a large Petri dish or other suitable container, filled with ice. The use of such a 'cold' stage greatly increases the survival of oocytes after *nuclear* injection.
2. Grasp the oocyte with watchmaker's forceps and rotate the oocyte until the apex of the pigmented half is facing the tip of the micropipette with the long axis of the micropipette at 90° to the equator between the pigmented and yolk hemispheres *(Figure 1a)*.
3. Insert the micropipette into the centre of the pigmented region of the oocyte, taking care not to insert it too far; the position of the oocyte nucleus and its relative dimensions are shown in *Figure 1a*.
4. If the whole unfractionated oocytes are to be analysed after the labelling period, inject up to 50 nl of DNA solution (but not more than this). However, if the oocytes are to be fractionated into nucleus and cytoplasm after labelling, inject no more than 10 nl of DNA (19). To label the RNA transcribed from the injected DNA, co-inject ^{32}P-labelled nucleotide along with the DNA (see Section 4.1). Alternatively, the radioisotope may be injected 24 h after the DNA (see step 6).
5. After injection[a], transfer the oocytes to Petri dishes (2.5 cm diameter) on ice containing ~5 ml of modified Barths' saline supplemented with 10 μg/ml penicillin and 10 μg/ml streptomycin[b]. Check that the oocytes are submerged and, after 30 min on ice, incubate them at 20°C.
6. If the RNA transcribed from the injected DNA is to be labelled and the ^{32}P-labelled nucleotide was not co-injected (step 4), then inject the nucleotide into the oocyte cytoplasm (Chapter 10, Section 3.3.3) 24 h after injection of the DNA.

   If the proteins encoded by the injected DNA are to be labelled, *either* inject the radioactive amino acid into the oocyte cytoplasm (Chapter 10, Section 3.3.3) 24 h after injection of the DNA *or* add it to the modified Barths' saline and culture the oocytes in microtitre wells 24 h after injection of the DNA (Chapter 10, Section 4).

Centrifugation Method

1. Fix nylon netting (0.8 mm mesh) to the bottom of a plastic Petri dish (5 cm diameter) with chloroform[c].
2. Fill the Petri dish with ~7 ml of modified Barths' saline.
3. Position each oocyte with its pigmented half uppermost.
4. Transfer the dish to the flat bottom of a 1 litre bucket of an MSE Coolspin 6 x 1 litre swing-out rotor or its equivalent.
5. Centrifuge at $450-600$ g for 10 min at 20°C. During this centrifugation *(Figure 1b)*, the pigment granules at the apex of the oocyte are displaced by the rising nucleus and offer an easily-identified target for microinjection *(Figure 2)*.
6. After centrifugation, flush the oocytes from the base of the Petri dish.
7. Transfer the oocytes to a microscope slide on the 'cold' stage of the stereomicroscope as described in step 1 of the 'blind' method.
8. Grasp each oocyte with watchmaker's forceps and position it relative to the micropipette as shown in *Figure 1b*.
9. Insert the micropipette in the nucleus, taking care not to go right through the nucleus.
10. If the whole oocytes are to be homogenised and analysed directly after the labelling period, inject up to 50 nl of DNA solution. However, if the oocytes are to be separated into nuclear and cytoplasmic fractions prior to analysis, inject only 10 nl of DNA solution. Note that the microinjection should be carried out within 45 min of centrifugation since the nucleus slowly settles back into the oocyte.

To label the RNA transcribed from the injected DNA, co-inject ^{32}P-labelled nucleotide along with the DNA (see Section 4.1). Alternatively, the radioisotope may be injected 24 h after the DNA (see step 12).

11. After injection[a], transfer the injected oocytes to modified Barths' saline and incubate them for 24 h as described in step 5 of the 'blind' method.

12. If the RNA transcribed from the injected DNA is to be labelled, and the ^{32}P-labelled nucleotide was not co-injected (step 10), then inject it into the oocyte cytoplasm (Chapter 10, Section 3.3.3) 24 h after injection of the DNA.

   If the proteins encoded by the injected DNA are to be labelled, *either* inject the radioactive amino acid into the oocyte cytoplasm (Chapter 10, Section 3.3.3) 24 h after injection of the DNA *or* add it to modified Barths' saline and culture the oocytes in microtitre wells 24 h after injection of the DNA (Chapter 10, Section 4).

---

[a]A recent report (11) recommends the immersion of oocytes in 90 mM potassium phosphate (pH 7.2), 10 mM NaCl, 1 mM $MgSO_4$ for 90 min following injection to promote wound healing. However the author has not noticed any improvement when using this procedure.
[b]The composition of modified Barths' saline is given in *Table 5* of Chapter 10.
[c]An alternative centrifugation support can be made by casting 1% agarose (in modified Barths' saline) in the Petri dishes to a depth of 0.2 cm and then creating pits by removal of plugs of agarose with a Pasteur pipette.

or III elicits no transcriptional response. The reasons for these phenomena are not clearly understood[2].

3.5 Loading Micropipettes with DNA

The procedure used for loading micropipettes with DNA is the same as that used to load mRNA (Chapter 10, Section 3.3.2). Briefly, a small volume $(1 - 2 \ \mu l)$ of the DNA solution is placed on a strip of 'Parafilm' or 'Nescofilm'. The micropipette is attached to the syringe via flexible polythene tubing and filled with paraffin oil. Using the stereomicroscope to monitor the operation, the micropipette is then filled by sucking up the DNA solution until the aqueous sample/paraffin oil interface is at the end of the field of view.

3.6 Microinjection

Two different methods have been generally used to microinject DNA into oocyte nuclei. The original ('blind') method *(Figure 1a)* involves injecting the DNA into the region of the oocyte normally occupied by the nucleus, the nucleus remaining unseen by the researcher throughout the procedure. The second method, developed by Birnstiel and his associates (16), involves the injection of centrifuged oocytes *(Figure 1b)*. During centrifugation, the nucleus approaches the surface of the oocyte and can be seen by virtue of the dilution of pigment in the surface region. The DNA is then injected directly into the nucleus. Both methods are described here.

3.6.1 *'Blind' Injection Method*

This method is described in *Table 2* whilst *Figure 1a* indicates the normal location of the nucleus during injection and its relative dimensions. Obviously the

[2]These observations regarding the necessity to inject such low amounts of rDNA have been disputed recently (R.Reeder, personal communication).

Figure 2. Appearance of centrifuged oocytes. During centrifugation the nucleus rises to the highest part of the oocyte. When the oocyte is positioned correctly for centrifugation, with its pigmented half uppermost *(Figure 1b)*, the position of the nucleus is indicated by a pale area at the apex of the pigmented surface. As the figure here shows, when the oocyte is *not* positioned correctly for centrifugation, the position of the pale area can vary.

novice will wish to determine the success of this method for injecting into the nucleus. The author has found the following approach to be very worthwhile for training purposes.

(i) Inject approximately 50 nl of 0.2% Trypan Blue [in phosphate-buffered saline (PBS)] into each oocyte, aiming for the nucleus.

(ii) After injection, transfer the oocytes to modified Barths' saline (Chapter 10, *Table 5*) in a Petri dish and, using two pairs of forceps, tear each oocyte apart, making the initial tear in the vegetal (unpigmented) half.

(iii) If the nucleus has been injected, then this will be disclosed by its dark blue staining. A miss will be denoted by a transparent nucleus and a diffusely-stained region of cytoplasm. With a little practice, a success rate of over 80% should be obtained.

3.6.2 *Centrifugation Method*

This procedure is described in *Table 2* and illustrated diagrammatically in *Figure 1b*. *Figure 2* shows the appearance of oocytes after centrifugation and prior to microinjection. Note that although this method facilitates injection of DNA into the oocyte nucleus rather than the cytoplasm, it has the disadvantage that it is easy to push the needle completely through the nucleus since it is considerably flattened by the centrifugation procedure.

4. RADIOACTIVE LABELLING OF INJECTED OOCYTES

4.1 **Labelling the RNA Transcribed from Injected DNA**

For many purposes it is most useful to label RNA transcripts with $[\alpha\text{-}^{32}P]$-nucleotides rather than $[^3H]$nucleosides or $[^3H]$nucleotides. This is because a greater variety of techniques are available for the analysis of ^{32}P-labelled transcripts. However, it is possible that the newly-available ^{35}S-labelled nucleotides will prove to be a desirable alternative to ^{32}P-labelled reagents because of the longer half-life of ^{35}S compared with ^{32}P. Unfortunately, although *Xenopus* oocytes are permeable to nucleosides, they are impermeable to nucleotides. Therefore, ^{32}P-labelled nucleotides (usually $[\alpha\text{-}^{32}P]GTP$ or $[\alpha\text{-}^{32}P]CTP$) must be injected into the oocyte for labelling purposes. Two alternative protocols are possible.

(i)   The ^{32}P-labelled nucleotide is co-injected with the DNA *(Table 2)*, usually approximately 50 nl at a concentration of $10-20$ mCi/ml (>400 Ci/mmol). This represents the injection of approximately 2.5 pmol of nucleotide per oocyte, an amount nearly 10-fold larger than the endogenous pool size (25). With this amount of injected label, about 50 000 d.p.m. should be incorporated into oocyte RNA over a 24 h period. Since the injected oocytes survive for several days and the injected DNA only becomes an efficient template for transcription after its assembly into chromatin, it is advisable to culture the oocytes for 24 h after the co-injection of DNA and ^{32}P-labelled nucleotide.

(ii)   The ^{32}P-labelled nucleotide is injected 24 h after injection *(Table 2)*. Although this makes the experiment more laborious, there is no need to inject the nucleotide into the nucleus since the cytoplasmic and nuclear pools equilibrate readily. Therefore, the second injection can be made into the cytoplasm (see Chapter 10, Section 3.3.3). This procedure has the advantage that the radioactive precursor is more efficiently incorporated compared with the co-injection protocol described above (see Section 4.2).

4.2 **Labelling the Proteins Encoded by Injected DNA**

The various aspects of labelling newly-synthesised proteins in *Xenopus* oocytes are discussed in Chapter 10, Section 4. When the gene injected codes for polyadenylated mRNA, more synthesis of the corresponding protein can be expected if labelling is delayed until 24 h after DNA injection. During this period an accumulation of polyadenylated mRNA will occur, whereas a corresponding accumulation of unstable, non-polyadenylated transcripts does not occur (see Chapter 10, Section 3.3).

As with labelling RNA transcripts (Section 4.1), two alternative labelling protocols are also possible when using amino acids to label the protein products:

(i)   since oocytes are permeable to amino acids, simply adding the radioactive amino acids to the incubation medium results in their incorpora-

Table 3. Isolation of RNA from Injected Oocytes.

1. *Either,* for whole oocytes, homogenise the oocytes at a concentration of $10-30$ oocytes/ml in
   1% SDS, 1.5 mM $MgCl_2$, 10 mM NaCl, 1 mg/ml Proteinase K (Calbiochem), 10 mM
   Tris-HCl, pH 7.6 (16). This homogenization buffer should be pre-incubated at 37°C
   before use;

   or, for separate analysis of nuclear and cytoplasmic RNA from individual oocytes [see
   Section 5.1 (iii)], enucleate each oocyte using method 1 described in Chapter 10, Section 6.1.1. Then homogenise each of the separate nuclear and cytoplasmic fractions in
   the above homogenisation buffer (0.1 ml/10 nuclei or 1.0 ml/30 cytoplasms) but containing 100 μg/ml of *E. coli* tRNA as carrier in the case of nuclei.
2. Incubate at room temperature for 20 min.
3. Add 4 M NaCl stock solution to bring the final concentration of NaCl to 0.3 M.
4. Add an equal volume of phenol-chloroform (1:1, v/v) previously saturated with homogenisation buffer.
5. Mix for 10 sec and then centrifuge at 10 000 g for 1 min.
6. Recover the upper (aqueous) phase and re-extract the organic phase plus interphase with 0.1
   volume of 0.3 M NaCl, 10 mM Tris-HCl, pH 7.6. Mix and centrifuge as in step 5.
7. Pool the two aqueous phases from steps 5 and 6 and re-extract this with an equal volume of
   phenol-chloroform as in steps 4 and 5.
8. Add 2.5 volumes of ethanol to the pooled aqueous phases and leave at -20°C overnight.
9. Recover the RNA by centrifugation (10 000 g for 2 min).
10. Dry the RNA precipitate under vacuum and dissolve it in the buffer of choice, depending on
    the analytical method to be used. If necessary, store the solution of RNA at -20°C prior to
    use.

---

tion into oocyte protein. Concentrations of $0.1-5.0$ mCi of radioactive
amino acid per ml of incubation medium are routine (Chapter 10, Section 4.2);

(ii) alternatively, the radioactive amino acid (50 nl of 10 mCi/ml) can be injected into the oocyte cytoplasm (Chapter 10, Section 4.5).

Analysis of the labelled proteins can be performed as described in Chapter
10 (Section 5).

5. ANALYSIS OF RNA TRANSCRIPTS

5.1 Isolation of RNA

(i) At the end of the incubation period, inspect the oocytes in the culture
dishes. Collect only healthy-looking oocytes (see Chapter 10, *Figure 9* for
photographs of healthy and unhealthy oocytes). Wash the oocytes in
modified Barths' saline.

Either

(ii) For the analysis of total oocyte RNA, homogenise the oocytes and isolate
the RNA as described in *Table 3*. The author prefers to use unfrozen
oocytes for this purpose although prior freezing on dry-ice has been
reported (17).

or

(iii) For the analysis of nuclear and cytoplasmic RNA separately, enucleate
each oocyte using the method described in Chapter 10, Section 6.1.1.
Since artifactual leakage of nuclear transcripts can occur into the

Table 4. Analysis of Specific Transcripts by Hybrid Release.

1. Dissolve complementary DNA (in double- or single-stranded form) in water and heat to 100°C for 10 min. Cool quickly on ice.
2. Filter the DNA through Millipore nitrocellulose filters (HAWP; 1.3 cm diameter) and then bake the filters at 80°C for 2 h.
3. Place each filter onto the flat bottom of a suitably-sized plastic scintillation vial insert. Add ~20−100 μg of oocyte RNA in 200 μl of 0.4 M NaCl, 0.2% SDS, 10 mM Pipes (pH 6.4). Incubate at 50°C for 2 h.
4. Wash each filter six times (1 min each) at 60°C in 1 x SSC[a], twice in 0.2 x SSC at 60°C and then once in 5 mM NaCl, 2 mM EDTA, 20 mM Tris-HCl (pH 7.7) at 50°C.
5. Elute the hybridised RNA transcripts by two successive 1 min incubations at 100°C in 0.2 ml of water. Pool these eluates. Add 10 μg of *E. coli* tRNA as carrier plus 2.5 volumes of ethanol to precipitate the nucleic acid (*Table 3*, step 8).
6. Recover the nucleic acid and dissolve it in buffer (*Table 3*, steps 9 and 10) before analysis by polyacrylamide gel electrophoresis (e.g. see Chapter 4, *Table 8*) or by translation (Chapter 8, Section 2.6.2).

[a]1 x SSC is 0.15 M NaCl, 15 mM trisodium citrate.

cytoplasm in some oocytes, Wickens and Gurdon (19) recommend that each oocyte should be individually analysed for nuclear and cytoplasmic RNA. Such leakages can subsequently be recognised by the presence of 40S ribosomal precursor RNA in the cytoplasmic extracts. RNA is isolated from the individual nuclei and cytoplasms as described in *Table 3*. In practice, these stringent precautions are realistic only if labelled RNA transcripts are being prepared.

5.2 Characterisation of RNA Transcripts

5.2.1 *Labelled RNA Transcripts*

RNA, labelled as described in Section 4.1 and isolated as described in *Table 3*, can be analysed by gel electrophoresis directly or after prior hybridisation to and elution from filters containing specific DNA fragments. This latter procedure (hybrid release) is described in *Table 4*. Some examples of this type of analysis are shown in *Figure 3*.

5.2.2 *Unlabelled RNA Transcripts*

An alternative approach, after the injection of DNA, is to incubate the oocytes in the absence of radioactive nucleotide. Following the extraction of RNA as described in *Table 3*, the unlabelled RNA transcripts can be analysed by a variety of techniques.

(i) *Northern blotting.* This technique, described in detail in *Table 5*, involves the electrophoresis of the RNA transcripts in an agarose gel followed by the transfer of the RNA to a nitrocellulose filter. The filter is then incubated in the presence of an appropriate labelled denatured DNA probe to which the RNA transcript of interest will hybridise. This technique allows the detection and sizing of transcripts containing regions complementary to either or both strands of the DNA probe. With care it can be used in a quantitative manner.

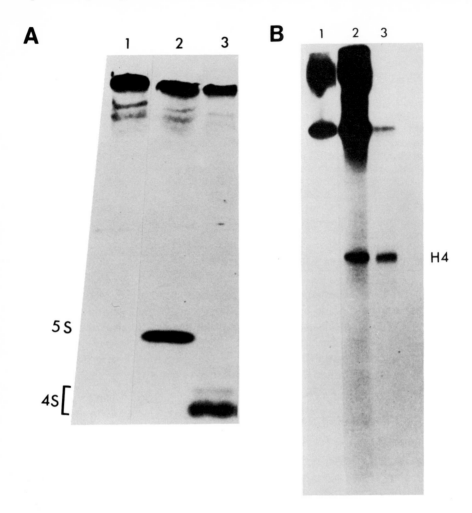

Figure 3. RNA synthesis in oocytes. **(A)** Oocytes were injected with distilled water **(lane 1)**, *Xenopus* DNA coding for 5S rRNA **(lane 2)** or *Xenopus* DNA coding for tRNA **(lane 3)**. [³H]GTP was injected at the same time. After incubation for 24 h, RNA was extracted as indicated in *Table 3* and analysed by electrophoresis in a 15% polyacrylamide gel before fluorography. **(B)** Oocytes were co-injected with *Xenopus* histone H4 DNA and [α-³²P]GTP. After incubation for 24 h, RNA was extracted as in *Table 3*. One aliquot of RNA **(lane 2)** was analysed directly by electrophoresis as in **A**, whilst a second aliquot **(lane 3)** was first hybridised to and then eluted from histone H4 DNA immobilised on a nitrocellulose filter. **Lane 1** shows unfractionated RNA (i.e., corresponding to **lane 2**) from oocytes injected with [α-³²P]GTP alone. These data have been kindly provided by R.Old.

An example of this kind of analysis is shown in *Figure 4*.

Single-stranded DNA probes must be used if analysis of the strand complementarity of the transcript is required. These can be obtained either by electrophoresis of the labelled DNA probe on a denaturing agarose gel and subsequent elution of the individual DNA strands or by using M13 coliphage

1 2 3

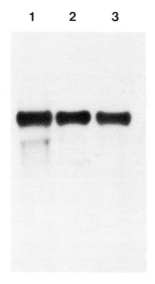

Figure 4. Northern blot analysis of oocyte RNA. Oocytes were each injected with 2 ng rabbit globin mRNA and then RNA was extracted at 0 h **(lane 1)**, 24 h **(lane 2)** and 48 h **(lane 3)** after injection. The RNA was then electrophoresed on a 1.5% formaldehyde-agarose gel, transferred to a nitrocellulose filter, and hybridised to ^{32}P-labelled globin DNA *(Table 5)*.

single-stranded recombinant DNA as a template for the preparation of labelled complementary DNA (28).

(ii) *Identification of the 5′ end of RNA transcripts.* Two methods can be used to identify the position on the injected gene corresponding to the 5′ end of the RNA transcript. These are nuclease S1 mapping and primer extension.

The principle of *nuclease S1 mapping* is described in detail in Chapter 5 Section 5.2.1. Two approaches are possible: the first uses an unlabelled DNA probe to analyse radiolabelled RNA transcripts (Chapter 3, Section 3.3), the second uses a radiolabelled DNA probe to analyse unlabelled RNA transcripts. The latter approach is more commonly used. Briefly, the procedure first involves hybridisation of the RNA transcripts to a radiolabelled DNA probe complementary to the 5′ end of the gene and some 5′-flanking sequence. Then the mixture is incubated with nuclease S1 which digests only single-stranded DNA but not DNA in hybrids so only that portion of the DNA probe hybridised to the RNA transcript will be resistant to degradation. The nuclease-resistant DNA fragment is subsequently sized by gel electrophoresis. By reference to the size of the original DNA probe, the size of the resistant DNA fragment then permits identification of the site within the DNA probe which corresponds to the 5′ end of the RNA transcript. Several practical variations of this basic protocol exist for nuclease S1 mapping using radiolabelled DNA

Table 5. Northern Blot Analysis.

1. Prepare the following stock solutions:

   Mops Gel Buffer

   The composition of this buffer is 1 mM EDTA, 5 mM sodium acetate, 20 mM Mops (pH 7.0). The buffer is required as 10- and 14.3-fold concentrated solutions. These are prepared as follows:

	10 x Mops Buffer	14.3 x Mops Buffer
Mops	41.9 g	5.93 g
sodium acetate	4.1 g	0.58 g
EDTA.Na$_2$	3.7 g	0.50 g
H$_2$O	to 1 litre	to 100 ml

   Loading Buffer

   This is prepared by mixing:
   500 μl deionised formamide[a]
   70 μl 14.3 x Mops buffer
   180 μl filtered formaldehyde[b]

   Hybridisation Buffer

   The composition of this buffer is:
   50% (v/v) deionised formamide[a]
   50 mM sodium phosphate buffer (pH 6.8)
   5 x SSC[c]
   0.02% bovine serum albumin
   0.02% polyvinylpyrrolidone
   0.02% Ficoll (mol. wt. 400 000)
   100 μg/ml sonicated, denatured salmon sperm DNA (Sigma)

   Wash Buffer

   The composition of this buffer is:
   50 mM Tris-HCl, pH 7.6
   2 mM EDTA
   0.5% tetrasodium pyrophosphate
   0.02% bovine serum albumin
   0.02% polyvinylpyrrolidone
   0.02% Ficoll (mol. wt. 400 00)

2. Dissolve 3 g of agarose in 20 ml of 10 x Mops buffer and 144 ml of distilled water by boiling. Cool to 60°C.
3. Add 36 ml of filtered formaldehyde[b] and pour the agarose into a standard horizontal gel mould.
4. Mix 5.5 μl of each RNA (<20 μg) dissolved in water with 16.5 μl of loading buffer.
5. Heat the RNA samples at 60°C for 5 min and cool quickly on ice.
6. Add 5 μl of glycerol containing 0.05% (w/v) bromophenol blue to each sample.
7. Load each sample into a separate well in the gel and electrophorese at 50 mA until the bromophenol blue reaches the end of the gel.
8. Remove the gel from the mould and soak it in 1% glycine for 60 min.
9. If a photographic record of the gel is required, add ethidium bromide to 2 μg/ml and leave at room temperature for 15 min for staining to occur. Destain the gel by three washes in distilled water (15 min each). Photograph the gel using u.v. illumination (see section 2.8 of ref. 27).
10. Soak the (stained) gel in 20 x SSC for 10 — 20 min.
11. Transfer the RNA from the gel onto a nitrocellulose filter (Schleicher and Schuell, BA85) using the standard technique for 'Southern transfer' (see ref. 26 and section 2.10 of ref. 27). The transfer buffer is 20 x SSC.

12. After transfer, bake the filter for 2 h at 80°C.
13. Pre-hybridise the filter in hybridisation buffer for at least 8 h at 42°C in a sealed plastic bag.
14. Remove the hybridisation buffer and replace it with 4 volumes of fresh buffer, 1 volume of 50% (w/v) dextran sulphate (mol. wt. 500 000, Sigma) and 0.1 volume of denatured (100°C, 5 min), labelled DNA probe. Hybridise for ~20 h at 42°C.
15. Wash the filter four times at room temperature with 2 x SSC, 0.1% SDS (5 min each wash), then twice at 50°C with 0.1 x SSC, 0.1% SDS (15 min each wash).
16. Wrap the *damp* filter in 'Saran' wrap or other suitable thin plastic film and expose it to X-ray film at −70°C in the presence of a calcium tungstate intensifying screen (e.g., Fuji Mach 2 or DuPont Cronex Lightning Plus).

The filter can be used repeatedly with different DNA probes to search for different RNA transcripts. To do this, remove the labelled probe from the filter by washing it in 0.05 − 0.1 x wash buffer at 65°C for 1 − 2 h. Then pre-hybridise the filter and incubate it with the next labelled DNA probe as described above.

---

[a]Deionised formamide is prepared as follows. Stir 100 ml formamide with 5 g of a mixed bed ion-exchange resin (e.g. Amberlite MB3) for 30 min at room temperature. Filter the formamide and store it in batches at −20°C.
[b]Filter formaldehyde through low-ash cellulose paper prior to use.
[c]1 x SSC is 0.15 M NaCl, 15 mM trisodium citrate.

probes. The procedure used by the author is described in *Table 6* and an example is shown in *Figure 5*. An alternative procedure is given in Chapter 5, *Table 10*.

In *primer extension*, a short fragment (30 − 80 bp) of DNA corresponding to the sequence near the anticipated 'cap' site of the gene of interest is labelled at the 5′ termini and then the single strand complementary to the RNA transcript is purified on a denaturing 'sequencing' gel. The purified strand is then hybridised to the RNA transcripts and extended in the presence of unlabelled deoxyribonucleotides and reverse transcriptase. The extension process continues until the 5′ end of the transcript is reached. The size of the labelled, extended primer is then accurately measured on a denaturing 'sequencing' gel. This identifies the exact 5′ end of the transcript. A practical protocol for primer extension is given in *Table 7*.

Primer extension is probably the better method for accurately locating the 'cap' site since a judicious choice of primer will ensure a relatively short extension and hence accurate size determination by subsequent gel electrophoresis. The extended primer can also be sequenced by Maxam-Gilbert sequencing (see ref. 29). However, if the position of the cap site is unknown, nuclease S1 mapping is probably the method of choice since strand separation, often a troublesome procedure, is not necessary.

(iii) *Identification of the 3′ end of RNA transcripts.* Primer extension, by virtue of its polarity, cannot be used to identify the 3′ end of a transcript. Nuclease S1 mapping is therefore the method of choice for this purpose. The method used is exactly the same as that described in *Table 7* except that the DNA probe should have a radioactively-labelled 3′ terminus. This can be accomplished using the Klenow fragment of DNA polymerase I and $[\alpha\text{-}^{32}P]$-deoxynucleoside triphosphates (see ref. 29). Examples of this procedure are described in reference 19.

Table 6. Mapping 5' and 3' ends of RNA Transcripts by Nuclease S1 Mapping.

1. Ethanol precipitate (*Table 3*, steps 8 and 9) both the labelled DNA probe to be used and the RNA transcript preparation.
2. Dissolve each precipitate in a small volume of S1 hybridisation buffer[a] (1 mM EDTA, 0.4 M NaCl, 80% deionised formamide, 40 mM Pipes, pH 6.4).
3. Mix together 5 μl (20 μg) RNA with 2 μl ($\sim$100 ng) of the DNA probe.
4. Transfer the mixture to a glass capillary and heat-seal this at both ends.
5. Incubate the capillary at 80°C for 10 min and then transfer it directly to 52°C *without* allowing it to cool. Incubate at 52°C for 5 h.
6. After hybridisation, mix 1 μl nuclease S1 ($\sim$1000 U/μl; Sigma) with 900 μl S1 assay buffer (0.28 M NaCl, 4.5 mM ZnSO$_4$, 20 μg/ml sonicated denatured salmon sperm DNA, 50 mM sodium acetate, pH 4.6). Dispense this solution into 150 μl aliquots.
7. Transfer the contents of the capillary tube (step 5) to one aliquot of 150 μl assay buffer containing nuclease S1 (step 6).
8. Incubate at 37°C for 30 min.
9. Stop the reaction by adding 6 μl of 0.2 M EDTA.
10. Extract with 1 volume of phenol-chloroform (1:1) and ethanol precipitate the nucleic acid from the aqueous phase (*Table 3*, steps 8 and 9).
11. Resuspend the precipitate in 2 μl of water and analyse the nucleic acid by electrophoresis on a polyacrylamide 'sequencing' gel (see section 3 of ref. 29 and Chapter 5, *Table 10*).

[a]This buffer should be prepared by dissolving 1.2 g Pipes and 2.3 g NaCl in 10 ml of 10 mM EDTA and 80 ml deionised formamide (see *Table 5*, first footnote). This takes about 15 min. Titrate the pH to pH 6.4 with concentrated NaOH and then adjust the final volume to 100 ml with distilled water.

5.3 Identification of RNA Polymerases Involved in Transcription

The three eukaryotic RNA polymerases are differentially inhibited by the fungal toxin α-amanitin. RNA polymerase II is inhibited by concentrations of 1 μg/ml and RNA polymerase III by 100 μg/ml of α-amanitin whilst RNA polymerase I is unaffected by concentrations up to 1 mg/ml. Therefore, this inhibitor can be used to identify the enzyme responsible in injected oocytes for transcription of a specific RNA species. Co-injection into oocytes of α-amanitin at 40 μg/ml in the DNA sample will completely inhibit transcription by RNA polymerase II whilst 100 μg/ml of α-amanitin in the injected sample will reduce tRNA and 5S rRNA synthesis by 90%.

6. ANALYSIS OF POST-TRANSCRIPTIONAL PROCESSING

Several features of the *Xenopus* oocyte lend themselves to sophisticated studies of post-transcriptional processing. Firstly, the oocyte nucleus is able to process primary transcripts correctly (17,19). Secondly, the large nucleus, which facilitates DNA injection, also allows one to inject RNA and this is processed by the oocyte. Thus transcription itself can be circumvented. Finally, rapid manual separation of the nucleus from the cytoplasm allows processing events which occur in these two cell compartments to be distinguished (17).

There are several types of post-transcriptional processing which are of general occurrence. These include the addition of 7-methylguanosine residues to the 5' termini of many primary transcripts ('capping'), polyadenylation, processive termination and splicing.

To date, using the *Xenopus* oocyte system to study post-transcriptional

1   2

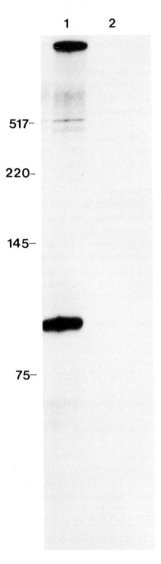

517-

220-

145-

75-

Figure 5. Nuclease S1 mapping of oocyte RNA. Recombinant DNA containing the *Herpes simplex* thymidine kinase *(tk)* gene was injected into oocytes and RNA was prepared from the oocytes 24 h later. The RNA was hybridised to a double-stranded DNA probe labelled at the 5′ end distal to the *tk* promoter. The resulting hybrid was digested with nuclease S1, denatured and electrophoresed on a 12% polyacrylamide 'sequencing' gel as described in *Table 6*. The major protected fragment **(lane 1)** had a length of 95 bases, thus locating the 5′ end of the major transcript at 95 bp upstream from the position of the labelled 5′ end of the probe. **Lane 2** shows a control experiment using RNA extracted from oocytes injected with water. The sizes (bp) of appropriate DNA fragment markers are also indicated.

events, it is in the analysis of RNA splicing that most progress has been made. The positions of the 5′ and 3′ splice sites used in the formation of a mature transcript can be determined by nuclease S1 mapping, utilising probes labelled at their 3′ or 5′ termini, respectively [see Section 5.2.2 (ii) and (iii)]. Wickens

Table 7. Mapping 5' and 3' ends of RNA Transcripts by Primer Extension.

1. Prepare a short fragment (30–80 bp) of DNA corresponding to the sequence near the anticipated cap site of the gene of interest. Label the 5' termini using T4 polynucleotide kinase and [γ-^{32}P]ATP and then purify the single strand complementary to the RNA transcript by electrophoresis on a denaturing 'sequencing' gel (see Chapter 5, *Table 10*). Ideally, to facilitate strand separation, the restriction enzymes used to prepare the fragment should be chosen so as to generate complementary strands of different length.

2. Mix together:
   1–3 μl   5' end-labelled DNA primer
   2 μl   5 x hybridisation buffer (2 M NaCl, 50 mM Pipes, pH 6.4)
   5–10 μg total RNA (from *Table 3*, step 9)
   distilled water to 10 μl.

3. Hybridise in a sealed glass capillary tube for at least 5 h. The exact temperature of hybridisation should be optimised by preliminary experiments, although 70°C is a reasonable compromise. Alternatively, 50% formamide (deionised as in *Table 5*) can be included in the hybridisation mixture and this allows the hybridisation temperature to be decreased to ~42°C. If formamide is used, the nucleic acids *must* be precipitated with ethanol (*Table 3* steps 8 and 9) before hybridisation to remove residual formamide which will be inhibitory in the subsequent primer extension reactions. Resuspend the precipitated nucleic acids in 10 μl of 1 x hybridisation buffer. Finally, SDS (1% final concentration) can be included in the hybridisation mixture, if desired, to inhibit RNase.

4. After hybridisation, dilute the 10 μl hybridisation mixture into 80 μl of 'master-mix' in a siliconised microcentrifuge tube. The composition of 'master-mix' is:
   10 mM DTT
   6 mM MgCl$_2$
   25 μg/ml actinomycin D
   0.5 mM each of dATP, dGTP, dTTP, dCTP
   50 mM Tris-HCl, pH 8.2.

5. Add 5 units of avian myeloblastosis virus (AMV) reverse transcriptase.

6. Incubate for 1 h at 42°C.

7. Add 0.1 volume of 3 M sodium acetate, pH 5.6. Mix and then extract the sample with 1 volume of phenol-chloroform (1:1).

8. Recover the aqueous phase and precipitate the nucleic acid by the addition of 2.5 volumes of ethanol followed by centrifugation and drying of the pellet (*Table 3*, steps 8–10).

9. Dissolve the precipitate in an appropriate volume of running buffer and electrophorese on a 12% polyacrylamide 'sequencing' gel (see section 3 of ref. 29).

and Gurdon (19) have quantitated the amount of faithful splicing that can occur on RNAs transcribed from injected templates. However, the events of transcription and processing must be separated to analyse the mechanisms of RNA splicing in detail. This is best done by injecting the precursor mRNAs into the oocyte nucleus. Green *et al.* (30), using an *in vitro* system for preparing large amounts of precursor mRNAs, have performed such an analysis on human β-globin precursor mRNA. The reader is referred to their work for a description of the methods since this author has no experience of the *in vitro* transcription systems involved.

Finally, with regard to termination, although it is possible to identify the 3' termini of RNA transcripts [Section 5.2.2 (iii)], this site probably does *not* correspond to the termination site for genes transcribed by RNA polymerase II. Studies of the authentic termination site are therefore needed. The oocyte will undoubtedly prove to be an excellent system for studying both sequence constraints and the involvement of other factors (see Section 8) on termination.

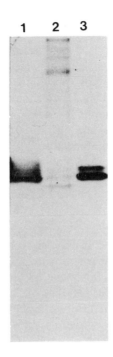

Figure 6. An example of a coupled transcription-translation assay. Oocytes were injected with recombinant DNA containing chicken ovalbumin cDNA sequences. After culture for 24 h in non-radioactive medium, [^{35}S]methionine was added and incubation continued for a further 24 h. The incubation medium **(lane 1)**, cytosol **(lane 2)** and membrane **(lane 3)** fractions were then prepared, immunoprecipitated with anti-ovalbumin antibody and electrophoresed as described in Chapter 10 (Section 5.5). The results clearly show the presence of ovalbumin in the membrane and secreted (incubation medium) material.

7. USE OF THE OOCYTE SYSTEM FOR COUPLED TRANSCRIPTION-TRANSLATION ASSAYS

The considerable virtues of the oocyte as a translational assay for injected mRNA are described in Chapter 10. There are also several reasons why its further use in a coupled transcription-translation assay could be of advantage. Firstly, in cases where a gene is expressed at a very low level *in vivo*, injection of the cloned DNA into *Xenopus* oocytes may circumvent the normal control mechanisms. This may allow the synthesis of larger amounts of protein and thus facilitate biochemical studies on the gene product. Secondly, using *in vitro* mutagenesis of DNA followed by its injection into *Xenopus* oocytes, it should be possible to study those DNA sequences important in intracellular compartmentation, enzyme activity, etc.

The methods for radioactive labelling of protein products in *Xenopus* oocytes and their subsequent analysis are described in Section 4.2 of this chapter and more comprehensively in Chapter 10 (Section 4). An example of this coupled system is shown in *Figure 6* where recombinant DNA containing a full-length chicken ovalbumin cDNA sequence was injected into oocyte nuclei.

Two potential problems in the use of the oocyte system for coupled transcription-translation should be borne in mind. Firstly, this author, in common with other investigators (11), has found the *translational* response to injected DNA to be extremely variable between different batches of oocytes. Therefore, whenever possible, frogs providing oocytes showing high levels of translation product from injected DNA should be re-used (see Chapter 10, Section 2.2.1). Secondly, accumulation of the *translation* product appears to be extremely dependent on the concentration of the injected DNA (J.Davey and A.Wilson, personal communication). Thus, whereas the author has always had success using the DNA concentrations recommended in Section 3.4, the use of higher concentrations (for example 1 mg/ml) can, *for some DNAs*, be counter-productive. The reason for this is not clear.

8. ASSAY OF REGULATORY PROTEINS

The *Xenopus* oocyte has proved an excellent vehicle for studying the transcriptional and post-transcriptional events which follow the injection of certain DNAs or RNAs. Very recently this work has been taken one stage further; the influence of proteins on transcriptional initiation (31) or termination (32) using injected DNA as template has been examined. In these experiments, co-injection of DNA and various protein fractions constitute an assay for the purification of the active regulatory protein. Clearly the *Xenopus* oocyte still has considerable potential as a transcriptional assay system.

9. REFERENCES

1. Mertz,J.E. and Gurdon,J.B. (1977) *Proc. Natl. Acad. Sci. USA,* **77,** 1502.
2. Colman,A. (1975) *Eur. J. Biochem.,* **113,** 339.
3. Gurdon,J.B. and Brown,D.D. (1978) *Dev. Biol.,* **67,** 346.
4. Kressmann,A. and Birnstiel,M.L. (1980) in *Transfer of Cell Constitutents into Eukaryotic Cells,* Celis,J., Graessmann,A. and Loyter,A. (eds.), Plenum Press, NY, p. 383.
5. Wickens,M.P. and Laskey,R.A. (1981) in *Genetic Engineering,* vol. **I,** Williamson,R. (ed.), Academic Press Inc., NY and London, p. 103.
6. Gurdon,J.B. and Melton,D.A. (1981) *Annu. Rev. Genet.,* **15,** 189.
7. Moss,T. (1982) *Cell,* **30,** 835.
8. Rungger,D. and Turler,H. (1978) *Proc. Natl. Acad. Sci. USA,* **75,** 6073.
9. De Robertis,E. and Mertz,J. (1978) *Cell,* **12,** 175.
10. Lasky,R.A., Honda,B.M., Mills,A.D., Morris,N.R., Wyllie,A.H., Mertz,J., De Robertis,E., and Gurdon,J.B. (1978) *Cold Spring Harbor Symp. Quant. Biol.,* **42,** 171.
11. Asselbergs,F.A.M., Smart,J.E. and Mathews,M.B. (1983) *J. Mol. Biol.,* **163,** 209.
12. Probst,E., Kressmann,A. and Birnstiel,M.L. (1979) *J. Mol. Biol.,* **135,** 709.
13. Old,R.W., Woodland,H.R., Ballantine,J.E.M., Aldridge,T.C., Newton,C.A., Bains,W.A. and Turner,P.C. (1982) *Nucleic Acids Res.* **10,** 7561.
14. McKnight,S.L. and Garvis,E.R. (1980) *Nucleic Acids Res.,* **8,** 5931.
15. Wickens,M.P., Woo,S., O'Malley,B.W. and Gurdon,J.B. (1980) *Nature,* **285,** 628.
16. Kressmann,A., Clarkson,S.G., Pirotta,V. and Birnstiel,M.L. (1978) *Proc. Natl. Acad. Sci. USA,* **75,** 1176.
17. Melton,D.A. and Cortese,R. (1979) *Cell,* **18,** 165.
18. Grosschedl,R. and Birnstiel,M.L. (1980) *Proc. Natl. Acad. Sci. USA,* **77,** 1432.
19. Wickens,M.P. and Gurdon,J.B. (1983) *J. Mol. Biol.,* **163,** 1.
20. McKnight,S. and Kingsbury,R. (1982) *Science (Wash.),* **217,** 316.
21. Grosschedl,R. and Birnstiel,M. (1980) *Proc. Natl. Acad. Sci. USA,* **77,** 7102.
22. Trendelenburg,M.F. and Gurdon,J.B. (1978) *Nature,* **276,** 292.
23. Bogenhagen,D.F. and Brown,D.D. (1981) *Cell,* **24,** 261.

24. Bogenhagen,D.F., Sakonju,S., Brown,D.D. (1980) *Cell,* **19**, 27.
25. Woodland,H. and Pestell,R. (1972) *Biochem. J.,* **127**, 597.
26. Thomas,P. (1980) *Proc. Natl. Acad. Sci. USA,* **77**, 5201.
27. Sealey,P.G. and Southern,E.M. (1982) in *Gel Electrophoresis of Nucleic Acids − A Practical Approach,* Rickwood,D. and Hames,B.D. (eds.), IRL Press, Oxford and Washington DC, p. 39.
28. Messing,J. (1981) in *Third Cleveland Symposium on Macromolecules: Recombinant DNA,* Walton,A. (ed.), Elsevier, Amsterdam, p. 143.
29. Davies,R.W. (1982) in *Gel Electrophoresis of Nucleic Acids − A Practical Approach,* Rickwood,D. and Hames,B.D. (eds.), IRL Press, Oxford and Washington DC, p. 117.
30. Green,M., Maniatis,T. and Melton,D. (1983) *Cell,* **32**, 681.
31. Jones,N., Richter,J., Weeks,D. and Smith,L. (1983) *J. Mol. Biol.,* in press.
32. Stunnenberg,H. and Birnstiel,M. (1982) *Proc. Natl. Acad. Sci. USA,* **79**, 6201.

CHAPTER 3

Transcription of Eukaryotic Genes in a Whole-cell Extract

JAMES L. MANLEY

1. INTRODUCTION

Three classes of nuclear DNA-dependent RNA polymerases have been identified in eukaryotic cells (1). RNA polymerase I catalyses the synthesis of ribosomal RNA (rRNA) precursors, RNA polymerase II transcribes primarily the genes which give rise to messenger RNA (mRNA), while RNA polymerase III transcription results in the production of transfer RNAs (tRNAs), 5S RNA and several other small RNAs of unknown function. It has been clear for many years that in order to study the mechanisms of transcription, as well as to identify the factors and nucleotide sequences which control gene expression, cell-free systems that accurately and specifically transcribe exogenous DNA are required. Early attempts at achieving this aim utilised purified RNA polymerases and were unsuccessful. In recent years, however, cell-free systems have been developed in which accurate transcription by all three types of RNA polymerases can be obtained. Two basic approaches have been successful. In one, purified RNA polymerase is supplemented with cell extracts which contain factors required for accurate transcription (2). This method is somewhat tedious because it is not a trivial matter to purify large amounts of active RNA polymerase. The other approach is to prepare concentrated cell extracts which contain not only the factors required for transcription, but also sufficient amounts of RNA polymerase so that addition of purified enzyme is not required. For RNA polymerase I (3) and III (4), such systems can be simply prepared from cytoplasmic extracts because sufficient amounts of these polymerases and their required factors leak out of the nucleus at hypotonic salt concentrations. RNA polymerase II, on the other hand, remains almost entirely within the nucleus. Thus, to obtain extracts containing this activity, extracts must be prepared from whole cells (5). The preparation and properties of such an extract, which contains all of the factors and enzymatic activities necessary for accurate and specific transcription by RNA polymerases I, II and III, and also for polyadenylation of mRNA precursors, are described here.

Most experiments to date have used extracts prepared from human cells which grow in suspension (HeLa or KB cells). Such extracts show quite broad species specificities for RNA polymerase II and III. Polymerase III genes from virtually all higher eukaryotes that have been tested are transcribed accurately in extracts from HeLa cells. Whole-cell extracts do not seem able to transcribe polymerase II genes from yeast accurately but have been shown to be capable of transcribing *Bombyx mori* silk fibroin (6) and chicken conalbumin (7) genes as well as many other genes from higher eukaryotes and their viruses.

Synthesis of mature RNA molecules, of course, requires additional enzymes and factors different from those needed to bring about accurate transcriptional initiation, for example, RNA processing enzymes. HeLa cell extracts appear to contain virtually all of the enzymes required for tRNA processing, including the pre-tRNA splicing enzymes (8). Similarly, although work with RNA polymerase I systems is just beginning, the extracts appear to contain at least one enzyme for processing rRNA precursor molecules (9,10). Finally, transcripts synthesised by RNA polymerase II in cell extracts are efficiently capped and methylated at their 5′ ends (2,5) and conditions have been developed recently that allow efficient and accurate polyadenylation of mRNA precursors (11). However, although several reports suggest that the newly synthesised RNA can be spliced (12 – 14), the splicing activities observed are low, and splicing is not reproducibly detected.

Clearly, soluble transcription systems can reproduce, under cell-free conditions, a number of the steps that are required *in vivo* to synthesise mature RNA (see ref. 15 for review). These systems are being used for the identification and characterisation of factors required for RNA synthesis, the elucidation of nucleotide sequences that control gene expression and an understanding of the mechanisms by which gene expression is regulated.

2. PREPARATION AND USE OF WHOLE-CELL EXTRACT FOR TRANSCRIPTION BY RNA POLYMERASE II

2.1 Preparation of Extract

Extracts are prepared by modification of a procedure originally described by Sugden and Keller (16) who used the method as a first step in RNA polymerase purification. The author has used HeLa cells almost exclusively. These cells are easily produced in large quantities and the resulting extracts are relatively free of nuclease activity when incubated at 30°C. Extracts with transcriptional activity have been prepared from a few other cell lines; a good rule is that cells which grow in spinner culture will give active extracts. Most other cell lines and tissues, however, have yielded extracts without detectable levels of transcription or with high levels of nuclease.

Details of the preparation procedure for HeLa cell extracts are given in *Table 1*. One litre of cells yields approximately 2 ml of extract (sufficient for 100 – 400 assays) containing 15 – 30 mg/ml of protein and up to 2 mg/ml of nucleic acid (primarily rRNA). Extracts as concentrated as this are necessary so that the volume of extract which must be added to each transcription reaction mixture is minimised and hence the final salt concentration is kept low (high salt severely inhibits transcription; see Section 4.3.2). Attempts to obtain more concentrated extracts by resuspending the pellet in a smaller volume after precipitation (*Table 1,* step 7), or by tying the dialysis bag tightly to reduce expansion during dialysis (*Table 1*, step 9), have not been reproducibly successful due to increased protein precipitation during dialysis. Likewise, dialysis

Table 1. Preparation of Whole-cell Extract.

1. Grow HeLa cells (2 − 25 litres) to a density of 4 − 8 x 10^5 cells/ml in suspension cultures in Eagle's minimal essential medium supplemented with 5% horse serum. The cell density appears not to be crucial, although slightly more active extracts have been obtained with cells harvested at the lower end of the range indicated.

The following operations are carried out at 0 − 4°C.

2. Harvest the cells by centrifugation (1000 g, 15 min) and wash them twice with phosphate buffered saline.

3. Resuspend the cells in four packed-cell volumes of 10 mM Tris-HCl (pH 7.9), 1 mM EDTA and 5 mM DTT (6 x 10^8 cells yield ~2 ml of packed cells). At this point, the cells should swell visibly.

4. After 20 min, lyse the cells by homogenisation in a Dounce homogeniser with eight strokes using a 'B' pestle.

5. Add four packed-cell volumes of 50 mM Tris-HCl (pH 7.9), 10 mM $MgCl_2$, 2 mM DTT, 25% (w/v) sucrose and 50% glycerol, and gently mix the suspension. With continued gentle stirring, add dropwise one packed-cell volume of a saturated (neutralised) solution of ammonium sulphate. After this addition, gently stir the highly viscous lysate for an additional 30 min. The stirring must be very gentle to prevent shearing of the DNA, which would interfere with its removal in the next step. Nuclear lysis can be detected by increased viscosity after approximately half of the required volume of ammonium sulphate has been added. Occasionally, lysates appear clumpy and only slightly viscous, rather than extremely viscous and uniform as usually observed. The author has obtained active extracts from both types of lysates, although more reproducibly from the latter. Approximately 80% of the extracts that are prepared are active.

6. Carefully pour the extract into polycarbonate tubes and centrifuge at 175 000 g for 3 h, for example using a Beckman 50.2 rotor run at 45 000 r.p.m.

7. Decant the supernatant so as not to disturb the pellet (leave the last 1 − 2 ml behind) and precipitate the protein and nucleic acid by adding solid ammonium sulphate (0.35 g/ml supernatant). After the ammonium sulphate has dissolved, add 1.0 μl of 1 M NaOH per gram of ammonium sulphate and stir the suspension for an additional 30 min.

8. Collect the precipitate by centrifugation at 15 000 g for 20 min. Drain off the supernatant completely and resuspend the pellet in 40 mM Tris-HCl (pH 7.9), 0.1 M KCl, 10 mM $MgCl_2$, 0.2 mM EDTA, 2 mM DTT and 15% glycerol equivalent to 1/20th the volume of the high speed supernatant (step 7).

9. Dialyse the suspension twice against the resuspension buffer (50 − 100 volumes each time) for a total of 8 − 12 h. The volume of the solution increases 30 − 50% during dialysis.

10. Centrifuge the dialysed suspension at 10 000 g for 10 min to remove insoluble material. Divide the resulting supernatant into small aliquots (0.2 − 0.5 ml), freeze these rapidly in liquid nitrogen or powdered dry ice and store at − 70°C. The extract retains full activity at − 70°C for at least a year and can be thawed and quick-frozen several times without loss of activity.

against buffer containing lower salt concentrations results in less active lysates, again as a result of increased protein precipitation.

2.2 The Transcription Reaction

Analytical reactions are conveniently carried out in disposable microcentrifuge tubes in 25 − 50 μl final volume. A typical 25 μl reaction mixture consists of:

7.5 − 15.0 μl of whole-cell extract

0.2 − 1.5 μg of template DNA

50 μM each of ATP, GTP, CTP, UTP

4 mM creatine phosphate

5 μCi [^{32}P]GTP ($\sim$10 mCi/ml aqueous commercial preparation;
>200 mCi/mmol).

Note that the salts required for transcription are all provided by the extract itself. The incubation is carried out at 30°C for 30−120 min.

Radioactive GTP is recommended as the radiolabelled nucleoside triphosphate to use when analysing RNA transcripts directed by weak promoters since it minimises problems of spurious labelling of other nucleic acids [Section 3.2 (ii)]. However, in other cases (e.g. see Section 4.1), radioactive UTP may be the tracer of choice. To label RNA transcripts uniformly, [α-^{32}P]-nucleoside triphosphate is required, whereas if only the 5' end of the transcript is to be labelled, a nucleoside triphosphate labelled in the β-position should be used. Label in the α-position is stable in whole-cell extracts but label in the β- and γ-positions may be labile due to various kinase and phosphatase activities present in the extract. This may result in significant exchange-labelling of other nucleoside triphosphates and hence results obtained using β- or γ-labelled triphosphates should be interpreted with caution.

2.3 Extraction of RNA

After the desired period of incubation, the reaction is terminated and the RNA is isolated by the procedure described in *Table 2*. Although somewhat tedious, this reproducibly yields intact RNA free of protein. The RNA solution obtained by this method is also essentially free of unincorporated nucleoside triphosphates so that the amount of ^{32}P incorporated into nucleic acid can be determined simply by measuring the Cerenkov radiation.

3. ANALYSIS OF RNA POLYMERASE II TRANSCRIPTION PRODUCTS

3.1 The Basic Procedure of Run-off Transcription

RNA polymerase II does not terminate transcription *in vitro*. However, discrete length RNA products can be generated by the 'run-off' assay. This method uses as template, DNA molecules which have been cleaved by a restriction enzyme that cuts downstream from a putative transcription start site. RNA polymerases which transcribe this DNA will stop or 'fall off' when they reach the end of the DNA molecule. If a substantial number of enzymes initiate transcription at the same site, then a population of RNA molecules of a discrete size will be produced. Such a population will migrate as a single band during subsequent gel electrophoresis. If DNA segments that have been cleaved by different restriction enzymes are used as templates in separate reaction mixtures, then the transcription start site can be deduced by comparison of the sizes of the RNAs produced. This technique has been widely used with cell-free transcription systems for mapping eukaryotic promoters.

To determine the size of a particular RNA transcript, the author has found the protocol of McMasters and Carmichael (17) to be simple and accurate. It involves denaturation of the RNA with glyoxal and then electrophoresis in an agarose gel. The method is suitable for analysing RNA over a wide range of

Table 2. Isolation of RNA after Transcription.

1. Terminate each transcription reaction (25 – 50 μl) by the addition of 250 μl of 7.5 M urea, 0.5% SDS, 10 mM EDTA and 10 mM Tris-HCl, pH 8.0.
2. Add an equal volume (275 – 300 μl) of phenol-chloroform-isoamyl alcohol (20:20:1) and vortex for 30 sec. Spin in a microcentrifuge (e.g., Eppendorf) for 5 min. Remove the aqueous phase (first aqueous phase) to a clean microcentrifuge tube (1.5 ml capacity) and leave the organic phase plus interphase.
3. Add to the organic phase plus interphase, 150 μl of 7 M urea, 0.35 M NaCl, 1% SDS, 10 mM EDTA, 10 mM Tris-HCl (pH 8.0) plus 25 μg of yeast tRNA. Mix by vortexing and then spin in a microcentrifuge for 20 sec. A very dense protein precipitate will form at the interphase making removal of the second aqueous phase impossible. Therefore, remove the *organic* phase with a Pasteur pipette, and re-extract the interphase plus second aqueous phase with 200 μl of chloroform. After brief vortexing and centrifugation (20 sec), the second aqueous phase can now be easily removed.
4. Pool the first and second aqueous phases; the volume now will be 0.45 ml. Add an equal volume of phenol-chloroform-isoamyl alcohol and mix by vortexing. Centrifuge for 5 min in a microcentrifuge, and then remove the aqueous phase, leaving the small interphase behind.
5. Re-extract the aqueous phase twice, each time with an equal volume of chloroform. After the first of these extractions, a small sticky protein precipitate can sometimes be detected; this should be left with the organic phase. After the second extraction, the interphase should be completely clear.
6. To the aqueous phase add 1 ml of ethanol, mix and then leave the mixture in a dry ice and ethanol bath for 5 min (or at least 2 h at − 20°C).
7. Spin for 5 min in a microcentrifuge in a cold room, and decant the supernatant.
8. Resuspend the pellet in 200 μl of 0.2% SDS, add 200 μl of 2 M ammonium acetate, and then precipitate the nucleic acid as described in step 6.
9. Repeat step 8.
10. Lyophilise the precipitated nucleic acid. Redissolve it in 60 μl of 0.2% Sarkosyl, 1 mM EDTA (pH 8.0) and store it at − 20°C.

sizes; there is a linear relationship between the distance migrated and $\log_{10}$ mol. wt. of the RNA from 0.2 kb to over 5 kb. The detailed procedure is described in *Table 3*. An example of the technique is shown in *Figure 1*. The RNAs were transcribed from recombinant plasmids containing the adenovirus late promoter and various segments of the long (30 kb) late transcription unit. Clearly, the whole-cell extract is capable of synthesising very long RNAs, although 7 – 8 kb is about the limit. Although the recombinant plasmids encode RNA processing signals, they are not detectably utilised in the whole-cell extract under the reaction conditions described in Section 2.2

The rate of elongation by RNA polymerase II in the reaction mixture is about 300 nucleotides/min. The rate *in vivo* is 10-fold higher, although this must be partially due to the higher temperature *in vivo* (37°C) compared with the *in vitro* reaction conditions (30°C). The rate of accumulation of 'run-off' transcripts *in vitro* is approximately linear for over an hour (5).

RNA polymerase II preferentially initiates transcription at the termini of DNA fragments and at internal nicks (18). The enzyme is also capable of end-labelling DNA fragments with [α-^{32}P]NTPs to yield full-length labelled molecules which are resistant to RNase digestion. These reactions are each sensitive to α-amanitin (0.5 μg/ml), although a distinct end-labelling activity

Table 3. Determination of the Size of RNA Transcripts.

1. Prepare glyoxalating solution by mixing:
    270 μl of dimethyl sulphoxide
    4.2 μl of 1 M sodium phosphate, pH 6.8
    60 μl of deionised glyoxal[a]
2. The large amount of rRNA in the whole-cell extract (Section 2.1) limits the amount of *in vitro* transcribed RNA which can be analysed by gel electrophoresis. To avoid overloading the gel, glyoxalate and electrophorese only one quarter of the RNA obtained from a standard 25 μl reaction mixture (Section 2.2); i.e., 15 μl of 60 μl of extracted RNA (*Table 1*, step 10). Mix:
    15 μl of RNA (in 0.2% Sarkosyl, 1 mM EDTA, pH 8.0)
    27 μl of glyoxalating solution
    Incubate at 50°C for 1 h.
3. Prepare a 1.4% agarose gel (20 cm long, 0.3 cm deep) in 10 mM sodium phosphate (pH 6.8). This buffer also serves as the reservoir buffer.
4. After adding bromophenol blue as tracking dye, load each glyoxalated RNA sample into a separate well (0.6 x 0.3 cm cross-section) in the agarose gel. Load suitable RNA mol. wt. markers[b] (which have been previously glyoxalated) into another well.
5. Electrophorese at 120 V for 2.5 h, rapidly recirculating the reservoir buffer throughout the run to prevent changes in pH. This voltage yields optimal results; lower voltages, and the consequently longer electrophoresis times required, lead to more diffuse bonds.
6. Visualise the endogenous rRNA and tRNA by staining with ethidium bromide; the amount of transcribed RNA will be too low for detection. This serves as a check on the recovery of RNA (from sample to sample) and its integrity[c]. Before the RNA can be stained with ethidium bromide it must be de-glyoxalated using alkali. The simplest procedure is as follows:
    Soak the gel in 10 volumes of 50 mM NaOH, 0.5 μg/ml ethidium bromide for 20 min. Transfer the gel to 10 volumes of 0.1 M Tris-HCl (pH 7.5) containing 0.5 μg/ml ethidium bromide for 40 min. Visualise the RNA by fluorescence under u.v. illumination (300 – 320 nm).
7. Dry the gel using a commercial gel dryer.
8. Expose the dried gel to X-ray film to detect the ^{32}P-labelled transcribed RNA.
9. Use the distances migrated by the molecular weight markers to construct a standard curve of $\log_{10}$ mol. wt. versus distance migrated and determine the molecular weights of the RNA transcripts by reference to this.

[a]Deionise glyoxal (Eastman) by stirring it with Amberlite resin (MB-1) (25 g/100 ml glyoxal) for 3 h. Filter off the resin and store the glyoxal in 0.5 ml aliquots at $-20°C$. It is stable for at least 6 months.
[b]The ^{32}P-labelled RNA size markers can be prepared by transcribing *in vitro* a DNA template containing a known promoter (e.g., the strong adenovirus 2 late promoter) that has been separately cleaved by different restriction endonucleases so that run-off RNAs of known size will be produced. After glyoxalation, such markers are stable for up to 2 months.
[c]If RNA degradation occurs during electrophoresis, it can be overcome by carrying out electrophoresis in the presence of SDS. This is most easily accomplished by adding SDS (BDH Chemicals Ltd.) to the reservoir buffer to a final concentration of 0.1% and then pre-electrophoresing the gel at 150 V for 45 min.

which is resistant to α-amanitin is present in some extract preparations.

The whole-cell extract contains the majority of the soluble proteins in the cell. Although most of these are of no concern, some can interfere with certain experiments. For example, most extracts have high levels of topoisomerase type I and II activities as well as DNA ligase. Thus, the topology of the DNA template can change rapidly in the reaction mix, preventing, for instance,

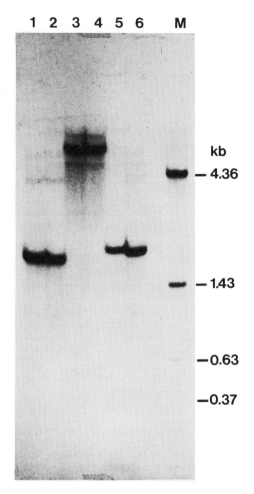

Figure 1. Analysis of RNA run-off products by glyoxalation and agarose gel electrophoresis. Recombinant plasmids containing the adenovirus 2 late promoter and various segments of the late transcription unit were constructed (R.Jove and J.L.Manley, unpublished), cleaved with a restriction enzyme, and used as templates for *in vitro* transcription. DNAs were cleaved so that run-off transcripts of 1.8 kb (**lanes 1** and **2**), 7.0 kb (**lanes 3** and **4**) and 1.95 kb (**lanes 5** and **6**) would be produced. Size markers (**lane M**) were also produced by *in vitro* transcription.

studies of supercoiled DNA. The extract also contains RNA polymerases I and III (16). Their contribution to the transcription obtained can be assessed by the degree of inhibition with α-amanitin; run-off transcription by RNA polymerase II is completely inhibited in the presence of 0.5 μg/ml of α-amanitin *(Figure 2)* whereas transcription by RNA polymerase I is resistant and RNA polymerase III is inhibited only by 200 μg/ml of α-amanitin. In fact, most template DNAs do not contain promoters for these enzymes and hence their contribution to the nucleotide incorporation is small. However, some genomic clones contain repetitive elements which do contain polymerase III promoters (see, for example, ref. 19).

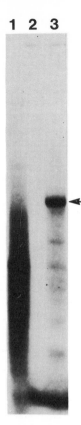

Figure 2. α-Amanitin sensitivity of transcription from the adenovirus 2 late promoter. RNA was synthesised in reaction mixtures (100 μl volume) each containing 400 μCi of [α-^{32}P]UTP (Section 2.2). The DNA template used (*Bal*I-E-pBR322) was a *Bal*I fragment of adenovirus serotype 2 DNA inserted into plasmid pBR322; this region of DNA contains the adenovirus 2 late promoter from which a specific 1750 nucleotide RNA (arrowed) is transcribed. After incubation, the RNA was isolated and glyoxalated and an aliquot (1%) was analysed by gel electrophoresis (Section 3.1). The reaction mixture analysed in **lane 1** contained 5 μl of whole-cell extract; **lane 2** contained 30 μl of extract plus α-amanitin (0.5 μg/ml); **lane 3** contained 30 μl of extract.

3.2 Limitations

The analysis of run-off transcripts as described in Section 3.1 is a simple, sensitive and accurate method for determining the origin of transcription for RNA synthesised in the whole-cell extract. However, it does have some limitations:

(i)   The whole-cell extract contains relatively high levels of nucleic acid. Since most of this is 18S and 28S rRNA, it is impossible to analyse more than 25 − 50% of the RNA obtained from a 25 μl reaction mixture in a single lane of a standard size gel without producing severe overloading in the regions finally occupied by these RNAs.

(ii) Specific transcripts produced by very weak promoters can sometimes be obscured by non-specific initiation or termination, or by exogenous

Table 4. Removal of DNA from RNA Transcript Preparations.

1. Dissolve the nucleic acid from step 9 of *Table 2* in 0.2 M ammonium acetate and then precipitate it with ethanol as in steps 6 and 7 of *Table 2*. Lyophilise.
2. Resuspend the dried pellet in 100 μl of 10 mM Tris-HCl (pH 7.5), 0.1 M NaCl.
3. Add RNase-free DNase (if necessary, treated with iodoacetate; ref. 20) to 50 μg/ml and MgCl$_2$ to 10 mM final concentration.
4. Incubate for 5 min at 37°C.
5. Add 100 μl of 10 mM EDTA (pH 8.0), 0.2% SDS.
6. Extract the sample with an equal volume of phenol-chloroform-isoamyl alcohol (20:20:1) and re-extract the aqueous phase with an equal volume of chloroform as in steps 4 and 5 of *Table 2*.
7. Add NaCl to a final concentration of 0.15 M and add 2 volumes of ethanol. Precipitate the RNA by incubation in dry ice-ethanol for 5 min (or at least 2 h at -20°C).
8. Recover the RNA by centrifugation for 5 min in a microcentrifuge in a cold-room (4°C).
9. Resuspend the RNA pellet in 60 μl of 0.2% Sarkosyl, 1 mM EDTA (pH 8.0) and store at -20°C.

nucleic acids (particularly DNA and rRNA and its breakdown products) labelled by end-labelling activities in the extract. This activity is insensitive to both α-amanitin and actinomycin D and can thus be distinguished from *de novo* RNA synthesis. The use of radioactive GTP as tracer produces the least end-labelling and UTP the most, whereas labelled CTP or ATP result in high levels of tRNA labelling by enzymes exchanging the 3'-terminal CCA.

(iii) The sites of initiation of RNA transcripts can be mapped only to within about 20 nucleotides by the usual procedure of run-off transcription, denaturation with glyoxal and sizing by agarose gel electrophoresis. To map the 5' ends of RNAs more precisely, short run-off transcripts can be analysed on polyacrylamide 'sequencing' gels. The structure of the 5' end of RNA synthesised by the whole-cell extract can also be studied by RNA 'fingerprinting' techniques (2,5). However this approach is not frequently used.

3.3 Analysis by Hybridisation and Nuclease S1 Digestion

By hybridising the ^{32}P-labelled RNA transcripts with a specific non-radioactive DNA probe, followed by digestion with nuclease S1, the problem of spurious (non-specific) RNA labelling sometimes associated with the basic method [Section 3.2 (ii)] can be eliminated since RNA not complementary to the DNA probe is destroyed by the nuclease. The first stage in the analysis is the removal of DNA from the crude RNA extract prepared after transcription by the whole-cell extract, since otherwise this will interfere with the hybridisation reaction. The DNA is removed as described in *Table 4*. Hybridisation and nuclease S1 digestion are then carried out as in *Table 5*. Double-stranded DNA can usually be used directly as the DNA probe, that is, without the need for prior purification of the DNA strand complementary to the RNA under investigation.

Table 5. Analysis of ^{32}P-labelled RNA Transcripts by Hybridisation and Nuclease S1 Digestion.[a]

1. Routinely use 0.25 μg of the DNA probe (~5 kb long) for the analysis of RNA transcripts from a single, standard transcription reaction or proportionately less (or more) for smaller (or larger) DNA fragments. This amount of DNA will certainly be in excess over the small amount of RNA synthesised in a standard transcription reaction and is also sufficient to drive the hybridisation reaction to completion in a short period of time. Add the DNA probe and the ^{32}P-labelled RNA preparation (free of DNA; *Table 4*) to 100 μl of 0.2 M sodium acetate (pH 5.2), 0.1% SDS in a microcentrifuge tube. Add 2.5 volumes of ethanol, mix, and leave in a dry ice-ethanol bath for 5 min (or −20°C for at least 2 h) to precipitate the nucleic acids.
2. Recover the nucleic acids by centrifugation for 5 min in a microcentrifuge in a cold-room (4°C). Decant the ethanol carefully and then allow the sample to air-dry for about 10 min. It is important that the precipitate does not become completely dry or it will be very difficult to dissolve.
3. Resuspend the sample in 15 μl of hybridisation buffer (80% deionised formamide[b], 0.4 M NaCl, 1 mM EDTA, 40 mM Pipes, pH 6.5).
4. Denature the DNA probe by heating at 68°C for 10 min and then quickly transfer the tube to 60°C. Incubate for 45 min at this temperature to allow hybridisation to occur.[c]
5. After hybridisation, rapidly dilute the reaction mixture by adding 200 μl of *ice-cold*, nuclease S1 digestion mixture [0.25 M NaCl, 1 mM ZnSO₄, 5% glycerol, 30 mM sodium acetate (pH 4.5), 2 x 10³ units/ml of nuclease S1 (Boehringer)]. When several samples are being analysed, they should be removed from the 60°C water-bath, diluted with the nuclease S1 mixture and mixed one at a time. This procedure prevents any DNA-DNA reannealing and possible displacement of hybridised RNA.
6. Allow the digestion with nuclease S1 to occur by incubating each sample at 45°C per 30 min.
7. After digestion, add 0.2 ml of phenol-chloroform (1:1, v:v) and *then* 10 μg of carrier tRNA to each sample. Mix. Spin in a microcentrifuge for 30 sec.
8. After centrifugation, remove the (upper) aqueous phase to a fresh microcentrifuge tube and precipitate the nucleic acids by adding 0.5 ml of ethanol and leaving in dry ice-ethanol for 5 min (or at −20°C for at least 2 h).
9. Recover the nucleic acids by centrifugation in a microcentrifuge for 5 min in a cold-room (4°C).
10. Dry each precipitate and dissolve it in 15−60 μl of 1 mM EDTA, 0.2% Sarkosyl. Store at −20°C if necessary.
11. The entire sample (or an aliquot) can be glyoxalated and the nuclease-protected RNA fragments can then be analysed by agarose gel electrophoresis as described in *Table 3*.

[a]Based on the procedure of Berk and Sharp (21).
[b]Deionise the formamide by shaking with Amberlite MB-3 mixed-bed resin (5 g per 100 ml of formamide) at room temperature for at least 30 min. Remove the resin by filtration and store the formamide at −20°C. Formamide prepared and stored in this manner is stable for at least 6 months.
[c]The hybridisation time and temperature given here are for adenovirus 2 DNA. The appropriate values to use for other DNA probes can be determined as described by Berk and Sharp (21).

For determination of the start point for *in vitro* transcription using nuclease S1, the DNA probe should be a genomic restriction fragment which extends from within the gene to some point upstream beyond the putative initiation site. After hybridisation and nuclease S1 digestion as usual, the RNA transcripts are glyoxalated and then sized by agarose gel electrophoresis (*Table 5*, step 11). Greater resolution can be obtained when this is carried out using a radiolabelled DNA probe and unlabelled RNA transcripts (fully described in Chapter 5, Section 5) since it is often easier to obtain the appropriate DNA size markers than the necessary RNA size markers (DNA and RNA molecules of the same chain length do not co-migrate precisely). Nevertheless, the use of radiolabelled RNA transcripts and an unlabelled DNA probe, as described in *Table 5*, has two advantages. First, it is as easy to synthesise radiolabelled RNA as it is to synthesise unlabelled RNA *in vitro*. An additional labelling step is therefore avoided. Secondly, in some instances it is more desirable to examine the RNA transcript directly. For example, if the RNA product of *in vitro* transcription is also present as an endogenous mRNA in the whole-cell extract, then radiolabelled RNA must be analysed.

4. OPTIMISATION OF TRANSCRIPTIONAL ACTIVITY BY RNA POLYMERASE II

4.1 Titration of DNA and Extract Concentrations

Titrations both of DNA and of extract concentration are non-linear. Measurement of run-off transcription as a function of DNA concentration at constant extract concentration shows firstly a threshold DNA concentration below which no transcription occurs, and secondly an inhibitory effect at high DNA concentrations (5). To exceed the minimal (threshold) concentration of DNA for transcription, carrier DNA unrelated to the DNA to be transcribed can be used. Thus, promoter-specific DNA, itself at a sub-threshold concentration, is actively transcribed in the presence of pBR322 or *Escherichia coli* DNA. Even the synthetic duplex, alternating co-polymers poly(dIC:dIC) and poly(dAT:dAT) can also act as carrier DNA, thereby demonstrating a total lack of sequence specificity in the bulk DNA requirement (22). An advantage of these co-polymers as carrier DNA is that the transcribed RNA products of poly(dIC:dIC) and poly(dAT:dAT) contain only two nucleotides. Thus, use of poly(dIC:dIC) as a carrier in a reaction containing $[\alpha\text{-}^{32}P]UTP$ yields no radioactive background product from non-specific copying of the carrier. The key aspect of this bulk DNA dependence is that, at a fixed total DNA concentration, the molar yield of transcripts per promoter is constant and independent of the source of carrier.

A critical dependence of transcription upon extract concentration is also observed (5). In fact, this dependence on DNA concentration and the protein concentration of the extract are related (23). Specific transcription can be obtained when the final concentration of extract protein in the reaction mixture is in the range 4 – 18 mg/ml. At the lower end of this range, the threshold DNA concentration is lower and transcription is less dependent on the absolute DNA concentration. The DNA optimum is approximately 10 μg/ml. At the higher extract concentrations, the threshold is increased so that it is sometimes necessary to use 60 μg/ml of DNA in order to obtain any transcription. Under these conditions, transcription is more dependent on the absolute DNA concentration. Clearly, for each new extract it is necessary to titrate the DNA and extract concentrations carefully to determine the optimal conditions. A typical study is shown in *Figure 3*. At very low extract concentrations, high levels of non-specific transcription may occur (compare lanes 1 and 4 of *Figure 3A*). This is probably because a required specificity factor is limiting under these

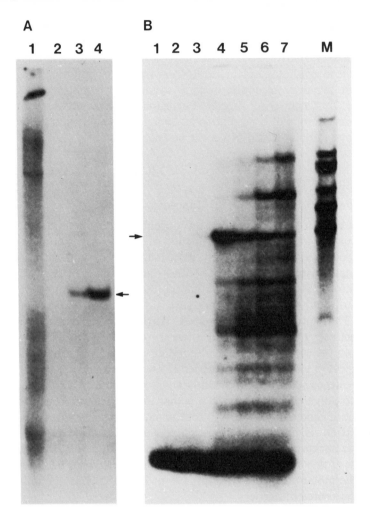

Figure 3. Determination of extract and DNA concentration optima for accurate transcription. RNA was synthesised in standard reaction mixtures using [α-^{32}P]UTP (Section 2.2) except that the concentrations of whole-cell extract (**A**) or DNA (**B**) were varied. The DNA template was *Bal*I-E-pBR322 (see legend of *Figure 2*). The positions of the specific transcript (1750 nucleotides) are indicated by the arrows. In the reactions analysed in **A**, the amounts of whole-cell extract used were 2.5 μl, 5.0 μl, 10.0 μl and 15.0 μl in **lanes 1 – 4**, respectively, with the DNA present at 50 μg/ml. In **B**, the concentrations of DNA used (μg/ml) were 0, 12.5, 25.0, 50.0, 75.0, 100.0 and 125.0 in **lanes 1 – 7**, respectively. **M** is an *Eco*RI digest of adenovirus 2 DNA labelled by nick-translation and used as DNA size markers.

conditions. At high DNA concentrations, the amount of specific initiation is reduced and transcripts both larger and smaller than the promoter-specific transcript (arrowed) are produced (compare lanes 4 – 7 of *Figure 3B*). The larger transcripts result from end-to-end transcription of the linear DNA template but the origin of the smaller transcripts is unknown (5).

4.2 **Variability between Extracts**

Extracts made from different cell preparations can vary in transcriptional activity over a 5- to 10-fold range, with about two thirds of these exhibiting activity within 2-fold of the observed maximum. Extracts should be compared for their activity using a run-off assay from a standard polymerase II promoter, such as the major late promoter of adenovirus 2. With optimal DNA and extract concentrations (Section 4.1) and $[\alpha$-$^{32}P]UTP$ at 100 Ci/mmol, a good extract (25 μl reaction volume) yields 10^6 d.p.m. in 1 h, equivalent to the synthesis of 20 ng of a 2200 nucleotide run-off transcript from the adenovirus 2 late promoter. This represents the synthesis of one RNA molecule per 10 DNA template molecules present, although the extract may actually be utilising only a fraction of the templates present with multiple rounds of initiation per active template.

4.3 **Reaction Conditions**

4.3.1 *Temperature*

One unusual feature of the whole-cell extract system is the temperature dependence of the reaction. Transcription is routinely carried out at 30°C and the RNA synthesised is stable for at least 8 h. Increasing the temperature to 37°C greatly enhances the rate of RNA degradation. Thus, RNA made at 30°C is degraded within 10 min at 37°C. Transcription assayed at 25°C yields the expected Arrhenius effect.

4.3.2 *Ionic Composition*

Specific transcription in the whole-cell extract is highly sensitive to ionic strength. Concentrations of KCl or NaCl above 60 mM significantly inhibit the reaction and concentrations of 120 mM or higher virtually eliminate specific transcription. Reactions can also be performed in $15 - 30$ mM $(NH_4)_2SO_4$. The divalent cations Zn^{2+}, Ca^{2+} or Mn^{2+} inhibit transcription. EDTA, 0.2 mM final concentration, is added routinely to control the possible inhibition by any heavy metal contaminants.

4.3.3 *pH*

Reactions are usually performed at pH 7.9, which is optimal for purified RNA polymerase II (1). However, there does not appear to be a sharp pH optimum for specific transcription by this enzyme in the whole-cell extract system.

4.3.4 *Nucleotides*

Even after extensive dialysis, most whole-cell extracts appear to contain a pool of free nucleotides (about 1 μM) (23). The routine addition of 4 mM creatine phosphate to the reaction mix ensures recharging of the nucleoside triphosphates, thus allowing a reduction in triphosphate concentrations and a consequent increase in the final specific radioactivity of the radiolabelled nucleotide added. In the presence of creatine phosphate, concentrations of

UTP, CTP and GTP as low as 5 μM are saturating for specific transcription, although higher concentrations (up to 500 μM) do not inhibit specific transcription (24). Because of endogenous pools, the transcription reaction is not fully dependent on addition of these three nucleotides. A higher concentration of ATP (50 μM) is required for optimal activity but ATP concentrations above 200 μM inhibit the reaction (23).

5. POST-TRANSCRIPTIONAL PROCESSING OF MESSENGER RNA PRECURSOR MOLECULES

5.1 Capping and Methylation

The dialysed whole-cell extract contains sufficient S-adenosylmethionine to methylate the 5′ ends of the mRNA precursor molecules synthesised by RNA polymerase II (5). The newly synthesised RNA is rapidly and quantitatively capped to form cap I structures [m^{7}GpppX(m)pY], although cap II structures [m^{7}GpppX(m)pY(m)pZ] have not been detected (25). Internal methylation has not been studied. Addition of exogenous S-adenosylmethionine does not affect the reaction (17,26), although S-adenosylhomocysteine inhibits specific transcription initiation by RNA polymerase II (26).

5.2. Splicing

Only low and somewhat irreproducible levels of mRNA splicing have been observed in whole-cell lysates similar (or identical) to the system described in this chapter (12 – 14). However, this is an active area of research and, therefore, this situation is likely to improve in the near future.

5.3 Polyadenylation

Transcription of exogenous DNA in whole-cell extracts under the conditions described in Section 2.2 results in the synthesis of RNA molecules that are not detectably modified by polyadenylation. However, conditions have been developed recently that allow for the efficient, accurate and specific polyadenylation of exogenous mRNA precursors in a whole-cell extract derived from HeLa cells (11). The mRNA precursors are synthesised using linear DNA templates under the standard reaction conditions (Section 2.2) and purified as described earlier *(Table 2)*. The 60 μl of RNA obtained *(Table 2, step 10)* from the standard transcription reaction is precipitated with ethanol once more, lyophilised and dissolved in 50 μl of sterile distilled water. There is no need to remove any contaminating DNA. The RNA is then incubated with a second aliquot of fresh whole-cell extract under conditions for polyadenylation. A standard polyadenylation reaction mixture (25 μl volume) contains:

    5 μl of whole-cell extract
    10 μl of RNA transcripts
    0.1 – 0.5 mM ATP
    4.0 mM creatine phosphate
    0.5 mM EDTA.

Each reaction mixture is incubated at 30°C for 60 min and the RNA is then extracted by the usual protocol *(Table 2)*. The extent of polyadenylation can be monitored either by affinity chromatography of the RNA on oligo(T)-cellulose (e.g. ref. 27) or by size analysis using glyoxalation followed by gel electrophoresis *(Table 3)*.

Figure 4 shows an example of polyadenylation of *in vitro* synthesised RNA analysed by the latter procedure. The recombinant plasmid originally used as template contained the strong adenovirus 2 late promoter fused to the T antigen-encoding sequences of SV40. The DNA had been cleaved with *Bam*HI, which cuts the DNA about 45 bp downstream from the poly(A) addition site utilised *in vivo*. The length of the poly(A) segment added (lane 2) was between 150 and 300 nucleotides.

The conditions required to obtain polyadenylation in the whole-cell extract are as precise as those required to obtain initiation of specific transcription although they are quite different. For example, polyadenylation is drastically inhibited at monovalent cation concentrations greater than 30 mM and divalent cation concentrations greater than 4 mM. There is also a requirement for a minimum concentration of RNA. As with transcription initiation, this requirement appears to be for total *mass* of RNA rather than for *specific* RNA sequences. Thus, when attempting to polyadenylate a pre-mRNA *in vitro*, the extract and RNA concentrations should be varied to optimise the reaction, using the values given above as a guide. Under optimal conditions, 70% or more of the precursor molecules become polyadenylated and the length of the poly(A) segment added is similar to that *in vivo*, giving rise to 150 – 300 nucleotide long stretches of poly(A). Furthermore, the elongation reaction appears to be processive, and the rate of elongation is relatively high. Under suboptimal conditions, both the *fraction* of precursor RNA that becomes polyadenylated and also the *length* of the poly(A) segment added are reduced.

The *in vitro* polyadenylation reaction is remarkably specific; only precursor mRNAs synthesised *in vitro* that contain a 3' end located at or slightly downstream from the authentic poly(A) addition site *in vivo* can be efficiently polyadenylated *in vitro*[1]. Thus, for future studies of the polyadenylation reaction, it may be fruitful to test the ability of RNAs produced from DNA templates that had been cleaved with different restriction endonucleases to serve as substrates for polyadenylation. These results are in marked contrast to those obtained when purified poly(A) polymerases are used to catalyse poly(A) addition. Then, no specificity is observed in primer selection, the reaction is inefficient, and the size of the poly(A) segment added is not controlled (see ref. 28 for review).

---

[1]The specific endonuclease activity implicated in the formation of mRNA 3' ends for *in vivo* polyadenylation (reviewed in ref. 15) has not yet been detected *in vitro*; AMP residues are added directly onto the end of the RNA created by run-off transcription.

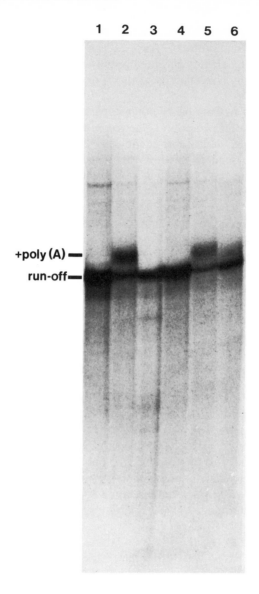

Figure 4. Characterisation of run-off RNA transcripts after polyadenylation. Run-off RNA was synthesised in the whole-cell extract system from a *Bam*HI-generated fragment of pϕ4-SVA, purified and added to a polyadenylation mixture as described in the text. Following extraction and glyoxalation, the RNA was analysed by electrophoresis through a 1.2% agarose gel. **Lane 1,** run-off RNA prior to polyadenylation; **lane 2,** RNA incubated in a polyadenylation reaction mixture as described in the text; **lane 3,** RNA incubated in a polyadenylation reaction mixture minus ATP; **lane 4,** RNA incubated in a polyadenylation reaction mixture minus extract but plus an equal volume of dialysis buffer; **lane 5,** identical to **lane 2** except that the amount of RNA added was reduced by 50%; **lane 6,** identical to **lane 3** except that the amount of RNA added was reduced by 80%. In **lanes 1 – 4,** the amount of RNA analysed was equivalent to 20% of the RNA in a polyadenylation reaction; in **lane 5,** 40% and in **lane 6,** 100% of the RNA was analysed.

6. TRANSCRIPTION BY RNA POLYMERASES I AND III IN WHOLE-CELL EXTRACTS

Whole-cell extracts contain significant levels of RNA polymerase I (R.Tjian, personal communication) and RNA polymerase III (e.g., ref. 19) activities. The ionic conditions required by these two polymerases may be different from those required by RNA polymerase II and, therefore, titrations of the salt concentration, as well as extract and DNA concentrations, should be carried out in preliminary studies. It is noteworthy that RNA polymerases I and III appear to transcribe exogenous DNA templates much more efficiently than does RNA polymerase II; each template can be transcribed 1 – 10 times by the former polymerases during the course of a standard incubation which is 10 – 100 times more efficient than transcription by RNA polymerase II (Section 4.2). Unlike RNA polymerases II and III, RNA polymerase I appears to display a distinct species specificity; RNA polymerase I contained in whole-cell extracts prepared from HeLa cells will transcribe human but not mouse rRNA genes (e.g. ref. 9).

Since RNA polymerase III, unlike RNA polymerase II, terminates transcription *in vitro*, and since the sizes of the RNAs produced are usually relatively small (however, see ref. 19), the products of RNA polymerase III transcription can be analysed directly. That is, the procedures of run-off transcription and nuclease S1 mapping are not usually required. In contrast, due to the large size of the transcripts from the rRNA genes, run-off and nuclease S1 procedures are normally necessary for analyses of transcription by RNA polymerase I.

Finally, whole-cell extracts, at least those prepared from HeLa cells, appear to contain most if not all of the enzymes involved in processing of tRNA precursor molecules (8) and at least one of the enzymes required for processing rRNA precursor molecules (9,10).

7. ACKNOWLEDGEMENTS

I thank R.Jove for carrying out the experiment shown in *Figure 1*. Work from the author's laboratory was supported by NIH grant GM-28983.

8. REFERENCES

1. Roeder,R.G. (1976) in *RNA Polymerase,* Losick,R. and Chamberlin,M. (eds.), Cold Spring Harbor Laboratory Press, New York, p. 285.
2. Weil,P.A., Luse,D.S., Segall,J. and Roeder,R.G. (1979) *Cell,* **18,** 469.
3. Grummt,I. (1981) *Proc. Natl. Acad. Sci. USA,* **78,** 727.
4. Wu,G.-J. (1978) *Proc. Natl. Acad. Sci. USA,* **75,** 2175.
5. Manley,J.L., Fire,A., Cano,A., Sharp,P.A. and Gefter,M.L. (1980) *Proc. Natl. Acad. Sci. USA,* **77,** 3855.
6. Tsujimoto,Y., Hirose,S., Tsuda,M. and Suzuki,Y. (1981) *Proc. Natl. Acad. Sci. USA,* **78,** 4838.
7. Wasylyk,B., Kedinger,C., Corden,J., Brison,O. and Chambon,P. (1980) *Nature,* **285,** 366.
8. Standring,D.N., Venegas,A. and Rutter,W.G. (1981) *Proc. Natl. Acad. Sci. USA,* **78,** 5963.
9. Grummt,I., Roth,E. and Paule,M.R. (1982) *Nature,* **296,** 173.
10. Miller,K.G. and Sollner-Webb,B. (1981) *Cell,* **27,** 165.
11. Manley,J.L. (1983) *Cell,* **33,** 595.

12. Weingartner,B. and Keller,W. (1981) *Proc. Natl. Acad. Sci. USA,* **78,** 4092.
13. Goldenberg,C.J. and Raskas,H.J. (1981) *Proc. Natl. Acad. Sci. USA,* **78,** 5430.
14. Kole,R. and Weissman,S.M. (1982) *Nucleic Acids Res.,* **10,** 5429.
15. Manley,J.L. (1983) *Prog. Nucleic Acid Res. Mol. Biol.,* **30,** in press.
16. Sugden,B. and Keller,W. (1973) *J. Biol. Chem.,* **248,** 3777.
17. McMaster,G.K. and Carmichael,G.C. (1977) *Proc. Natl. Acad. Sci. USA,* **74,** 4835.
18. Lewis,M.K. and Burgess,R.R. (1980) *J. Biol. Chem.,* **255,** 4928.
19. Manley,J.L. and Colozzo,M.T. (1982) *Nature,* **300,** 376.
20. Zimmerman,S.B. and Sandeen,G. (1966) *Anal. Biochem.,* **14,** 269.
21. Berk,A.J. and Sharp,P.A. (1977) *Cell,* **12,** 721.
22. Hansen,U., Tenen,D.J., Livingston,D.M. and Sharp,P.A. (1981) *Cell,* **27,** 603.
23. Fire,A., Baker,C.C., Manley,J.L., Ziff,E.B. and Sharp,P.A. (1981) *J. Virol.,* **40,** 703.
24. Handa,H., Kaufman,R.J., Manley,J.L., Gefter,M.L. and Sharp,P.A. (1981) *J. Biol. Chem.,* **256,** 478.
25. Jove,R. and Manley,J.L. (1983) *J. Biol. Chem.,* in press.
26. Jove,R. and Manley,J.L. (1982) *Proc. Natl. Acad. Sci. USA,* **78,** 5842.
27. Manley,J.L., Sharp,P.A. and Gefter,M.L. (1979) *J. Mol. Biol.,* **135,** 171.
28. Edmonds,M. and Winters,M.A. (1976) *Prog. Nucleic Acid Res. Mol. Biol.,* **17,** 149.

CHAPTER 4

Transcription of RNA in Isolated Nuclei

WILLIAM F. MARZLUFF and RU CHIH C. HUANG

1. INTRODUCTION

The establishment of faithful *in vitro* transcription systems is an essential prerequisite for studies of the mechanism of transcription of specific genes and the processing of primary transcripts in eukaryotic cells. Two major approaches have been utilised. One approach starts with isolated nuclei and uses the endogenous chromatin as the template for transcription. The other is a DNA-dependent system, which uses specific cloned DNA fragments as templates for transcription. The former approach emphasizes the preservation of cellular activities *in vitro* whereas the latter has focussed on determining the DNA sequences and protein factors required for the accurate expression of cloned genes in the test tube. Both approaches have their advantages and should be considered in concert for a comprehensive analysis of RNA biosynthesis. This chapter describes the first approach while the second approach is discussed in Chapter 3.

The principal advantage of using isolated nuclei for transcriptional studies is that it is a system which is as close to the intact cell as possible. Within the isolated nucleus the chromatin is maintained in its native state. The integrity of the nucleus is maintained during the cell-free synthesis of RNA and the newly-made RNA remains associated with the nucleus, as it does in the intact cell. Thus, to a first approximation, the activity one measures reflects the activity of that nucleus in the cell; the same genes are being expressed in the same relative amounts in the isolated nucleus as in the intact cell.

The principal disadvantages are:

(i) The complexity of the RNAs being synthesised. All of the $10^4 - 10^5$ different genes active in the intact cell are presumably also transcribed in isolated nuclei. Thus, to study the synthesis of a particular RNA one requires extremely sensitive (hybridisation) assays.

(ii) When the nuclei are isolated there are many RNA polymerases already engaged in transcribing RNA so that much of the initial RNA biosynthetic activity observed *in vitro* reflects the completion of these RNA chains. Therefore sensitive assays for *initiation* of RNA synthesis are required to distinguish molecules which have been initiated *in vitro* from molecules which have simply been *completed*.

(iii) There is a large amount of endogenous RNA in the isolated nuclei. If this is not separated from the RNA which is synthesised *in vitro*, this RNA can interfere with the hybridisation assays designed to quantitatively measure newly-made RNA.

Several useful methods have been developed in the authors' laboratories to resolve these problems. It is possible to measure relative rates of transcription and changes in these rates. Sensitive assays for measuring chain initiation have been developed using thionucleoside triphosphates as substrates and appropriate cloned DNA sequences as hybridisation probes. Finally, since the RNA products are presumably assembled in native ribonucleoprotein particles after transcription, they may provide substrates for the study of RNA processing and transport. Most of these methods may be of general use in many eukaryotic systems, although the authors have used them only with cultured mouse myeloma cells and sea urchin embryos.

2. PREPARATION OF NUCLEI

2.1. **Introduction**

The purpose is to prepare a cell-free system which is active in RNA synthesis. The RNA synthesised should reflect both qualitatively and quantitatively the RNA synthesised in the intact cell. In particular, only the correct DNA strand should be transcribed, and initiation, termination and processing should occur at the same sites *in vitro* as *in vivo*. In addition, only those genes which are expressed *in vivo* should be expressed *in vitro*. To obtain nuclei active in faithful RNA synthesis, it is necessary to preserve RNA polymerase activity and nuclear structure during the isolation and to exclude degradative activities (nucleases, proteases, triphosphatases) from the preparation. As long as faithful RNA synthesis is maintained, the nuclei may vary from relatively crude nuclei still contaminated with substantial amounts of cytoplasmic material to highly purified nuclei.

The choice of starting material for the preparation obviously depends on the precise aims of the subsequent experiments. If possible, it is best to start with cultured cells (1,2) especially since most mammalian cell cultures are low in endogenous nuclease and protease activities. However, if one is studying a gene which is expressed only in a particular tissue or is properly regulated only in a particular tissue in the intact animal then obviously one must start with that tissue itself. This introduces a variety of potential problems, including cell heterogeneity. Nevertheless, with care, it is possible to prepare nuclei with good RNA synthetic activity from a range of tissues (3,4). This section describes a basic procedure and its modifications which are applicable to a wide range of cell types.

2.2 **Basic Procedure**

2.2.1 *Tissue Culture Cells*

Nuclei highly active in RNA synthesis can be easily prepared from HeLa cells simply by lysing the cells in Nonidet P-40 (NP-40) and collecting the nuclei by centrifugation at low speed (5). However, HeLa cells are an exception. For all other cultured cells tested, the most reliable preparations of active nuclei are

Table 1. Preparation of Nuclei from Tissue Culture Cells[a].

---

1. Prepare the following three buffers, each lacking DTT and Triton X-100. Autoclave the buffers and store at 4°C. Just prior to use, add DTT and Triton X-100 from stock solutions (1.0 M DTT, 10% Triton X-100).

   Buffer I
     0.32 M sucrose
     3.0 mM $CaCl_2$
     2.0 mM magnesium acetate
     0.1 mM EDTA
     0.1% Triton X-100
     1.0 mM DTT
     10.0 mM Tris-HCl, pH 8.0

   Buffer II
     2.0 M sucrose
     5.0 mM magnesium acetate
     0.1 mM EDTA
     1.0 mM DTT
     10.0 mM Tris-HCl, pH 8.0

   Storage Buffer
     25% glycerol
     5.0 mM magnesium acetate
     0.1 mM EDTA
     5.0 mM DTT
     50.0 mM Tris-HCl, pH 8.0

2. Harvest the cells by centrifugation (500 g for 5 min at 25°C) and resuspend them in buffer I at a density of $5 \times 10^6 - 2 \times 10^7$ cells/ml.[b]

3. Homogenise the cells (10 − 15 strokes) in a Dounce homogeniser fitted with a tight (B) pestle.

4. Dilute the homogenate with 1 − 2 volumes of buffer II and then layer this over a cushion of buffer II occupying one third the volume of a centrifuge tube.

5. Centrifuge at 30 000 g for 45 min at 4°C[c].

6. Resuspend the nuclei at $5 \times 10^7 - 2 \times 10^8$/ml (0.5 − 2.0 mg DNA/ml) in storage buffer. Nuclei from organisms with a lower DNA content per nucleus are stored at the same DNA concentration.

7. Quick-freeze the nuclei in small aliquots and store these in liquid nitrogen.

---

[a]For cells where endogenous nucleases or proteases are a potential probem, add inhibitors to all buffers as specified in Section 2.3.
[b]The cells may be washed with PBS prior to resuspension in buffer I but the authors have not found this to be necessary with the cultured cells which have been used.
[c]A preliminary low speed separation of nuclei from the homogenate prior to this centrifugation does not give satisfactory results. A great deal of other material pellets with the nuclei and is subsequently very difficult to separate.

made by lysing the cells in an iso-osmotic solution and then cleaning the nuclei by sedimentation through a dense solution of sucrose (1,2). The non-ionic detergent causes the cells to lyse easily without prior swelling and greatly reduces the amount of cytoplasmic contamination. This basic protocol is described in *Table 1*. Since the density of the nuclei is characteristic of the cell-type from which they are isolated, the correct concentration of sucrose in buffer II of the basic isolation procedure (*Table 1*) should be determined by preliminary experimentation since it is buffer II which forms the sucrose cushion through which the nuclei sediment (*Table 1*, steps 4 and 5). For example, nuclei from myeloma cells will not sediment through sucrose concentrations greater than 2.05 M and nuclei from early stages of sea urchin embryos will not sediment through sucrose concentrations greater than 1.95 M. The correct sucrose concentration which permits nuclei to pellet is usually in the range 1.8 − 2.2 M. The concentration of sucrose in the homogenate which is layered over the sucrose cushion (*Table 1*, step 4) is also important. It must be high enough to prevent the accumulation of intracellular membranes and organelles at the interface; otherwise the yield of nuclei is substantially reduced. The final sucrose concentration of the homogenate is governed by the volume of buffer II added to it (*Table 1*, step 4). Therefore, this must also be determined by preliminary experimentation. For example, to obtain nuclei from sea urchin embryos it is necessary to increase the sucrose concentration to 1.4 M in the homogenate prior to centrifugation otherwise the interface becomes clogged with membranous material (6). Finally, it is very important to use dilute homogenates since this reduces both the amount of material which accumulates at the interface and the material which adventitiously adsorbs to the nuclei. Once frozen in liquid nitrogen the nuclei retain activity for several months but should not be re-frozen for future use after thawing.

The authors have found this method to work well with several cultured cell lines and sea urchin embryos. Other investigators have used similar methods on other cell types.

2.2.2 *Whole Tissues*

The basic protocol described above for the preparation of nuclei from cultured cells can also be applied to the isolation of nuclei from whole tissues, once these have been homogenised. A suitable procedure is described in *Table 2*.

Table 2. Preparation of Nuclei from Whole Tissues.

1. Prepare the buffers described in *Table 1*, step 1, but add inhibitors to all the buffers as specified in Section 2.3.
2. Cut the tissue into small pieces[a]. Rinse these in ice-cold 0.14 M NaCl, 10 mM Tris-HCl, pH 8.0.
3. Homogenise the tissue pieces in buffer I[b] using a suitable homogeniser[c]. Use 10 ml of buffer I per gram of tissue.
4. Filter the homogenate rapidly through several layers of cheesecloth to remove connective tissue.
5. Homogenise the filtrate (10 − 20 strokes) using a Dounce homogeniser fitted with a tight (B) pestle. Monitor cell lysis by microscopy.
6. Proceed to isolate the nuclei as described in *Table 1*, steps 4 − 7.

[a]For the preparation of nuclei from liver, perfuse the liver with 0.14 M NaCl, 10 mM Tris-HCl, pH 8.0, prior to step 2.
[b]The composition of buffer I is given in *Table 1*.
[c]The homogeniser used will depend on the nature of the tissue. For soft tissues such as liver, a Dounce homogeniser should suffice.

2.3 **Precautions against Nucleases and Proteases**

For cells where endogenous proteases are a potential problem, PMSF should be added to all buffers used for the isolation of nuclei (*Table 1*), including the storage buffer, to a final concentration of 0.1 mM from a 0.1 M stock solution in isopropanol. This will effectively inhibit serine proteases without affecting RNA synthesis. Endogenous nuclease activity is more of a problem. However, the removal of all divalent ions from buffer I (*Table 1*) and inclusion of 1 mM EGTA and 1 mM spermidine will inhibit nucleases which require divalent ions, including many DNases (6,10).

Since an intact chromatin structure is required for accurate transcription, it is important to assay nuclei for intactness of both the DNA and histones even if the above precautions are taken. Suitable tests are described in *Table 3*. In addition, the nuclei can be analysed for RNase by performing a 'chase' experiment (*Table 3*). The nuclei are incubated under the standard conditions for transcription but an excess of unlabelled ribonucleotides (2 mM) is added to the reaction after 5 min and incubation is continued. If the size of the RNA product, as judged by sucrose gradient centrifugation or gel electrophoresis, does not change and is relatively large, then there is now extensive RNase activity.

2.4 **Variations of the Nuclear Isolation Procedure**

2.4.1 *Cells Rich in Pigment Granules or Lysosomes*

In cases where there is a high content of pigment granules or lysosomes in the tissue or cultured cells, the use of the standard method (Section 2.2) involving non-ionic detergent will result in lysis of these organelles, the adsorption of released pigments to the nuclei and potential degradation of some nuclear constituents by the released lysosomal enzymes. To avoid these problems, nuclei can be prepared from these sources by a similar method without using detergent. The cells are swollen in 25 mM KCl, 2 mM magnesium acetate, 10 mM Tris-HCl (pH 7.5) for 5 min at 4°C and then homogenised as usual (*Tables 1* and *2*). The pigment granules and lysosomes remain intact and are separated from the nuclei during the subsequent centrifugation procedure.

2.4.2 *Non-aqueous Methods for Isolation of Nuclei*

During aqueous isolation procedures, a large amount of material is lost from the nuclei, including all the small molecules, many enzymes and possibly some RNAs. Gurney (11) has pioneered the use of non-aqueous methods which yield nuclei containing all the small molecules normally present in nuclei *in vivo*. These nuclei are active in both RNA and DNA synthesis.

2.5 **Isolation of Chromatin with Endogenous RNA Polymerase Activity**

It is possible to prepare chromatin from isolated nuclei which retains endogenous RNA polymerase activity (12); see *Table 4*. Chromatin prepared in this way is nearly as active in RNA synthesis as are intact nuclei. However, it loses activity on freezing whereas nuclei can be stored for several months in liquid

Table 3. Assays for DNase, RNase and Protease in Isolated Nuclei.

DNase

To assay DNase activity in isolated nuclei, one can examine either the endogenous DNA or exogenously-added DNA for degradation[a]. The latter assay is more sensitive in detecting small amounts of DNase.

Endogenous DNA

1. Incubate the nuclei under identical conditions used for transcription (*Table 5*) for 30 min but omit the radioactive nucleoside triphosphates.

2. Lyse the nuclei by adding 10 volumes of 1% SDS, 0.5 M NaCl, 10 mM EDTA, pH 8.0. Add Proteinase K to 50 μg/ml and incubate at 37°C for 30 min.

3. Add an equal volume of phenol saturated with 50 mM Tris-HCl, pH 8.5. Mix thoroughly and then separate the phases by low-speed centrifugation.

4. Recover the aqueous phase and then dialyse it against 10 mM NaCl, 1 mM EDTA, 10 mM Tris-HCl, pH 8.0.

5. Determine the size of the isolated endogenous DNA by agarose gel electrophoresis, comparing it with DNA prepared from nuclei which have not been incubated. Use a 0.8% agarose gel prepared and run in 1 mM EDTA, 50 mM Tris-borate, pH 8.3. Add Bromophenol blue to the DNA samples to 0.001% final concentration then load these onto the gel. In a separate lane, load phage λ DNA restricted with *Hind*III together with intact phage λ DNA as size markers. Electrophorese at 100 V until the Bromophenol blue reaches the end of the gel. Visualise the DNA by staining the gel with 0.5 μg/ml ethidium bromide for 15 min. View the DNA bands by u.v. illumination. DNA from mouse myeloma cell nuclei, for example, should be larger than 50 kb and DNA from sea urchin nuclei should be larger than 20 kb after incubation for 30 min.

Exogenous DNA

1. Incubate the nuclei under identical conditions used for transcription (*Table 5*) for 30 min but omit the radioactive nucleoside triphosphates and include supercoiled plasmid DNA (2 μg per 0.2 ml reaction mixture).

2. Isolate the DNA and determine its size by agarose gel electrophoresis as described above. The disappearance of supercoils and appearance of open circular DNA is evidence of DNA nicking which could be due to topoisomerase activity.

RNase

1. Incubate the nuclei as described in *Table 5*.

2. After 5 min, prepare RNA from a portion of the reaction mixture using the protocol described in *Table 6*.

3. Add a 40-fold excess (2 mM) of unlabelled ribonucleoside triphosphate. This stops the incorporation of radioactivity into RNA without inhibiting RNA synthesis. Incubate for an additional 25 min and then prepare RNA from the reaction mixture as described in *Table 6*.

4. Determine the size of the radiolabelled RNA in the two samples (steps 2 and 3 above) either by sucrose gradient centrifugation (*Table 7*) or agarose gel electrophoresis (*Table 8*). Monitor the labelled RNA by liquid scintillation counting or autoradiography, respectively. If the nuclei lack significant RNase activity, the two samples will contain RNA of about the same size and the RNA will be large. If desired, the gels can be stained with ethidium bromide (*Table 8*) to monitor the size of the endogenous RNA.

Protease

1. Incubate the nuclei as described in *Table 5* for 30 min but omit the radioactive ribonucleoside triphosphate.

2. After the incubation, add an equal volume of 0.4 M H_2SO_4. Incubate at 4°C for 30 min.

3. Centrifuge at 10 000 *g* for 10 min. Recover the supernatant and add four volumes of ethanol. Mix and store at − 20°C overnight or − 70°C for 2 h.

4. Recover the precipitated histones by centrifugation at 10 000 *g* for 10 min.

5. Wash the histone pellet with 80% ethanol and then dry it under vacuum.

6. Analyse the histones by gel electrophoresis. Three methods are available and are described in references 7 − 9. Protease activity in the isolated nuclei is manifest by the detection of degraded histones when compared with those isolated from nuclei which have not been incubated. Histone H1 is particularly sensitive to degradation by protease.

[a]A rapid test of DNA size is to lyse the nuclei with 10 volumes of 1% SDS, 10 mM EDTA. If the DNA is intact, the lysed nuclei form a gel, whereas if the size of the DNA has been reduced even only slightly (to 20 kb or less), a relatively non-viscous solution is obtained.

Table 4. Preparation of Chromatin Active in Transcription[a].

1. Prepare nuclei as described in *Tables 1* or *2*.

2. Suspend the nuclei in 10% glycerol, 1 mM DTT, 1 mM EDTA, 10 mM Tris-HCl, pH 8.0 at 5 x 10[7] nuclei/ml. Centrifuge at 1000 *g* for 2 min.

3. Repeat step 2 until the nuclei lyse, yielding a gelatinous pellet (chromatin).

4. Resuspend the chromatin in 1 ml of 0.12 M KCl, 10% glycerol, 1 mM DTT, 10 mM Tris-HCl, pH 8.0 and centrifuge at 1000 *g* for 2 min[b].

5. Resuspend the chromatin at 1 mg DNA per ml (10[8] nuclei per ml for cultured mouse cells) in 10% glycerol, 1 mM DTT, 10 mM Tris-HCl, pH 8.0 by shearing first through an 18-gauge needle, and then through a 22-gauge needle.

6. Use the chromatin immediately for transcription, incubating it under the same conditions used for RNA synthesis in isolated nuclei (*Table 5*).

[a]See also *Table 1* of Chapter 5.
[b]This wash step, to remove non-chromatin components, is optional. The non-chromatin components are soluble in 0.12 M KCl whereas chromatin itself is insoluble.

nitrogen. Methods for isolation of chromatin and transcription by *exogenous* RNA polymerase are described in detail in Chapter 5.

3. RNA SYNTHESIS IN ISOLATED NUCLEI

3.1 Conditions for RNA Synthesis

For RNA synthesis, nuclei are incubated with all four ribonucleoside triphosphates, one of which is radiolabelled. The concentrations must be at least 50 μM for CTP, GTP and UTP and 100 μM for ATP for maximal activity. Routinely, 500 μM of each of the three unlabelled triphosphates are used and 50 μM of the labelled triphosphate (100 μM if the label is in ATP).

When studied *in vitro* using DNA templates, RNA polymerases I, II and III each have their own salt optima. However, all of these enzymes must function under similar conditions in the cell. Not surprisingly, therefore, the optimal conditions for RNA synthesis in mammalian nuclei for all three polymerases are 80 − 120 mM KCl, 5 − 8 mM magnesium acetate, 25 mM Tris-HCl, pH 7.5 − 8.0. There is no significant change in activity when conditions are varied within these limits. At low salt concentrations (<25 mM KCl) the activities of RNA polymerases II and III are reduced and the major activity is due to RNA polymerase I. RNA polymerase II is greatly stimulated by Mn^{2+} ions when studied using DNA as template *in vitro*, but even a low concentration of Mn^{2+} ions (1 mM) inhibits transcription by all three polymerases by about 50%. Addition of the polyamines spermine or spermidine to a final concentration up to 1 mM has no effect on the rate of total RNA synthesis.

The detailed procedure for transcription by isolated nuclei is described in *Table 5*. The reaction mixture is assembled by preparing a 2-fold concentrated solution of the various salts and ribonucleoside triphosphates and then adding an equal volume of nuclei. Any of the four $[\alpha-^{32}P]$NTPs can be used as the radiolabelled nucleotide precursor. However, the authors have observed that some nuclei (e.g., sea urchin nuclei) transfer radiolabel from $[\alpha-^{32}P]$UTP to the 3' end of some RNAs. Reactions are also known in which $[\alpha-^{32}P]$ATP (polyadenylation or tRNA maturation), or $[\alpha-^{32}P]$CTP (tRNA maturation) are added to the end of intact RNAs. In addition, $[\alpha-^{32}P]$GTP is the substrate in the capping reaction. Therefore, although similar rates are obtained regardless of which radiolabelled triphosphate is used, $[\alpha-^{32}P]$GTP is the best choice for most experiments since there is less incorporation of this nucleotide into RNA by reactions which are independent of transcription. The reaction mixture is incubated at a temperature in the range $25-37°C$ for mammalian nuclei or $15-25°C$ for sea urchin nuclei. The temperature usually chosen is $25°C$ since RNA synthesis continues for a much longer time at this temperature. Whilst a systematic investigation has not been carried out, there does not seem to be a significant difference in the types of RNA made at different temperatures in the ranges quoted. The reaction is allowed to continue for $30-60$ min and then total RNA is isolated (see Section 3.3) for subsequent analysis.

If desired, the time course of transcription can be monitored by removing aliquots at various times during the incubation and determining TCA-precipitable radioactivity (*Table 5*). The amount of RNA synthesised is expressed as pmol GMP incorporated per microgram DNA per unit time. The DNA concentration can be determined by counting an appropriate dilution of the nuclei using a haemocytometer (if the DNA content per nucleus is known) or by a standard colorimetric assay for DNA. The initial rate of RNA synthesis is typically $0.1-0.2$ pmol GMP incorporated/μg DNA/min. Synthesis of RNA usually continues for at least one hour although often at a somewhat reduced rate of $0.05-0.1$ pmol GMP incorporated/μg DNA/min.

3.2 Determination of the Activities of Different RNA Polymerases

The three eukaryotic RNA polymerases may be readily distinguished using the fungal cyclic peptide, α-amanitin (*Figure 1*). At a concentration of 0.01 μg/ml, this inhibits mammalian RNA polymerase II activity by 50% whereas RNA polymerase III is inhibited to this extent only at much higher concentrations (25 μg/ml) (13). RNA polymerase I is not inhibited by this peptide. Therefore, to determine the amount of RNA synthesis due to each polymerase, perform three separate incubations (*Table 5*):

Table 5. Transcription by Isolated Nuclei.

1. Prepare the following three solutions using autoclaved distilled water and store these at −20°C.

    Solution A

    2 mM each of ATP, CTP, UTP
    0.2 mM [α-^{32}P]GTP (10−40 Ci/mmol; 500−1000 μCi/ml)[a]
    0.1 mM S-adenosylmethionine

    To prepare this solution, mix one-tenth volumes of separate stock solutions (20 mM each) of ATP, CTP and UTP and one-tenth volume of 1 mM S-adenosylmethionine in 1 mM H$_2$SO$_4$. Freeze-dry the [α-^{32}P]GTP and re-dissolve it in solution A containing sufficient GTP to bring the final concentration of [α-^{32}P]GTP plus unlabelled GTP to 0.1−0.2 mM. It is important that the pH of the solution is pH 7.0. Therefore, if the nucleotides used are in the free acid form, adjust the pH to pH 7.0 using Tris base. Add water to the correct final volume.

    Solution B

    0.6 M KCl
    12.5 mM magnesium acetate

    Solution C

    1% SDS
    10 mM EDTA, pH 7.0

2. For a reaction mixture of 0.2 ml final volume, mix 50 μl of solution A, 40 μl of solution B and 10 μl of autoclaved distilled water. Then start the reaction by adding 0.1 ml of nuclei suspended in storage buffer (*Table 1*)[b]. Incubate at 25°C.

3. To monitor transcription, remove aliquots (1−5 μl) at the start of the incubation and at 10 min intervals, spotting each onto a square of Whatman 1MM filter paper labelled with pencil. Place these into a beaker containing ice-cold 10% TCA, 1% sodium pyrophosphate (15−30 ml per filter). Incubate on ice for 10 min with occasional mixing. Discard the solution and then wash the filters twice for 5 min each with 5% TCA, 1% sodium pyrophosphate. Finally, rinse the filters with absolute ethanol or acetone, air-dry them and determine the retained radioactivity in a liquid scintillation counter using a toluene-based scintillation fluid.

    In a typical reaction the amount of radioactivity incorporated into TCA-precipitable material will increase 20- to 100-fold during the initial 30 min incubation period. There is a significant background radioactivity (at the start of the incubation) due to non-specific binding of the radionuclide to the nuclei.

4. Stop the reaction by adding 10 volumes of solution C and mixing well. This lyses the nuclei and results in a clear viscous solution.

5. Extract the RNA as described in *Table 6*.

---

[a]Solution A described here contains [α-^{32}P]GTP as the radioactive ribonucleoside triphosphate. Any of the other radioactive nucleotides may be used (see the text). In all cases, the final concentration of the labelled nucleotide in the reaction mixture should be 0.05−0.1 mM.
[b]The Tris-HCl buffer component of the reaction mixture (see Section 3.1) is provided by the storage buffer in which the nuclei are suspended.

(i)   without α-amanitin,
(ii) with 0.5 μg/ml α-amanitin,
(iii) with 100 μg/ml α-amanitin.

Add the α-amanitin to the relevant reaction mixtures prior to addition of the nuclei. Next add the nuclei and leave the mixtures on ice for 5 min. Finally, start the reactions by transferring the mixtures to 25°C.

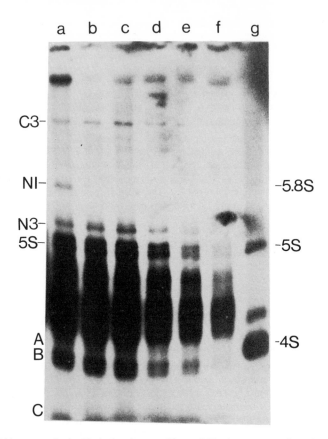

Figure 1. RNA was synthesised in isolated sea urchin nuclei in the presence of varying amounts of α-amanitin. The RNA prepared from the extranuclear fraction was analysed by polyacrylamide gel electrophoresis as described in *Table 8*. The gel was stained with ethidium bromide to locate the position of the 5.8S, 5S and 4S RNA markers and the radioactive RNA transcripts were then detected by autoradiography. The concentrations of α-amanitin used were: **lane a**, no α-amanitin; **lane b**, 1 μg/ml; **lane c**, 10 μg/ml; **lane d**, 100 μg/ml; **lane e**, 250 μg/ml; **lane f**, 500 μg/ml. **Lane g** contained 4S tRNA and 5S rRNA. The N1 RNA transcript is a product of RNA polymerase II and the other RNAs are all products of RNA polymerase III. The RNAs migrating between 4S and 5S rRNA are the tRNA precursor RNAs. N1 and N3 are small nuclear RNAs (6). C3 is the major small cytoplasmic RNA. RNAs A, B and C are unidentified RNAs which have not been detected *in vivo*.

The difference in the amount of RNA made in reactions (i) and (ii) represents RNA polymerase II activity. The difference between (ii) and (iii) represents RNA polymerase III activity and the activity in reaction (iii) is due to RNA polymerase I. In nuclei isolated from cultured cells, about 45% of the RNA synthesis is due to polymerase I, 50% to RNA polymerase II and 5% to RNA polymerase III. In sea urchin embryo (blastula) nuclei, 90−95% of the activity is due to RNA polymerase II and the remainder to RNA polymerases I and III. In each case, the proportion of the total activity due to each polymerase is similar to that found *in vivo*, an indication that under the conditions for RNA synthesis described above, the isolated nuclei reflect the pattern of RNA synthesis found *in vivo*.

Table 6. Isolation of RNA.

---

1. After the transcription reaction has been stopped by addition of 10 volumes of solution C (*Table 5*, step 4), add one-tenth volume of 2 M sodium acetate, pH 5.0. Mix well.

2. Add an equal volulme of water-saturated phenol-chloroform (2:1) and mix well.

3. Incubate the mixture at 55°C for 5 min and then cool on ice for 5 min.

4. Centrifuge the mixture at 10 000 *g* for 10 min at 4°C.

5. Carefully recover the (upper) aqueous phase, avoiding the interface.

6. Add 2.5 volumes of absolute ethanol to the aqueous phase and incubate at −70°C for 1 h (or −20°C for 16 h) to precipitate the RNA.

7. Recover the RNA by centrifugation at 10 000 *g* for 10 min at 4°C.

8. Wash the RNA by resuspending the pellet in cold (4°C) 70% ethanol and centrifugation as in step 7.

9. Repeat step 8.

10. Dry the RNA pellet under vacuum and then dissolve the RNA in 0.5 ml of 0.3 M NaCl, 0.1% SDS, 1 mM EDTA, 10 mM Tris-HCl, pH 7.5.

11. Remove residual unincorporated, labelled nucleotide by chromatography on a column (0.5 cm x 10.0 cm) of Sephadex G50 or Biogel P6 in the above buffer.

12. Recover the RNA (in the void volume) by ethanol precipitation as described in steps 6 and 7.

13. Dissolve the RNA in sterile water and store at −20°C.

3.3 Isolation of RNA

After the reaction, the RNA may be isolated by any of the standard procedures used for nuclear RNA:
(i)   extraction with phenol in the presence of SDS at pH 5.0 and 55°C (14);
(ii) treatment with DNase I followed by extraction with phenol (15);
(iii) extraction with guanidinium isothiocyanate (16).

The guanidinium isothiocyanate procedure was developed for tissues rich in endogenous RNase. Since the nuclei must not contain significant amounts of RNase if transcription is to work well (Section 2.3), there is usually no need for this procedure. The DNase I procedure (15) leaves the RNA contaminated with small DNA fragments. Therefore, of the three methods available, the authors prefer the hot phenol-SDS procedure. This is described in *Table 6*.

4. GENERAL ANALYSIS OF RNA SYNTHESIS

The method used for analysis of transcription by isolated nuclei depends upon the aims of the experiment. Sections 4.1, 4.2 and 4.3 describe general procedures for the analysis of total RNA products, determination of the rate of elongation of RNA chains by nuclei and assays for initiation of transcription, respectively. Suitable techniques for the analysis of RNA transcripts of particular genes are described in Section 5.

4.1 Analysis of Total RNA Products

The size range of the total RNA products can be analysed by sucrose density gradient centrifugation (*Table 7* and *Figure 2*), monitoring the RNA

Table 7. Sucrose Gradient Centrifugation of RNA.

1. Prepare 10 – 40% (w/v) sucrose gradients in 0.1 M NaCl, 1 mM EDTA, 0.1% SDS, 10 mM Tris-HCl, pH 7.5. This range of sucrose concentrations allows all the RNA species of interest (4S – 45S) to be displayed.

2. Dissolve the RNA sample in 0.1% SDS, 1 mM EDTA, pH 7.5. Load 0.5 ml of sample onto a 17 ml gradient or 1.5 ml onto a 38 ml gradient.

3. Centrifuge the gradients at 90 000 *g* for 16 h. Suitable rotors are either the Sorvall AH627 or the Beckman SW27. Under these conditions the 28S RNA sediments ~60% of the distance down the centrifuge tube.

4. Fractionate each gradient through a u.v. monitor (set to read at 254 nm) into 25 fractions. There is sufficient rRNA in isolated nuclei to serve as size markers (28S and 18S) when monitored by u.v. absorbance in this manner. Locate the labelled RNA transcripts by spotting an aliquot (5 – 20 μl) of each fraction onto a square of Whatman filter paper and determining the TCA-precipitable radioactivity as described in *Table 5*, step 3.

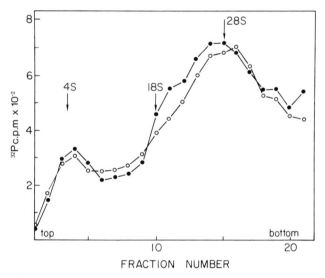

Figure 2. RNA was synthesised in isolated mouse myeloma nuclei for 30 min using [α-^{32}P]GTP as a radiolabelled precursor and was then fractionated on a sucrose gradient as described in *Table 7*. The gradient profiles show RNA synthesised as described in *Table 5* (○) and RNA synthesised in the presence of a cell extract (●) as described in Section 6.4. The positions of 28S and 18S rRNA and 4S tRNA are indicated as markers.

transcripts by scintillation counting of fractions. Alternatively, small RNA transcripts can be examined by electrophoresis in denaturing polyacrylamide gels (*Table 8* and *Figure 1*) whilst large RNAs are best resolved by electrophoresis in denaturing agarose gels (*Table 8*). The size of the RNA product is relatively large if the nuclei are free of ribonuclease (see ref. 1 and *Figure 2*).

Although particular small RNAs, incuding 5S rRNA and tRNA, are readily detected and quantitated by the techniques of gel electrophoresis, it is usually necessary to employ sensitive hybridisation assays to study individual mRNAs. These techniques are described in Section 5.

Table 8. Gel Electrophoresis of RNA.

Small RNA Transcripts

Small RNA (<8S) is easily analysed by electrophoresis in 10% polyacrylamide gels in the presence of 7 M urea.

1. Prepare the acrylamide:bisacrylamide (30:1) stock solution as follows.

acrylamide	30 g
bisacrylamide	1 g
urea	42 g
20 x TBE buffer[a]	5 ml
Distilled water to 100 ml final volume	

2. Degas a suitable volume of acrylamide:bisacrylamide stock solution for the electrophoresis apparatus being used. This is conveniently achieved using a vacuum line for 10 min. Add 30 μl TEMED and 0.7 ml l of freshly-prepared 10% ammonium persulphate per 100 ml of gel mixture, mix well and pour the gel.

3. Dissolve the radioactive RNA sample in a small volume of distilled water. Mix 10 μl of the RNA with 20 μl of 10 M urea, 0.1 x TBE[a], 0.001% Bromophenol blue, 0.001% xylene cyanol. Heat at 65°C for 10 min and then load the sample onto the gel.

4. Electrophorese at 900 – 1000 V using 1 x TBE[a] buffer as electrode buffer until the xylene cyanol reaches the bottom of the gel[b]. The gel plates should be hot to the touch during electrophoresis.

5. Transfer the gel to Whatman 3MM filter paper and dry it using a commercial gel dryer. Detect the radioactive RNA by autoradiography. Alternatively, the gel may be stained with 0.5 μg/ml ethidium bromide for 15 min prior to drying and endogenous nuclear RNAs visualised by u.v. illumination.

Large RNA Transcripts

1. Prepare a 0.8% agarose gel in 10 mM sodium phosphate, pH 7.0.

2. Dissolve the radioactive RNA in 1 M deionised glyoxal[c], 50% DMSO, 10 mM sodium phosphate, pH 7.0 and incubate at 50°C for 1 h[d]. This glyoxalation step denatures the RNA.

3. Load the glyoxalated RNA onto the gel and electrophorese at 100 V with Bromophenol blue as marker. Continuously recirculate the electrode buffer between the anodic and cathodic reservoirs during electrophoresis to avoid the pH increasing above pH 8.0 or the RNA will be de-glyoxylated. Continue electrophoresis until the dye has migrated two-thirds of the way down the gel[e].

4. Detect the radioactive RNA by autoradiography, If marker RNAs (e.g., 28S and 18S rRNA) have also been electrophoresed these can be visualised by staining with ethidium bromide.

[a]TBE buffer: 50 mM Tris, 50 mM boric acid, 50 μM EDTA.
[b]tRNA migrates ~ 80% as fast as the xylene cyanol.
[c]Prepare 3 M glyoxal (Sigma) in water and deionise this by passage through a 1 ml column of Bio-Rad resin AG501-X8. Add 3 M glyoxal to 67% DMSO, 10 mM sodium phosphate, pH 7.0 to prepare the glyoxalating solution.
[d]Glyoxalation of RNA as described by McMaster and Carmichael (17).
[e] tRNA runs ahead of the Bromophenol blue.

4.2 Rate of Elongation

The average rate of elongation can be determined relatively easily. The detailed protocol is described in *Table 9*. In summary, the nuclei are incubated for a short period of time (0.5 – 1.0 min) with [³H]GTP or another ³H-labelled nucleoside triphosphate. The RNA is isolated, digested with RNases T1 and T2 and the resulting nucleotides and nucleosides are then separated by paper

Table 9. Measurement of Rate of Elongation of RNA.

1. Add the nuclei to the standard reaction mixture for transcription (*Table 5*) containing all four unlabelled ribonucleoside triphosphates at the normal concentrations but lacking the radio-labelled GTP.

2. Incubate for 5 min at the desired temperature.

3. Add the required amount of [α-^{32}P]GTP. After $0.5 - 1.0$ min, stop the reaction by adding 10 volumes of 1% SDS, 10 mM EDTA, pH 7.0.

4. Isolate the RNA as described in steps $1 - 6$ of *Table 6*.

5. Dissolve the RNA in 0.5 ml of 1% SDS, 1 mM EDTA, pH 7.0 and analyse it by sucrose gradient centrifugation as described in *Table 7*. This sucrose gradient centrifugation step is designed to separate large RNA (which is still being elongated) from small RNA (see Section 4.1). Therefore, pool those gradient fractions containing RNA larger than 15S.

6. Precipitate the pooled RNA with ethanol (*Table 6*, steps 6 and 7).

7. Dissolve the RNA in 20 μl of 10 mM sodium acetate, pH 5.0. Add 10 units each of RNase T1 (Sigma) and RNase T2 (Sigma). Incubate at 37°C for 3 h.

8. Spot the digest onto Whatman 3MM chromatography paper (10 cm x 45 cm). Also spot radiolabelled GMP and guanosine as markers onto a separate lane. Dry the spots using a hair-dryer.

9. Fractionate the samples by ascending chromatography in butan-1-ol:0.1 M ammonium chloride (6:1) for 16 h at room temperature.

10. Hang the chromatography paper in a fume cupboard to dry. Then cut each lane transversely into 0.5 cm wide strips.

11. Soak each strip in 1 ml of water for 1 h.

12. Recover the water eluates, add 10 ml of Triton-toluene scintillation fluid[a] to each and count in a liquid scintillation counter. Having located the positions of the GMP and guanosine markers, determine the radioactivity of each of these molecules in the RNA transcript samples.

The length of RNA made in the incubation is given by the ratio of radioactivity in GMP compared with guanosine. This value is divided by the time period of incubation to give the rate of elongation.

[a]13.5 g Omnifluor (NEN):1 litre Triton X-100:2 litres toluene.

chromatography (18). The ratio of radioactivity in GMP compared with free guanosine is a measure of the length of the RNA transcript formed during the time period of incubation and hence this gives an estimate of the rate of elongation. Small RNAs must be removed by sucrose gradient centrifugation prior to digestion of the RNA since these will not be representative of nascent transcripts. A typical elongation rate is approximately 5 nucleotides/sec for mouse myeloma nuclei (18). At this rate, a 45S rRNA precursor molecule (10 000 nucleotides long) would be completely synthesised in 30 min *in vitro*.

The rate of RNA synthesis, as measured by incorporation of [α-^{32}P]GTP into TCA-precipitable material (*Table 5,* step 3), is about $0.1 - 0.2$ pmol GMP incorporated/μg DNA/min which represents only $5 - 10\%$ of the *in vivo* rate. On the other hand, the rate of elongation *in vitro* (5 nucleotides/sec) is 20% of the estimated *in vivo* rate of 25 nucleotides/sec. Thus it is not obvious that all RNA polymerase molecules which initiated their RNA transcripts *in vivo* simultaneously elongate these *in vitro*.

4.3 Initiation of RNA Synthesis

4.3.1 *Use of [γ-³²P]Nucleoside Triphosphates*

An unambiguous definition of a primary transcript is that it must contain the pppXp structure at the 5' end of the initiated RNA molecule. Incorporation of a [γ-³²P]ribonucleoside triphosphate into RNA and retention of a radiolabelled γ-phosphate at the 5' end has commonly been used for measuring the initiation of RNA chains *in vitro* (2,12). A suitable protocol is given in *Table 10A*. This method, which results in the incorporation of a single radioactive atom per initiated transcript, works well in the analysis of transcripts which are short and are re-initiated frequently (2,12), for example 5S rRNA and tRNA. It is only applicable to situations where the RNA product retains its triphosphate terminus, that is, it is not capped.

4.3.2 *Use of γ-Thionucleoside Triphosphates*

The use of [γ-³²P]ribonucleoside triphosphates to assay for RNA initiation as described above depends on the detection of the single γ-labelled nucleotide in the nascent RNA. Analysis of initiation of transcription becomes impractical by this method when the RNA of interest is very long or is initiated infrequently *in vitro*, since the level of detection is decreased for these RNAs. In addition, the γ-phosphate is usually rapidly removed from the primary transcript in complex systems such as nuclei due to capping or phosphatase activities.

Eckstein introduced the use of thionucleotides for studies involving nucleic acid metabolism (19), and has made extensive use of these to investigate the stereochemistry of such enzymatic reactions (20). We have adopted the use of [γ-S]ribonucleoside triphosphates to specifically label RNA initiated *in vitro*, which can subsequently be isolated by affinity chromatography on an organomercurial matrix, namely Hg-agarose or Hg-cellulose (21,22). The use of such an affinity label in conjunction with [α-³²P]ribonucleoside triphosphates permits the RNA transcribed *in vitro* to be isolated with a much higher specific activity than that allowed by utilising [γ-³²P]ribonucleoside triphosphates. In addition, it eliminates the competition from RNA that was simply elongated during the transcription reaction. Furthermore, [γ-S]nucleotides are resistant to phosphatase cleavage and, when incorporated into RNA, also prevent capping from occurring (23). Inhibition studies with α-amanitin show that all three RNA polymerases in mouse myeloma nuclei can utilise the γ-thio analogues (22). In addition, the rate of RNA synthesis is usually greater when using [γ-S]nucleotides than with unsubstituted nucleotides, presumably because of the higher stability of the former, although this remains to be investigated. Thus, this method can be utilised to study the transcription products of all three classes of eukaryotic RNA polymerases, including messenger RNAs and viral RNAs (22,24).

The use of [γ-S]ribonucleoside triphosphates to analyse RNA initiation is described in *Table 10B*. Either Hg-agarose or Hg-cellulose can be used as the affinity matrix for purification of the RNA transcripts. Mercury-agarose is prepared according to the method of Cuatrecasas (25) using Bio-Gel A-15M. A

Table 10. Assays for Initiation of RNA Synthesis.

A. Use of [γ-32P]Nucleoside Trisphosphates[a]

1. Incubate the nuclei as described in *Table 5* but use [γ-32P]GTP or [γ-32P]ATP (1 – 5 mCi/ml; 200 Ci/mmol).

2. Isolate RNA as described in *Table 6*.

3. Dissolve an aliquot of the RNA in 20 μl of 10 mM sodium acetate, pH 5.0. Add 10 units each of RNase T1 (Sigma) and RNase T2 (Sigma). Incubate at 37°C for 3 h.

4. If initiation of RNA synthesis has occurred, the 32P label will be present in either pppGp or pppAp after digestion with RNases T1 and T2. These radioactive tetraphosphates can be detected by t.l.c. on PEI-cellulose using 1.5 M KH_2PO_4, pH 3.5 as the developing solvent.

B. Use of γ-Thionucleoside Triphosphates

1. Incubate the nuclei as described in *Table 5* but with [γ-S]GTP or [γ-S]ATP (Boehringer) in place of GTP or ATP[b]. One or more of the other three nucleoside triphosphates can be radioactively labelled with α-32P. Alternatively use [γ-S][β-32P]GTP or [γ-S][β-32P]ATP (19,20).

2. Isolate RNA as described in *Table 6* but use 0.1% SDS, 0.1 M NaCl, 20 mM EDTA, 10 mM Tris-HCl, pH 7.9, for the chromatographic separation on Sephadex G-50 or Biogel P6 (*Table 6*, step 11).

3. Pool the RNA fractions from the Sephadex G50 fractionation which correspond to the void volume and load this directly onto a column consisting of 1 – 2 ml of Hg-agarose or Hg-cellulose pre-equilibrated in the buffer described in step 2.

4. Wash the Hg-agarose or Hg-cellulose column with 10 volumes of the above buffer.

5. Elute the specifically-bound RNA (γ-S-RNA) with 10 mM DTT or 1% 2-mercaptoethanol in the above buffer. RNA which elutes under these conditions must be the result of initiation *in vitro*.

C. Use of β-Thionucleoside Triphosphates

This procedure is carried out exactly as described by Eckstein and his colleagues (32,33).

---

[a]Since the radioactive phosphate is incorporated into the RNA transcript only at the 5′ end, this assay of initiation is applicable only where the transcripts comprise a large proportion of the final RNA product (e.g., 5S rRNA or tRNA) and normally has a triphosphate terminus (i.e., is not capped).

[b]To show that the sulphur is indeed incorporated at the 5′ terminus of the RNA, use [γ-S][β-32P]-GTP or [γ-S][β-32P]ATP in the *in vitro* incubation (*Table 5*), then isolate the RNA (*Table 6*) and analyse the distribution of incorporated label as follows. Add NaOH to the RNA to 0.3 M final concentration and incubate at 37°C for 18 h to digest the RNA. Isolate the sulphur-containing nucleotides by chromatography on Hg-agarose as described in steps 3 – 5 of *Table 10B*. Add 1 mg of yeast tRNA (digested previously with RNase) to the eluted sulphur-containing nucleotides and chromatograph this mixture on a column of DEAE-cellulose in 1 mM 2-mercaptoethanol, 7 M urea, 50 mM Tris-HCl (pH 8.0). Elute the nucleotides with a linear gradient of 0.05 – 0.30 M NaCl. Measure the radioactivity of collected fractions. Nucleoside tetraphosphate has a net charge of − 6 under these conditions and therefore the 32P-labelled, sulphur-containing nucleotides should elute in this position.

---

low cyanogen bromide concentration (< 10 mg/ml) must be used to minimise extensive cross-linking during the gel activation. A Hg-agarose gel prepared by activation with 2.5 mg/ml of cyanogen bromide shows high binding efficiency and selects all sizes of RNA equally well (21). Preparation of Hg-cellulose is described in *Table 11*.

Table 11. Preparation of Mercury-cellulose.

1. Incubate cellulose powder with 5 M NaOH overnight at 4°C under a nitrogen atmosphere.

2. Wash the cellulose by filtration with 100 volumes of water and then 100 volumes of 0.75 M potassium carbonate, pH 11.0.

3. Resuspend the washed cellulose in 0.75 M potassium carbonate, pH 11.0 to a final concentration of 2.5 g/100 ml. Add cyanogen bromide (0.5 g/ml in dioxane) to a final concentration of 25 mg/ml.

4. Allow the activation to proceed for 6 min at room temperature, maintaining the pH at pH 11.0 by titration with 10 M KOH.

5. Filter the activated cellulose rapidly and wash it with 100 volumes of cold water, then 100 volumes of cold 50% DMSO, and finally 50 volumes of (100%) DMSO at room temperature.

6. Resuspend the cellulose at 0.1 g/ml in DMSO and add *p*-aminophenyl mercuric acetate to 20 mM (using a 1 M stock of this reagent in DMSO). Stir overnight at room temperature.

7. Filter the Hg-cellulose. Resuspend it in 0.4 M 2-amino-2-methyl-1,3-propanediol in DMSO and stir overnight at room temperature.

8. Wash the Hg-cellulose with 100 volumes of 50% DMSO then 100 volumes of water. Store at 4°C in 0.01% sodium azide.

4.3.3 *Use of β-Thionucleoside Triphosphates*

[γ-S]ATP and [γ-S]GTP have been successfully used to study the initiation of transcription in nuclei from a variety of sources such as mouse myeloma (26,27), yeast (28), *Xenopus* (29), *Physarum* (30) and virus-infected mouse cells (24). However, RNase contamination of crude preparations of nuclei can cause gross degradation of nuclear RNA. This may introduce artifacts into initiation assays by providing RNA fragments with free 5'-hydroxyl groups that can be subsequently labelled with [γ-S]nucleotides acting as phosphate donors for endogenous polynucleotide kinase (31). In these cases, [β-S]ribonucleoside triphosphates (32,33) can be used instead of [γ-S]ribonucleoside triphosphates since the former are not substrates for polynucleotide kinase (34). However, other aspects of RNA synthesis (in particular, capping) have not yet been studied using [β-S]nucleotides. Furthermore, it is not clear to what extent the different classes of eukaryotic RNA polymerases are able to use [β-S]nucleotides as substrates. Finally, unlike [γ-S]nucleotides (Section 4.3.2), [β-S]nucleotides are not yet commercially available and hence must be synthesised in the laboratory (32,33).

4.3.4 *Capping of RNA In Vitro*

Capping of RNA polymerase II transcripts *in vivo* normally occurs very soon after the initiation of RNA synthesis. Several cap structures exist. These are classified according to the number of ribose moieties which are methylated in the structure ^{m}GpppXpYpZpN..... The cap 0 structure (found naturally in yeast) contains no O-methyl groups. Cap I structures (O-methylated ribose only in the X nucleoside) and cap II structures (O-methylated ribose in both

Table 12. Characterisation of Caps in RNA: Method A[a].

1. Incubate the nuclei for *in vitro* transcription as described in *Table 5* except include S-adenosylmethionine in the reaction mixture at a final concentration of 20 μM. Use 500 μCi/ml of [α-^{32}P]GTP, [β-^{32}P]nucleoside triphosphate or [^{3}H]S-adenosylmethionine as the radiolabelled precursor.

2. Isolate the RNA as described in *Table 6*.

3. Dissolve the RNA in 50 μl of 50 mM Tris-HCl, pH 7.5.

4. Add 10 units each of RNase T1 and RNase T2 (Sigma) and incubate at 37°C for 3 h. The products of this digestion will be mononucleotides, oligonucleotides due to 2'-0 methylation of ribose residues and a variety of cap structures (see Section 4.3.4).

5. Add 100 μg of carrier RNA (e.g., rRNA or tRNA), digested with pancreatic RNase (10 μg, 37°C, 1 h).

6. Mix the RNA digestion products with 1 ml of 7.0 M urea, 1 mM EDTA, 10 mM Tris-HCl, pH 7.5 and apply this to a column (0.6 cm x 30.0 cm) of DEAE-cellulose pre-equilibrated in this buffer. Elute the column with a 0 − 0.35 M gradient (100 ml total volume) of NaCl in the same buffer. Monitor the A_{254} of the eluate continuously but also collect 2 ml fractions and determine their radioactivity by Cerenkov counting. The absorbance profile, due to the carrier RNA digestion mixture (step 5), indicates the elution position of individual size classes of oligonucleotides, separated on the basis of charge (*Figure 3*). Radioactivity eluting with trinucleotides or larger can be due to cap structures in the RNA synthesised *in vitro*; cap 0 structures elute with trinucleotides, cap I structures elute with tetranucleotides and cap II structures with pentanucleotides (*Figure 3*).

7. Pool those column fractions corresponding to material eluting at charge −4 to −7.

8. Dilute the pooled fractions by adding 3 volumes of water. Load the samples onto a column of DEAE-cellulose prepared in a Pasteur pipette and pre-equilibrated with water.

9. Wash the column with 4 volumes of water to remove salts and urea and then elute the nucleotides with a minimal volume of 2.0 M triethylammonium bicarbonate, pH 7.5.

10. Freeze dry.

11. Dissolve the nucleotide residue in 200 μl of water.

12. Repeat steps 11 and 12 until all the triethylammonium bicarbonate is removed (3 − 4 times).

13. To confirm the presence of caps, digest the oligonucleotides with nuclease P1 and alkaline phosphatase (Sigma). To do this, dissolve the freeze-dried nucleotides in 10 μl of 10 mM sodium acetate, pH 5.8, add 5 μg of nuclease P1 and incubate at 37°C for 1 h. Then adjust the pH of the mixture of pH 7.0 − 8.0 using 1 M Tris-HCl pH 9.0 and add alkaline phosphatase to a final concentration of 2 μg/ml. Incubate at 37°C for 3 h. These reactions convert cap structures to cap 0 structures (Section 4.3.4) and oligonucleotides to nucleosides plus inorganic phosphate.

14. Resolve the caps from the nucleosides by thin-layer electrophoresis on a cellulose plate (20 cm x 20 cm; Kodak Ltd.) run at 200 V for 3 h in 5% acetic acid, 0.5% pyridine, 1 mM EDTA until a xylene cyanol marker has migrated 9 − 10 cm. Alternatively, isolate the caps by ascending t.l.c. on PEI-cellulose in 0.5 M KH$_2$PO$_4$ adjusted to pH 3.5 with phosphoric acid. In this case, continue chromatography until the buffer reaches the top of the plate. Suitable cap markers can be obtained from P-L Biochemicals and are visualised by u.v. illumination after electrophoresis or chromatography.

15. Since there are two possible types of capping nucleotide [7-methylguanosine in mRNA (8) and 2,2,7-trimethylguanosine in SnRNAs (8)], it is necessary to separate these. Therefore, elute the cap structures from the cellulose thin-layer plate (after step 15 electrophoresis) using water or from the PEI-cellulose plate (after step 15 chromatography) using 2 M triethylammonium bicarbonate, pH 7.5.

16. Freeze dry the eluted material.

17. Dissolve the caps in 5 μl of 10 mM sodium acetate pH 5.0 and add 10 units of acid pyrophosphatase from tobacco (BRL). Incubate at 37°C for 2 h.

18. Freeze dry. Dissolve in $1-2$ μl of water.

19. Separate 7-methylguanosine monophosphate from 2,2,7-trimethylguanosine monophosphate by two-dimensional t.l.c. as follows. Spot the digest onto a cellulose thin-layer plate (20 cm x 20 cm; Kodak Ltd.) and carry out chromatography with isobutyric acid:ammonium hydroxide:H_2O (66:1:33) in the first dimension and 0.1 M sodium phosphate buffer, pH 6.8:ammonium sulphate:propan-1-ol (100:60:2 v/w/v) in the second dimension (37). Chromatograph the four ribonucleoside 5'-monophosphates and 7-methyl guanosine monophosphate (P-L Biochemicals) as markers.

A typical separation is shown in *Figure 3*.

---

[a]Adapted from reference 35.

the X and Y nucleosides) occur in mammalian mRNAs. The first O-methylation reaction occurs in the nucleus and cap I structures are found on hnRNA but the second O-methylation reaction may occur in the cytoplasm since cap II structures are not found in hnRNA. In the formation of the cap, two phosphates are derived from the α- and β-phosphates of the initiating ribonucleoside triphosphate and one from the α-phosphate of GTP used in the capping reaction. Following formation of the cap, the guanosine is methylated at the 7 position. In addition to mRNAs, mammalian small nuclear RNAs (the U series) also contain cap II structures. In this case the guanosine is methylated twice on the 2-amino group as well as on the 7 position, yielding 2,2,7-trimethyl guanosine.

If [α-^{32}P]GTP is used as a precursor, then caps formed *in vitro* will be labelled. Alternatively the cap can be labelled by using ATP, GTP, UTP or CTP containing ^{32}P in the β position. The label in the cap is then derived from the first nucleotide in the RNA chain. Therefore, incorporation of the α-phosphate of labelled GTP into the cap structure of RNA synthesised *in vitro* indicates that the RNA was capped *in vitro* and incorporation of a [β-^{32}P]nucleoside triphosphate into a cap structure *in vitro* implies that the transcription was initiated *in vitro*. Thus, although capping is formally a post-transcriptional modification, it is discussed here as one of the potential assays which may be used for initiation of RNA synthesised *in vitro*. Whichever radioactive precursor is used, it is necessary to separate the large amount of nucleotides incorporated into RNA during elongation from the cap itself so that the latter can be analysed. This can be done as described in *Table 12*. In summary, the RNA is digested by RNases T1 and T2 followed by chromatography on a column of DEAE-cellulose. This resolves cap 0, cap I and cap II structures produced by the RNase digestion [GpppXp, GpppX(m)pYp, GpppX(m)pY(m)pZp, respectively]. However, internal RNA fragments containing O-methyl groups [e.g., X(m)pY(m)pZp] may contaminate the caps as well as incompletely digested RNA. Thus the material eluting from the DEAE-cellulose with charge -4 to -7 is digested with nuclease P1 and alkaline phosphatase to generate core cap structures (GpppX$_{OH}$). The core cap structures are then characterised by digestion with nucleotide pyrophosphatase

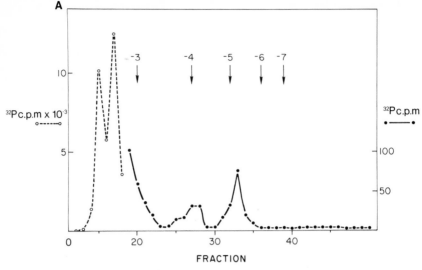

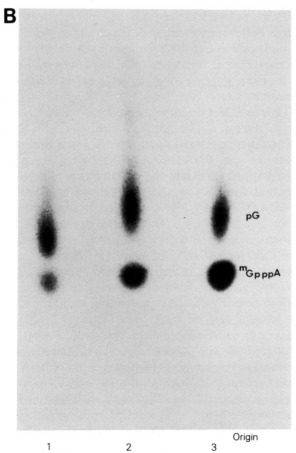

C

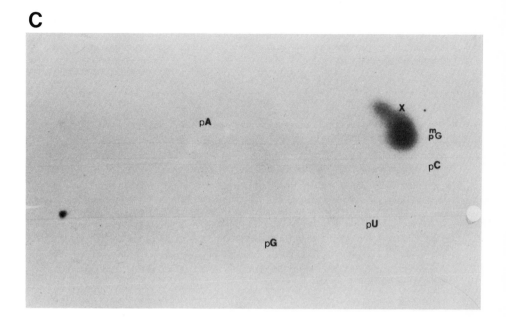

Figure 3. Capping of RNA *in vitro*. (**A**) RNA was synthesised *in vitro* by nuclei in the presence of [α-³²P]GTP (*Table 5*). The RNA was digested with RNases T1 and T2 and applied to a DEAE-cellulose column in 7.0 M urea (*Table 12*, steps 1 – 6). Ribosomal RNA digested with pancreatic RNase was included to provide charge markers. The oligonucleotides were eluted with a gradient of 0 – 0.35 M NaCl in 7.0 M urea, 1 mM EDTA, 10 mM Tris-HCl, pH 7.5. The position of oligo-nucleotides of defined size (indicated by their net charges of − 3, − 4, − 5, etc.) was determined by continuous monitoring of the eluate at 254 nm. The eluate was also collected into fractions and the radioactivity of each fraction was determined by Cerenkov counting. Since 98% of the eluted radioactivity is present in mononucleotides, there is a change in scale as the data is plotted; (○- - -○) ³²P c.p.m. x 10⁻³, (●——●) ³²P c.p.m. (**B**) Column fractions corresponding to the material eluting at charge − 4, − 5 and − 6 plus − 7 were pooled, de-salted and digested with nuclease P1 followed by alkaline phosphatase (*Table 12*, steps 7 – 13). The nucleotides were resolved by thin-layer electrophoresis at pH 3.5 (*Table 12*, step 14). The GMP (pG) detected in the samples could result from O-methylguanosine in cap I and cap II structures or from incomplete digestion of the original RNA. **Lane 1**, material eluting with charge − 4; **lane 2**, material eluting with charge − 5; **lane 3**, material eluting with charge − 6 and − 7. (**C**) The cap structures (ᵐGpppA) shown in **B** were eluted with water, digested with acid pyrophosphatase from tobacco and the nucleotides resolved by two-dimensional t.l.c. as described in *Table 12* (steps 15 – 19). Spot X is probably 2,2,7-trimethylguanosine monophosphate. The positions of AMP (pA), GMP (pG), CMP (pC) and UMP (pU) markers, detected by u.v. illumination, are indicated as well as 7-CH₃-GMP (ᵐₚG). The cloned DNAs were a gift of Drs. Jack Ho and David Stafford.

(Sigma) or acid pyrophosphatase from tobacco (BRL) followed by separation of the resulting mononucleotides by two-dimensional t.l.c. *Figure 3* shows a typical analysis of capping of RNA *in vitro* according to this protocol.

An alternative approach developed by Contreras and Fiers (36) allows one to detect the cap core structures in a one-step procedure (*Table 13*). The RNA synthesised *in vitro* is digested with nuclease P1 and the resulting nucleotides are separated by electrophoresis on cellulose acetate followed by chroma-

Table 13. Characterisation of Caps in RNA: Method B[a].

1. Incubate the nuclei for transcription *in vitro* as described in *Table 5* except include S-adenosylmethionine in the reaction mixture at a final concentration of 20 μM. Use 500 μCi of [α-^{32}P]GTP, [β-^{32}P]nucleoside triphosphate or [^{3}H]S-adenosylmethionine as the radiolabelled precursor.

2. Isolate the RNA as described in *Table 6*.

3. Dissolve the RNA in 10 μl of 0.1 M formic acid-pyridine, pH 6.0 and add 10 μg of nuclease P1. Incubate at 37°C for 3 h.

4. Dry. Dissolve in 10 μl of 0.1 M ammonium bicarbonate pH 8.0 and incubate with 0.5 units of alkaline phosphatase (Worthington) at 37°C for 1 h.

5. Dry the digest under vacuum. Dissolve the residue in a small volume of water and dry again. Repeat 2 – 3 times to remove all traces of ammonium bicarbonate.

6. Dissolve the nucleotides in a minimal volume of water and spot the solution onto pre-wetted cellulose acetate strips. Use xylene cyanol (blue) and acid fuchsin (pink) as markers. Electrophorese at 2000 – 3000 V in 5% acetic acid – 0.5% pyridine – 5 M urea, pH 3.5, until the acid fuchsin has migrated 20 – 25 cm. Then transfer the strip directly to a PEI-cellulose plate (20 cm x 20 cm). Arrange the strip so that the origin of the cellulose acetate strip is at one edge of the PEI plate and the acid fuchsin is off the PEI plate.

7. Chromatograph the PEI plate using 1.3 M formic acid-pyridine, pH 4.2, until the buffer front reaches the top of the plate.

8. Dry the plate and expose it to X-ray film using a Dupont Lightning-Plus intensifying screen. The identity of the caps can be determined using the appropriate markers available from P-L Biochemicals.

9. Elute the caps with 2.0 M triethylammonium bicarbonate, pH 7.5, and analyse these as described in steps 16 – 19 of *Table 12*.

[a]After Contreras and Fiers (ref. 36).

tography on PEI-cellulose. Most of the label incorporated into RNA is released as nucleoside monophosphates and the caps [GpppX(m)p] can be resolved and identified using appropriate markers (available from P-L Biochemicals). The incorporation of [β-^{32}P]ATP and [β-^{32}P]GTP into caps of SV40 RNA during transcription by isolated nuclei has been conclusively demonstrated using this technique (36).

In general, approximately 10% of RNA polymerase II transcripts are capped following transcription in nuclei (35).

5. ANALYSIS OF SPECIFIC RNA TRANSCRIPTS

5.1 **Purification of RNA Transcripts**

Although the use of nucleic acid hybridisation allows the identification of specific RNA transcripts, one of the major difficulties in using isolated nuclei as a transcription system is how to distinguish RNA which has been transcribed *in vitro* from the vast excess of endogenous RNA present in nuclei. In particular, it is desirable to separate RNA chains *initiated in vitro* from those which have simply been *elongated*. To overcome this problem, several protocols have been developed which allow the physical separation of *in vitro* RNA transcripts from the endogenous RNA. This physical separation is useful under two conditions:

(i) when the assay depends on the use of a labelled probe rather than direct analysis of labelled RNA, for example, mapping with nuclease S1 (see Chapter 5, Section 5.2);

(ii) when the large amount of endogenous RNA interferes with the use of a hybridisation assay, for example for rRNA or small nuclear RNAs.

RNA-dependent RNA synthesis can be a major problem during transcription studies of isolated chromatin using exogenous RNA polymerase (Chapter 5). However, RNA-dependent RNA synthesis does not occur in isolated nuclei which rely upon endogenous RNA polymerase for transcription. Therefore any RNA transcripts which are isolated must be the products of authentic gene transcription. Nevertheless, one should routinely check that the observed incorporation is due to transcription using α-amanitin (for RNA polymerase II and III) (Section 3.2) or actinomycin D (25 μg/ml) for RNA polymerase I.

5.1.1 *Use of Mercurated Nucleoside Triphosphates*

Dale, Livingston and Ward (37) first demonstrated that 5-mercurated UTP (Hg-UTP) could be utilised as substrate for RNA synthesis by *Escherichia coli* RNA polymerase. Since mercury-substituted RNA (Hg-RNA), like other organomercurials, exhibits a high affinity for mercaptans, the use of a sulph-hydryl-containing agarose affinity column (for example, Bio-Rad Affi-Gel 501) allows the separation of newly-synthesised Hg-RNA from endogenous RNA.

The conditions for *in vitro* transcription using Hg-UTP are as described in *Table 5* except that 12 mM 2-mercaptoethanol should be used in place of DTT in the nuclei storage buffer (*Table 1*). Following synthesis, the newly-synthesised Hg-RNA is heated and then chromatographed on thiolated Sepharose (see *Table 6* of Chapter 5) at 55 – 60°C. The elevated temperature is necessary to obtain optimal separation of Hg-RNA from endogenous RNA. Several drawbacks have been observed with this method. The rate of RNA synthesis in nuclei is reduced by 30 – 35% with Hg-UTP as substrate even when the Hg group is blocked by 2-mercaptoethanol. The chain length of the Hg-RNA is also shorter when compared with RNA made with unsubstituted nucleotides (37). Nevertheless, the authors have used this procedure to show that template restriction for mRNA coding for immunoglobulin *k* light chain is retained in nuclei isolated from myeloma cells. The RNA transcribed *in vitro* in nuclei from mouse myeloma 66.2 cells, a *k*-chain producing cell line, contained 0.02% *k*-chain mRNA. Nuclei from the cell line MOPC 315, which produces λ light chains, synthesised less than 0.0008% *k*-chain mRNA. Subsequently, this method has been used successfully in the analysis of globin (38), ovalbumin (39) and viral gene transcription (40).

5.1.2 *Use of α-Thionucleoside Triphosphates*

The authors have recently utilised [α-S]ribonucleoside triphosphates as an alternative to Hg-UTP in transcription studies where separation of the newly-synthesised RNA from endogenous RNA is desired (26). The conditions for *in*

Table 14. Quantitation of RNA by Hybridisation to Immobilised DNA.

1. If necessary (see Section 5.2), linearise the DNA probe by digestion with an appropriate restriction endonuclease.

2. Extract the DNA with phenol saturated with 50 mM Tris-HCl, pH 8.0.

3. Add 2.5 volumes of absolute ethanol and precipitate the DNA at $-70°C$ (for 1 h) or at $-20°C$ (for 16 h).

4. Recover the DNA by centrifugation (10 000 g, 10 min, 4°C) and wash the pellet with ice-cold 70% ethanol.

5. Dry the DNA pellet under vacuum.

6. Dissolve the DNA in 0.2 M NH_4OH, 2.0 M NaCl at a final concentration of 50 $\mu g/ml$..

7. Denature the DNA by boiling for 2 min followed by quick-cooling in ice.

8. Spot the DNA onto a nitrocellulose filter (Schleicher and Schuell, BA85) pre-soaked in 2 x SSC[a]. This is most conveniently carried out using commercial equipment, for example, a Hybridot apparatus (BRL Inc.). With this apparatus, up to 100 μl (5 μg DNA) can be applied per dot. A convenient method of analysing several RNA samples each with several DNA probes is to prepare a single row of dots with each sample. The filter can then be cut into strips, each containing one dot of each DNA probe to be analysed.

9. Following application of the DNA, wash the bound DNA by passing 50 μl of 2 x SSC through each Hybridot well.

10. Bake the filter at 80°C for 2 h in a vacuum oven and then cut it into strips ready for hybridisation.

11. To reduce the background radioactivity eventually bound to the filter, radioactive RNA preparations should be freed of unincorporated radiolabelled nucleotides before hybridisation, by chromatography on Sephadex G50 as described in *Table 6*, steps 10 and 11.

12. Pool the fractions corresponding to the void volume of the column, add 2.5 volumes of ethanol and keep at $-70°C$ (for 1 h) or $-20°C$ (for 16 h) to precipitate the RNA.

13. Centrifuge the RNA (10 000 g, 10 min, 4°C) and then dry the pellet under vacuum.

14. Pre-hybridise the filter strip at 52°C for 30 min by sealing it in a small plastic bag (use a commercial electric bag sealer) along with 0.5 ml of degassed prehybridisation buffer prepared by mixing:
    0.25 ml deionised formamide[b]
    0.125 ml 20 x SSC
    5 μl 10% SDS
    5 μl 0.1 M EDTA, pH 7.5
    5 μl 1.0 M Tris-HCl, pH 7.5
    10 μl 0.5 $\mu g/ml$ poly(A) (Miles)
    20 μl 100 x Denhardt's solution[c]
    5 μl 1 mg/ml denatured *E. coli* DNA
    75 μl H_2O

15. After the pre-hybridisation, carefully open the bag and discard the buffer. Re-dissolve the RNA (step 13) in 75 μl of water and mix this with the reagents listed in step 14, omitting the water. Transfer the RNA solution to the bag and re-seal it, avoiding air bubbles. Incubate at 52°C for 72 h.

17. After hybridisation, open the bag and place the filter in a tray with 20 ml of 5 x SSC, 0.1% SDS, 1 mM EDTA, 10 mM Tris-HCl, pH 7.5. Agitate the filter gently for 30 min at 52°C. Repeat this wash procedure a total of four times.

18. *Either*
    Wash the filter once with 20 ml of 0.1 x SSC at room temperature. If an even more stringent wash is desired, this wash can be carried out at a higher temperature (up to 65°C),

or

to minimise non-specific binding of RNA to the filter, incubate the filter with 10 μg/ml of RNase A in 0.3 M for 10 min at 37°C. Then wash the filter with 20 ml of 0.3 M NaCl at 25°C.

19. Allow the filter to air-dry and then autoradiograph it using a calcium tungstate intensifying screen if necessary.

20. To quantitate the amount of RNA bound to each dot of DNA,

   either, scan the autoradiographic image of the dot using a densitometer, having first calibrated the film response to a known amount of the same radioisotope,

   or, punch out each dot from the filter strip and determine the radioactivity by liquid scintillation counting. In this case, the background is determined by counting a similar sized dot from a portion of the filter where no DNA is bound or by determining the amount of radioactive RNA bound to a dot of non-homologous DNA (e.g., pBR322).

---

[a]The composition of 1 x SSC is 0.15 M NaCl, 15 mM trisodium citrate.
[b]Formamide is deionised as follows. Stir 100 ml of formamide with 10 g of mixed bed resin [e.g. Bio-Rad AG501-X8(D)] for 2 h at room temperature. Filter off the resin and store the formamide at −20°C.
[c]100 x Denhardt's solution consists of 10% nuclease-free BSA (e.g., from BRL), 10% polyvinylpyrrolidine, 10% Ficoll (mol. wt. 400 000). The BSA is omitted if the hybridised RNA is to be eluted subsequently from the filter (see *Table 15*).

vitro transcription are as given in *Table 5*. The synthesis of such thiolated RNA in myeloma nuclei is linear with time for at least one hour, and the RNA transcribed in the presence of 5 μM [α-^{35}S]ATP is identical to that transcribed with 5 μM ATP plus [α-^{32}P]nucleotides, as assayed by gel electrophoresis under denaturing conditions. The [α-^{35}S]ATP may be used as sole radiolabelled precursor to yield [α-^{35}S]RNA or in conjunction with [α-^{32}P]GTP to yield ^{32}P-labelled thiolated RNA.

Separation of newly-synthesised thiolated RNA (α-S-RNA) from endogenous RNA is achieved by affinity chromatography on an organomercurial matrix such as Hg-cellulose or Hg-agarose. The preparation of these matrices has been described earlier in this chapter (Section 4.3.2). Isolation of γ-S-RNA by chromatography on Hg-cellulose or Hg-agarose is described in *Table 10B*. However, in contrast to γ-S-RNA (*Table 10B*), binding of α-S-RNA to Hg-cellulose is very inefficient at pH 7.9, though it occurs readily at pH 4.9. Therefore the separation is carried out using 0.1 M NaCl, 40 mM EDTA, 0.1% SDS, 50 mM sodium acetate, pH 4.9. For the separation to be effective at this pH, the mercury titre of the Hg-cellulose must not exceed 10 μmol Hg/g of cellulose since higher titre Hg-celluloses exhibit considerable non-specific binding. The low organomercurial content required limits the use of this matrix to the fractionation of relatively small quantities of thiolated RNAs. No size discrimination of the RNA bound by Hg-cellulose is observed and α-S-RNA as large as 45S can be recovered from this matrix with yields approaching 75%.

5.2 Quantitation of Specific RNA Transcripts

The amount of a particular RNA synthesised can be quantitated by RNA-DNA hybridisation. The DNA probe used is either double-stranded DNA cloned in a plasmid or single-stranded DNA cloned in coliphage M13. The most convenient method is to immobilise the DNA probe as a series of 'dots' (each ~0.5 cm in diameter) on a nitrocellulose membrane filter. The filter is then incubated in a solution of the labelled RNA transcripts for hybridisation to occur. After washing to remove unhybridised RNA, the amount of hybridised RNA can be determined either by autoradiography followed by densitometry of the photographic images or by punching out the dots and counting these in a liquid scintillation counter. The protocol is described in *Table 14*. However, certain points require further discussion.

(i)   Spotting of the DNA probes onto the nitrocellulose membrane is most conveniently carried out using a commercial apparatus which filters multiple samples (usually 96 samples in an 8 x 12 typical microtitre plate format) onto one filter simultaneously. Such equipment, for example the 'Hybridot' apparatus marketed by BRL Inc. (*Table 14*), relies upon vacuum filtration and thus the DNA must bind to the filter as the sample is filtered through it. Since supercoiled plasmid DNA will not bind to nitrocellulose, plasmid DNA to be used as the probe must first be linearised using a restriction enzyme.

(ii) The DNA probe should be at least 1.0 kb in size to ensure quantitative binding to the nitrocellulose filter.

(iii) The RNA must be freed of all ^{32}P-labelled nucleotides by gel filtration prior to hybridisation. If this is not carried out, there is usually an uneven 'speckled' background on the resulting autoradiographs.

(iv) Under the hybridisation conditions specified in *Table 14*, the plasmid DNA is generally in excess (except for genes for structural RNAs which are found in large amounts in the nucleus, e.g., rRNA and small nuclear RNAs) and the hybridisation is quantitative. To confirm this, three experiments should be carried out:

    (a) Vary the amount of input RNA; the amount of RNA hybridised should be proportional to the RNA input.

    (b) Vary the amount of DNA on the filter; there should be no effect on the amount of RNA hybridised if the DNA is in excess.

    (c) Rehybridise that RNA which does not bind to the filter to the same DNA; none of the RNA should bind.

(v)   Plasmid DNA lacking the DNA sequence of interest should be used as a control DNA to monitor for non-specific binding of RNA. Routinely the background is 0.0001 − 0.001% of the input radioactivity. To reduce the background further and to digest any RNA fragments which are not completely complementary to the DNA probe, the filters can be treated with 10 µg/ml of RNase A or 10 units/ml of RNase T1 in 0.3 M NaCl for 10 min at 37°C, followed by a wash in 0.3 M NaCl alone (*Table 14,* step 18) prior to drying and autoradiography.

An example of the use of dot hybridisation is shown in *Figure 4*.

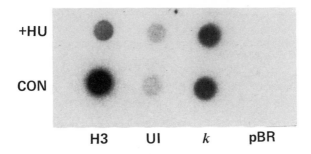

Figure 4. Hybridisation of labelled RNA to DNA 'dots'. Nuclei were prepared from exponentially growing myeloma cells (CON) or from cells treated for 1 h with 5 mM hydroxyurea (+HU). Transcription *in vitro* was carried out with [α-^{32}P]GTP as radiolabelled precursor. Equal amounts of RNA (4 x 10^6 c.p.m.) were used for each hybridisation to 'dots' of various DNA probes. The DNAs used were H3 (a mouse H3 gene), U1 (a mouse U1 RNA gene), *k* (a mouse *k* light chain immunoglobulin gene) and pBR (plasmid pBR322). The filters were washed and the hybridised RNA detected by autoradiography.

5.3 Determination of the Size of Specific RNA Transcripts

Two approaches can be used to determine the size of a specific RNA synthesised *in vitro*:

(i)   the RNA can be hybridised to a specific cloned DNA, the hybridised RNA eluted and then its size determined *or*

(ii) the RNA can be fractionated by size prior to hybridisation.

The latter is probably a better procedure for the analysis of large RNA transcripts which might be degraded during hybridisation.

5.3.1 *Hybridisation Prior to Size Determination*

To determine the size of the complete RNA transcript, the RNA is hybridised to cloned DNA immobilised on a nitrocellulose filter, eluted and then analysed by electrophoresis either in a polyacrylamide gel (RNA molecules <500 nucleotides long) or an agarose gel (for larger RNAs). This procedure is described in *Table 15A*.

As a variation of this method, the immobilised DNA:RNA hybrids can be treated with RNase T1 which will trim the RNA if the molecule extends beyond sequences represented in the cloned DNA. The RNase T1 is effectively removed by washing the filters with SDS prior to elution of the RNA (*Table 15B*). An example of this technique is shown in *Figure 5* where intact sea urchin N1 RNA was bound to a filter which contained DNA complementary to one-half of the gene. After digestion with RNase T1, the RNA was reduced to one-half the size corresponding to the region which was present in the DNA. One important application of this method is that cloned cDNAs bind nascent RNAs containing intervening sequences but on RNase digestion the intervening sequences are degraded leaving RNA fragments the size of the exons which comprise the cDNA. Thus an analysis of the RNA transcripts with and without RNase T1

Table 15. Elution of Hybridised RNA from Immobilised DNA.

A. Standard Procedure

1. If the nitrocellulose filters have been used for liquid scintillation counting (*Table 14*, step 20), first wash them with chloroform.

2. Allow each nitrocellulose disc to air-dry and then place it in a microcentrifuge tube. Add 75 μl of 98% formamide (deionised)[a], 0.2% SDS, 10 μg/ml yeast tRNA[b].

3. Incubate the filter discs at 60°C for 10 min and then recover the solution containing the eluted RNA.

4. Elute the filters once more as in steps 2 and 3 and combine the two eluates.

5. To the pooled eluates, add 0.2 ml of 0.6 M NaCl and then 1.0 ml of absolute ethanol. Leave at −70°C for 30 min.

6. Centrifuge at 10 000 g at 4°C for 10 min to recover the RNA.

B. Elution After Digestion with RNase T1[c]

1. If necessary, remove contaminating scintillation fluid from the filter discs as in step 1 of Section A above.

2. Incubate each dried filter disc in 100 μl of 50 mM sodium acetate, pH 5.5 containing 10 units of RNase T1 at 37°C for 10 min.

3. Wash each filter disc three times with 100 μl of 0.1% SDS, 2 x SSC[d], for 10 min each time. This inactivates residual RNase T1.

4. Elute the RNase-resistant RNA as described in steps 2 − 6 of Section A above.

---

[a]Formamide is deionised as described in footnote b of *Table 14*.
[b]Alternatively the RNA can be eluted by heating each filter with 100 μl of 10 μg/ml yeast tRNA in water at 100°C for 2 min. Essentially identical results are obtained using either procedure.
[c]RNase A cannot be used in place of RNase T1 since it is not completely inactivated by the washing procedure.
[d]See *Table 14*, footnote a.

digestion yields information on both the size of the nascent RNA transcript and the size of exons within the corresponding gene.

5.3.2 *Size Determination Prior to Hybridisation*

The procedure of Schibler *et al.* (42) allows one to hybridise agarose gel fractions directly to DNA dots (*Table 16*). The RNA is fractionated on a 0.8% agarose gel under denaturing conditions either with formaldehyde (43), glyoxal (17) or methyl mercuric hydroxide (44). The gel is then sliced and each slice dissolved in 0.6 M sodium perchlorate. The solution can be hybridised directly to filters containing immobilised cloned DNA and the amount of hybrid measured as usual. This allows accurate determination of the sizes of transcripts containing specific sequences and is the preferred procedure for the analysis of large RNA transcripts.

5.4 **Determination of the Termini of RNA Transcripts**

Important questions concerning RNA synthesis *in vitro* are whether the RNA starts and stops at unique sites and the relationship of these sites to the initiation and termination sites used *in vivo*. These questions can be answered using

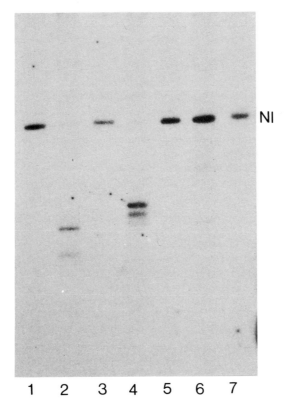

Figure 5. Determination of the size of RNA transcripts. The DNA probe was a plasmid containing a gene for sea urchin small nuclear N1 RNA (6) which is 163 nucleotides long. *Ava*I cleaves the gene at position 116 whilst *Hinf*I cleaves at position 87. Plasmids containing either the intact gene, or the gene cleaved with *Ava*I or *Hinf*I were immobilised on nitrocellulose. The filters were incubated with ^{32}P-labelled RNA transcribed by sea urchin nuclei *in vitro*. Pairs of filters were analysed for each plasmid DNA probe. From one filter the RNA was eluted directly and the other was treated with RNase T1 prior to elution of the RNA. The eluted RNA was analysed by electrophoresis on a 10% polyacrylamide gel in 7.0 M urea. **Lane 1**, untreated RNA eluted from unrestricted N1 DNA; **lane 2**, RNA bound to *Hinf*I cleaved N1 DNA and treated with RNase T1 prior to elution; **lane 3**, RNA bound to *Hinf*I N1 DNA and eluted without RNase treatment; **lane 4**, RNA bound to *Ava*I cleaved N1 DNA and treated with RNase T1 prior to elution; **lane 5**, RNA bound to *Ava*I cleaved N1 DNA and eluted without RNase treatment; **lane 6**, as **lane 1**; **lane 7**, RNA bound to uncut N1 DNA but then treated with RNase T1 before elution. In the *Hinf*I digest **(lane 2)** bands from the 5′ end (90 nucleotides) and from the 3′ end (75 nucleotides) are present. In the *Ava*I digest **(lane 4)** there are two bands each ~120 nucleotides long. Both are derived from the 5′ end. The presence of *two* bands may be due to heterogeneity of cleavage by RNase T1. The 3′ end fragment has been run off the gel.

the techniques described above. Cloned genomic DNA is used. Specific subcloned fragments which contain either the 5′ or the 3′ end of the gene are prepared and these are bound to nitrocellulose. These are then hybridised with the labelled RNA transcripts and the hybrids treated with RNase T1. The RNA is then eluted and analysed by gel electrophoresis. If appropriately small cloned fragments are chosen, the RNA fragments protected will be small, even

Table 16. Size Determination of RNA Prior to Hybridisation[a].

The procedure involves fractionation of the RNA by electrophoresis under denaturing conditions using methyl mercuric hydroxide followed by hybridisation of eluates from gel slices to locate the RNA of interest.

Caution: methyl mercuric hydroxide is extremely toxic and therefore all procedures which involve this reagent should be carried out in a well-ventilated fume hood.

1. Prepare a 1% agarose gel in 7.5 mM methyl mercuric hydroxide, 50 mM boric acid, 10 mM sodium sulphate, 1 mM EDTA, pH 8.3. Load this with the radioactive RNA sample and electrophorese using the above buffer (without methyl mercuric hydroxide) as described by Bailey and Davidson (44).

2. After electrophoresis, cut the gel into 1 – 2 mm slices. Dissolve each slice in a small volume of 2 M sodium perchlorate, 0.1% SDS, 10 mM 2-mercaptoethanol, 4-fold concentrated Denhardt's solution[b], 0.25 mg/ml yeast tRNA, 50 mM Tris-HCl, pH 7.4 at 68°C for 2 h.

3. Remove an aliquot of each sample and determine the total radioactivity by liquid scintillation counting.

4. Prepare nitrocellulose filter discs (Schleicher and Schuell, BA85, 0.9 cm diameter) onto each of which has been baked (*Table 14*) 5 µg of a DNA probe against the RNA under study. Place one filter into each RNA sample and cover the solution with a layer of paraffin oil. Hybridise at 65°C for 18 h.

5. After hybridisation, wash the filters with 2 M sodium perchlorate, 0.1% SDS, 20 mM Tris-HCl, pH 7.4 at 68°C for 1 h, then with 0.1% SDS, 2 x SSC[c] at 68°C for 1 h, and then with 2 x SSC at room temperature for 1 h. In each case use 10 ml of wash buffer for each filter.

6. Wash the filters with 0.1 x SSC at room temperature for 15 min. Alternatively, treat the filters with RNase to reduce the background binding of RNA to the filter, as described in *Table 14*, footnote d.

7. Quantitate the hybridised, labelled RNA by liquid scintillation counting (*Table 14*, step 20).

[a]From reference 42.
[b]The composition of Denhardt's solution is given in footnote c of *Table 14*.
[c]See *Table 14,* footnote a.

those derived from large transcripts. This is important so that their size can be measured accurately. When coupled with DNA sequence information this approach allows one to determine the termini of RNA transcribed *in vitro*. An example of such an analysis for the ribosomal RNA genes of *Lytechinus variegatus* is shown in *Figure 6*. *Figure 6A* compares the termini of the rRNA synthesised in isolated nuclei from blastula embryos, and the termini of the RNA synthesised *in vivo*. *Figure 6B* shows that the RNA synthesised *in vitro* is copied only from the sense strand of the DNA. Fragments corresponding to the 5′ or 3′ regions of the gene were subcloned in coliphage M13. The various DNAs (four clones each containing one strand of one end) were hybridised with the RNA transcripts and then the length of the bound RNA was determined. Note that this also allows one to determine the *amount* of transcription in the 5′ and 3′ regions and therefore to assess the amount of RNA chain completion (which will label the 3′ end preferentially) in contrast to *de novo* transcription which will label both ends equally. In the case shown (*Figure 6*), the 300 nucleotides at the 5′ end of the 8-kb transcript are labelled significantly suggesting that initiation is occurring.

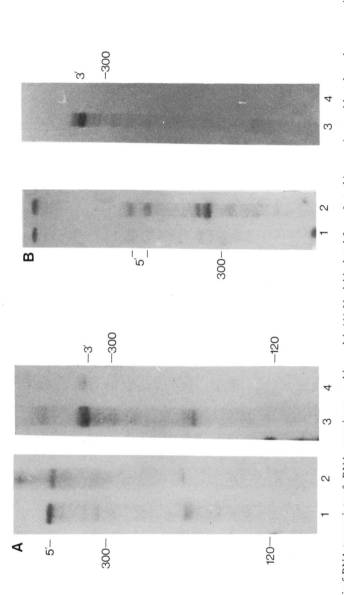

Figure 6. Termini of RNA transcripts of rRNA genes in sea urchin nuclei. (**A**) Nuclei isolated from *Lytechinus variegatus* blastula embryos were incubated with [α-³²P]GTP. The labelled RNA was hybridised to cloned DNA containing either the 500 bp of the 5′ end of the rRNA gene or the 500 bp of the 3′ end of the rRNA gene. In a parallel experiment, RNA labelled *in vivo* with ³²PO₄ was hybridised to the same DNAs. The hybrids were treated with RNase T1, the RNA was eluted and then analysed by electrophoresis on a 10% polyacrylamide gel. The positions of marker 7S RNA (300 nucleotides) and 5S RNA (120 nucleotides) are indicated. **Lane 1**, *in vitro* synthesised RNA hybridised to 5′ gene fragment; **lane 2**, *in vivo* synthesised RNA hybridised to 5′ gene fragment; **lane 3** *in vitro* synthesised RNA hybridised to 3′ gene fragment; **lane 4**, *in vivo* synthesised RNA hybridised to 3′ gene fragment. Multiple bands are resolved at the 5′ end (**lane 2**). Whether this is due to heterogeneity of cleavage by RNase T1 or variable initiation sites is not known. Likewise, the origin of the doublet in **lane 3** is unknown. (**B**) The RNA made *in vitro* was hybridised to separated strands of the DNAs cloned in coliphage M13. The hybrids were treated with RNase T1 and analysed on either a 6% polyacrylamide gel (**lanes 1 and 2**) or a 10% polyacrylamide gel (**lanes 3 and 4**). The position of 7S RNA (300 nucleotides) is indicated as a marker in each case. **Lanes 1 – 4** show the RNA which hybridised to the 5′ end of the non-coding strand, the 5′ end of the coding strand, the 3′ end of the coding strand and the 3′ end of the non-coding strand, respectively.

119

6. APPLICATIONS OF NUCLEAR TRANSCRIPTION SYSTEMS

Transcription in isolated nuclei provides one of the best methods to answer several important questions. The most useful experiments so far have involved measuring the relative rates of transcription of a gene and determining the structure of the primary transcript. Potentially this type of system can be used to detect regulatory factors and factors required for RNA processing.

6.1 Measurement of the Relative Rate of Transcription of a Gene

The relative rate of transcription of a particular gene (that is, relative to the transcription of other genes) can be determined by measuring the amount of that RNA synthesised *in vitro* using DNA excess hybridisation (dot hybridisation; *Table 14*) and comparing this with the total amount of RNA synthesised. Generally, 5 μg of DNA per dot (*Table 14*) provides sufficient DNA excess for quantitative hybridisation unless one is dealing with a structural RNA (e.g., rRNA or small nuclear RNAs). The advantage of this technique is that the data are not complicated by degradation of RNA in the isolated nuclei. However, the assumption made is that the transcription rate of all genes *in vitro* reflects the activity *in vivo*. This may not be the case for small transcripts which may initiate inefficiently while large transcripts are being elongated after *in vivo* initiation. In this situation, the large transcripts will be a larger proportion of the RNA labelled *in vitro* than they would be after a pulse label *in vivo*. To compensate for this, one can carry out the transcription for a short period of time (5 min) when most transcription should represent this elongation compared with longer times (30 or 60 min) when variations in the efficiency of initiation of different genes should be more apparent. The relative rates of transcription of the gene are measured for each time period. If the relative rates of transcription are independent of the incubation period then the measured rate is probably an accurate reflection of the rate of transcription *in vivo*. An application of this technique to measure changes in histone gene transcription is shown in *Table 17* (45). The relative rates of transcription *in vitro* agree well with those *in vivo* as determined by pulse-labelling cells for 5 min with [³H]uridine (*Table 17*). In this experiment, the proportion of RNA that was histone mRNA was independent of the time period of transcription, despite the fact that these transcripts are only 500 nucleotides long. This implies either that these genes are efficiently initiated *in vitro* or that a number of RNA polymerase II molecules are already bound to these genes *in vivo* and begin transcription at different times during the *in vitro* incubation.

6.2 Definition of the Primary Transcript

Because of the low level of RNA degradation in nuclei isolated from many sources, it is possible to detect primary transcripts which are not readily detected *in vivo* due to rapid processing of the RNA, especially at the 3' end. However, termination does not necessarily occur faithfully *in vitro*. To ensure that the transcripts observed *in vitro* are also found *in vivo*, Weintraub *et al.* (46) and Hofer *et al.* (47) labelled nuclei from cells expressing globin genes for

Table 17. Relative Rate of Histone Gene Transcription.

Isolated Nuclei

Nuclei isolated from exponentially growing mouse myeloma cells or cells treated with cytosine arabinoside (ara-C; 40 μg/ml) were incubated with [^{32}P]GTP under conditions of *in vitro* transcription (*Table 5*). For each set of nuclei, the labelled RNA (6.4 x 10^6 c.p.m.) was hybridised to dots of various DNA probes[a] immobilised on nitrocellulose filters (*Table 14*) and the amount of RNA bound to each dot of immobilised DNA was determined by liquid scintillation counting.

Source of RNA	DNA probe[a]			
	H3.2	H2a	k	pBR322
Control cells (c.p.m.)	1265	1007	680	30
% input	0.020	0.016	0.012	0.0005
Ara-C treated cells (c.p.m.)	226	149	624	25
% of control	21.0	14.8	91.8	–

Cells Pulse-labelled in vivo

Parallel cultures of cells were pulse-labelled with [^{3}H]uridine for 10 min and the amount of labelled RNA hybridising to the various DNA probes was determined as described above.

Source of RNA	DNA probe[a]			
	H3.2	H2a	k	pBR322
Control cells (c.p.m.)	1137	976	2244	40
% of input	0.016	0.014	0.032	0.0006
Ara-C treated cells (c.p.m.)	200	170	1847	35
% of control	17.6	17.4	85.0	–

[a]The DNA probes used were: H3.2, a mouse histone H3 gene (45); H2a, a mouse histone H2A gene (45); *k*, a light chain immunoglobulin gene (42); pBR322, plasmid control.

1 min with [^{32}P]GTP. They then hybridised the RNA transcripts to a genomic DNA fragment located 1.0 kb downstream from the 3' end of the globin gene. Since RNA polymerase II can only transcribe about 300 nucleotides/min *in vitro* (18), any labelled RNA complementary to a region more than 500 bases from the 3' end of the gene must have been due to transcripts which had been initiated *in vivo* and extended into that region *in vitro*. They concluded that the globin primary transcript extended downstream of the last exon of globin mRNA. It is possible to define the 3' region of the primary transcript using a 1-min labelling period for *in vitro* transcription using isolated nuclei and appropriate probes against the 3' end of the gene. It has not been possible to detect the transcription of these 3' sequences readily *in vivo* presumably because they are rapidly degraded (47).

6.3 Isolation and Study of Primary Transcripts

This section describes a method for the isolation of primary RNA transcripts following their initiation with a [γ-S]ribonucleoside triphosphate. This technique can be used for the identification and isolation of transcripts with a unique initiating nucleotide. The transcripts can then be sequenced to detect the exact sites of initiation within a cloned DNA segment.

Transcription is carried out by incubating the nuclei as described in *Tables 5* or *10* using [γ-^{35}S]ATP or [γ-^{35}S]GTP (New England Nuclear) as the radio-labelled nucleoside triphosphate. Confirmation that the [γ-^{35}S]nucleotide has been incorporated at the 5′ end of the RNA transcript can be obtained as described in footnote b of *Table 10*. [γ-^{35}S]RNA transcripts are hybridised to a cloned DNA probe of the gene of interest using DNA bound to a nitrocellulose filter (*Table 14*) and hybridised RNA is subsequently eluted (*Table 15*). Next, these specific transcripts are cleaved enzymatically at specific bases (see Section 7 of ref. 48 for detailed procedures). The 5′-[γ-^{35}S]oligonucleotides can be separated from oligonucleotides generated by internal cleavage by electrophoresis through a polyacrylamide gel prepared using *p*-acrylamidophenyl mercury as the monomer (41). The thiol-containing 5′-oligonucleotides remain bound to the mercurated gel whilst oligonucleotides generated by internal cleavage migrate through the gel. The former can then be eluted from the matrix by continuing electrophoresis with a buffer containing a negatively-charged thiol compound. These oligonucleotides will all contain the initiating sequence [^{35}S]pppXp but will be of different lengths depending on the location of the cleavage sites. After electrophoresis, therefore, a 'ladder' of different-sized fragments will have been produced which is detected by autoradiography. The RNA sequence at the 5′ end of the transcript can then be read directly from the ladder as with DNA sequencing gels. In fact, the method works even without the elimination of internally-cleaved oligonucleotides since these are unlabelled and so fail to be detected by autoradiography. Therefore one can determine the sequence at the 5′ end of the primary transcript simply by analysing the enzymatic cleavage products (see above) on a typical sequencing gel (48). By comparison of this RNA sequence with that of the gene from which the transcripts were derived, one can accurately map the start sites of transcripts initiated *in vitro*.

To demonstrate the effectiveness of γ-thionucleotides as indicators of initiation, the authors have isolated *in vitro* RNA transcripts coded by phage λCb$_2$ DNA. Transcripts which initiated with either A or G were selected by using [γ-S]ATP or [γ-S]GTP, respectively (49). The relative efficiency of transcription was unaffected by replacement of ATP or GTP with the corresponding γ-thionucleotides and no errors in the fidelity of transcription were detected as a result of the substitution (49). The same principle can be used for isolating RNA transcripts specifically initiated with A or G in isolated nuclei. In addition, since the pyrimidine analogues [γ-S]UTP and [γ-S]CTP are now available, these may be used to test whether eukaryotic RNA polymerases can use pyrimidine nucleotides to initiate RNA chains instead of ATP or GTP.

6.4 Study of Post-transcriptional Modifications

Nuclei isolated from cultured cells by the method described in Section 2.2 are deficient in the reactions of RNA processing, polyadenylation and methyl-

Table 18. Preparation of Extracts for Post-transcriptional Modification.

Preparation of Whole Cell Extracts[a,b]

1. Harvest the cells by centrifugation (500 g, 5 min, 25°C) and wash them twice with 0.14 M NaCl, 10 mM Tris-HCl, pH 7.5.

2. Wash the cells once in 5 volumes of 25 mM KCl, 2 mM magnesium acetate, 1 mM DTT, 10 mM Tris-HCl, pH 7.5.

3. Resuspend the cell pellet in one volume of this buffer (routinely 5 x 10^8 cells are resuspended in 1 ml of buffer) and allow the cells to swell for 5 min at 4°C.

4. Lyse the cells by homogenisation (10 strokes with a B pestle in a Dounce homogeniser).

5. Adjust the homogenate to 0.35 M KCl with 4.0 M KCl.

6. Centrifuge the homogenate at 10 000 g for 5 min to pellet the nuclei and mitochondria.

7. Recover the supernatant and centrifuge it at 100 000 g for 1 h at 4°C.

8. Recover the supernatant and store it frozen in small aliquots in liquid nitrogen. The extracts are stable for at least 3 months.

Preparation of Extracts from Nuclei[a,b]

1. Prepare a cell homogenate as described in steps 1 – 4 above.

2. Centrifuge the homogenate at 1000 g for 3 min at 4°C to pellet the nuclei.

3. Homogenise the crude nuclei in 1 ml of 0.35 M KCl, 2 mM magnesium acetate, 1 mM DTT, 10 mM Tris-HCl, pH 7.5. The increased salt concentration causes the nuclei to lyse.

4. Prepare the nuclear extract by centrifugation as described in steps 6 – 8 of the first part of this table. This crude nuclear extract contains all the polyadenylation and methylation activities of whole cell extracts but without the contaminating cytoplasmic proteins and RNA.

[a]Endogenous RNAs can be removed by passing the freshly prepared extract through a DEAE-cellulose column (3 ml volume) equilibrated in 0.3 M KCl, 2 mM magnesium acetate, 1 mM DTT, 10 mM Tris-HCl, pH 7.5. The column is eluted with the same buffer and the fractions containing protein (which is not retarded by the column) are pooled.
[b]If desired, the small molecules (ribonucleoside triphosphates and salts) can be removed by chromatography of the extract on a column of Sephadex G25 (10 ml volume) equilibrated in 10% glycerol, 0.12 M KCl, 5 mM magnesium acetate, 1 mM DTT, 10 mM Tris-HCl, pH 7.5. The void volume is pooled and stored in liquid N_2. If the extract is not chromatographed on Sephadex G25, there is a significant amount of ribonucleoside triphosphates in the extract which reduces the specific activity of the radiolabelled nucleoside triphosphate during subsequent use.

ation. In addition, they can be used to produce primary transcripts lacking caps. Thus this system provides a potential starting point to reconstitute the reactions of RNA maturation.

6.4.1 *Polyadenylation*

Although the isolated nuclei exhibit little polyadenylation of RNA polymerase II transcripts, this activity can be restored by the addition of extracts prepared either from whole cells or nuclei (50). The preparation of suitable extracts is described in *Table 18*.

Transcription is carried out as usual using [α-^{32}P]GTP (*Table 5*) except that each reaction mixture also contains 80 μl of whole cell or nuclei extract. Care should be taken to adjust the concentration of salts and other components to maintain these at the concentrations in the normal transcription reaction mixture (*Table 5,* step 1). After transcription, the RNA transcripts are isolated as

Table 19. Isolation of Polyadenylated RNA by Chromatography on Poly(U)-Sepharose.

1.	Dissolve the RNA sample (prepared as in *Table 6*) in application buffer (0.4 M NaCl, 1 mM EDTA, 0.5% SDS, 10 mM Tris-HCl, pH 7.5). Heat at 80°C for 2 min to denature the RNA and then chill on ice.
2.	Load the RNA sample onto a 1 ml column of poly(U)-Sepharose pre-equilibrated in application buffer at room temperature and collect the unbound RNA.
3.	Re-load the unbound RNA onto the column as in step 2. Repeat this loading a total of three times to ensure complete binding of polyadenylated RNA. Only ~10% of the total input radioactivity is usually bound to the column.
4.	Wash the column with application buffer until no more RNA passes through (monitor the radioactivity). Then elute the column with 5% formamide (v/v) in elution buffer (1 mM EDTA, 0.5% SDS, 10 mM Tris-HCl, pH 7.5). This releases weakly-bound RNA which may contain only short oligo(A) stretches.
5.	Elute the column with a 10 ml linear gradient of 5 – 60% formamide in elution buffer. Collect 0.5 ml fractions. Polyadenylated RNA containing 3′ poly(A) ~200 residues long elutes in fractions 11 – 15.
7.	Finally, elute the column with 90% formamide in elution buffer. Less than 0.1% of the input radioactivity should be eluted in this fraction.
8.	Re-equilibrate the column in application buffer ready for re-use.

Table 20. Determination of the Length of Poly(A) Tracts.

1.	Incubate the nuclei for *in vitro* transcription as described in *Table 5* but using 1 mCi/ml of [^{3}H]ATP or [α-^{32}P]ATP as precursor and with each incubation mixture containing 80 μl of whole cell or nuclei extract. The total ATP concentration should be 100 μM. The concentrations of other components should be adjusted to be the same as in the normal transcription reaction mixture (*Table 5*, step 1).
2.	Isolate the RNA as described in *Table 6* and then purify polyadenylated RNA by affinity chromatography on poly(U)-Sepharose (*Table 19*).
3.	Pool the fractions containing polyadenylated RNA transcripts. Add 40 μg of carrier tRNA, NaCl to 0.3 M and then 2.5 volumes of ethanol. Leave the RNA at −20°C overnight to precipitate.
4.	Recover the RNA pellet by centrifugation (10 000 g for 15 min) and then dissolve it in 50 μl of 0.4 M NaCl, 50 mM Tris-HCl, pH 7.5. Add 0.1 g of RNase A and 1 unit of RNase T1. Incubate at 25°C for 30 min. This digests all RNA sequences other than poly(A) (which is protected from digestion by the high salt concentration).
5.	Chromatograph the RNA digest on poly(U)-Sepharose and isolate the radioactive poly(A) (*Table 19*).
6.	Add 10 μg of carrier tRNA to the radioactive poly(A), NaCl to 0.3 M and then 2.5 volumes of ethanol and leave the mixture overnight at −20°C to allow the poly(A) to precipitate.
7.	Recover the poly(A) by centrifugation at 10 000 g for 15 min.
8.	Determine the size of the radioactive poly(A) by polyacrylamide gel electrophoresis as described in *Table 8*. Convenient size markers are the small nuclear RNAs (SnRNAs) of mammalian cells.

usual (*Table 6*) and then polyadenylated transcripts are detected by chromatography on poly(U)-Sepharose and elution with a formamide gradient (50) (*Table 19*). This procedure allows one to separate RNAs on the basis of

Table 21. Analysis of Poly(A) by Alkaline Hydrolysis.

1. After transcription (*Table 5*), isolate the polyadenylated RNA transcripts by affinity chromatography on poly(U)-Sepharose (*Table 19*) and digest these with RNases A and T1 (*Table 20*, steps 1 – 4) to release the radioactive poly(A).

2. Dry the sample under vacuum and dissolve it in 25 μl of 0.3 M KOH. Seal this in a glass capillary tube and incubate at 37°C for 18 h.

3. Recover the hydrolysed poly(A) into a microcentrifuge tube by breaking the capillary. Adjust the pH of the contents to neutrality using 3 M perchloric acid and then centrifuge in a microcentrifuge for 10 min to remove the precipitated potassium perchlorate.

4. Dry the supernatant under vacuum and then dissolve the residue in 20 μl of water containing 100 μg of adenosine.

5. Spot the sample onto a strip of Whatman No.1 paper. Chromatograph for 18 h at room temperature using 1-butanol:0.1 M ammonium chloride (6:1, v/v) as solvent.

6. Visualise the adenosine marker by illumination with u.v. light.

7. Cut the chromatogram into strips (1 cm wide) and elute each separately with a small volume of 0.3 M NaOH. Neutralise the pH of each fraction with 3 M acetic acid and determine the radioactivity of each by scintillation counting using a Triton X-100-toluene fluor[a]. AMP remains at the origin and adenosine migrates 10 – 20 cm.

[a]13 g Omnifluor (NEN), 2 litres toluene, 1 litre Triton X-100.

poly(A) length while oligo(dT)-cellulose will bind all polyadenylated RNA with similar affinity regardless of poly(A) length. Thus chromatography on poly(U)-Sepharose allows one to distinguish the true poly(A)$^+$ RNA from RNA containing adventitious stretches of deoxyadenosines coded for in the genome.

To determine the size of the poly(A) tracts of polyadenylated RNA transcripts, *in vitro* transcription is carried out using [³H]ATP or [α-³²P]ATP as precursor and the length of the poly(A) determined directly by gel electrophoresis after digestion of the rest of the RNA by RNases A and T1 (*Table 20*).

Finally, one can show that the poly(A) has been synthesised *in vitro* by measuring the ratio of [³H]adenosine to [³H]AMP after digestion with alkali (50) (*Table 21*). This distinguishes true synthesis from turnover of the ends of the molecule. The ratio of [³H]AMP to [³H]adenosine should be equal to the length of the poly(A) tract. If turnover of the end of the poly(A) or extension of pre-existing poly(A) is taking place, then the ratio of [³H]AMP to [³H]adenosine will be much lower than the length of the poly(A).

6.4.2 *Methylation*

The RNA methylation activity of isolated nuclei can be restored (35) by addition of the same cell extract which restores polyadenylation (Section 6.4.1). The methylation of RNA can be followed using [³H]CH$_3$-S-adenosyl-methionine as the radiolabelled precursor (*Table 22*). The methylated bases are identified after digestion of the RNA with RNase T1, nuclease P1 and alkaline phosphatase, followed by thin-layer electrophoresis and chromatography. Methylation occurs on the RNA transcription products of all three eukaryotic RNA polymerases. There is also extensive methylation of endogenous rRNA

Table 22. Methylation of RNA Transcripts.

1.	Incubate nuclei as described in *Table 5* but using [³H]CH₃-S-adenosylmethionine as the radiolabelled precursor at a concentration of 10 μM. The specific radioactivity of this radio-isotope should be at least 10 Ci/mM. Also include 80 μl of cell extract (prepared as in *Table 18*) in each reaction mixture. The other components should be at the same concentration as in the normal transcription reaction mixture (*Table 5, step 1*).
2.	Isolate the RNA transcripts as described in *Table 6*. Only 20 − 30% of the [³H]CH₃-S-adenosylmethionine is incorporated into RNA. The rest is incorporated into DNA and protein.
3.	Digest the methylated RNA with RNases T1 and T2 and fractionate the oligonucleotides by chromatography on DEAE-cellulose in 7 M urea (*Table 12, steps 3 − 6*).
4.	Pool the eluted oligonucleotides then de-salt the sample and treat it with nuclease P1 and alkaline phosphatase (*Table 12, steps 8 − 13*).
5.	Analyse the nucleosides by t.l.c. to identify the methylated bases as described in reference 36.

precursors, but the methylation of tRNA precursors and mRNA occurs primarily on RNA molecules synthesized *in vitro* (35).

6.4.3 *Capping of Messenger RNA Precursor Molecules*

Procedures for the analysis of capping of primary RNA transcripts synthesised *in vitro* are described in Section 4.3.4.

Shatkin and his colleagues have established that reovirus mRNAs are initiated with GTP and the nascent chains are immediately capped and form a methylated 5′-terminal cap structure, m⁷GpppGᵐ, in the presence of S-adenosylmethionine (51). Under appropriate conditions, cap formation can also occur post-transcriptionally which suggests that the activities of the polymerase and capping enzymes are not tightly coupled. The authors have studied the synthesis of reovirus RNA by viral core enzymes using thionucleotides as substrates, in an effort to isolate the primary transcription product prior to the processing events. Using [β-³²P][γ-S]GTP as a probe for initiation, primary transcripts of reovirus mRNA were synthesised *in vitro* by the core enzyme associated with the viral preparation (23). The bulk of the transcripts contained triphosphorylated 5′ termini which had not been capped. Furthermore, the virion-associated phosphohydrolase could hydrolyse GTP to GDP, but could not hydrolyse [γ-S]GTP. It was thus shown that capping was a post-transcriptional event. Since the γ-thiophosphate group is acid-labile, it can be removed by weak acid, such as 7% formic acid (23), without cleaving the RNA phosphodiester linkage. This allows one to study the regulation of the capping reaction using *in vitro* initiated RNA.

6.4.4 *Splicing of Messenger RNA Precursor Molecules*

There has been little success in analysing the splicing of RNA polymerase II primary transcripts to yield mature mRNA molecules *in vitro* using isolated nuclei from mammalian cells. Blanchard *et al.* (52) and Flint *et al.* (53) have reported that adenovirus RNA is spliced in a crude preparation of nuclei from HeLa cells but it is not clear how efficient this is. Nevertheless, it seems likely that mammalian cell nuclei will prove useful in preparing substrates for splicing and detecting the spliced products by the hybridisation assays described above. The advantage of this approach of using whole nuclei may be that the RNAs will be in their 'native' conformation complexed with appropriate proteins.

6.4.5 *Processing of tRNA Precursors*

Isolated nuclei are able to synthesise tRNA precursor molecules but these are not processed to mature tRNA (2). The absent activity is *not* restored by cell extracts prepared in 0.35 M KCl (*Table 18*) which are able to restore polyadenylation (Section 6.4.1) and methylation (Section 6.4.2) activities. However, it can be restored by extracts prepared from crude nuclei in 0.15 M KCl (2). The processing of tRNA precursors to mature tRNA is most easily monitored by polyacrylamide gel electrophoresis (*Table 8*) using 10% gels.

6.4.6 *Processing of Ribosomal RNA Precursors*

At the time of writing, there have been no reports of successful processing of the eukaryotic rRNA precursors in a cell-free system.

6.5 **Study of Nuclear Transport of Mature RNA**

To assay for RNA released from the nuclei, the transcription reaction (*Table 5*) is diluted with one volume of 0.1 M KCl, 5 mM magnesium acetate, 10 mM Tris-HCl, pH 7.5 and then layered over an equal volume of glycerol storage buffer (*Table 1*) and centrifuged at 1000 *g* for 2 min. The supernatant contains the released RNA and the pellet contains the RNA retained within the nuclei. When transcription is carried out in the absence of cell extracts, there is no release of RNA except for the small RNAs (5S RNA, tRNA precursors, small nuclear RNAs) (6). On inclusion of an extract (*Table 18*) prepared from myeloma cells, there is release of some additional larger RNA molecules including poly(A)$^+$ RNA (50). The released RNAs can be analysed for specific transcripts by the hybridisation assays described above.

6.6 **Assays for Regulatory Factors**

A potentially important application of isolated nuclei is in the detection of factors involved in regulating transcription and in studying how these factors interact with chromatin. The authors have used nuclei isolated from sea urchin embryos to detect and partially purify an inhibitor of RNA polymerase III present in sea urchin eggs (6). Stoute and Marzluff (54) have shown that extraction of nuclei with 0.35 M KCl greatly reduces transcription by all three RNA polymerases and that transcription can be restored by adding high mobility group (HMG) proteins 1 and 2. Thus it is possible to alter transcription reversibly in isolated nuclei. In the future this approach should yield considerable insight into the regulation of transcription.

7. ACKNOWLEDGEMENTS

The methods described in this chapter were developed in the authors' laboratories over the past decade. We particularly acknowledge the contributions of Drs. Anthony E. Reeve, Mitchell Smith, Gilbert Fragoso, Jeffrey Moshier, David Cooper, Anne Brown, Gil Morris and Don Sittman in developing many techniques related to the work described in this chapter. This work has been supported by grants from the NIH.

8. REFERENCES

1. Marzluff,W.F., Murphy,E.C. and Huang,R.C.C. (1973) *Biochemistry (Wash.)*, **12**, 340.
2. Marzluff,W.F., Murphy,E.C. and Huang,R.C.C. (1974) *Biochemistry (Wash.)*, **13**, 3689.
3. Tilghman,S.M. and Belayew,A. (1982) *Proc. Natl. Acad. Sci. USA*, **79**, 5254.
4. Ernest,M.M., Schultz,G. and Feigelsen,P. (1976) *Biochemistry (Wash.)*, **15**, 824.
5. Price,R. and Penman,S. (1972) *J. Mol. Biol.*, **70**, 435.
6. Morris,G.M. and Marzluff,W.F. (1983) *Biochemistry (Wash.)*, **22**, 645.
7. Panyim,S. and Chalkley,R. (1969) *Arch. Biochem. Biophys.*, **130**, 337.
8. Zweidler,A. (1978) *Methods Cell Biol.*, **17**, 223.
9. Thomas,J.O. and Kornberg,R.D. (1978) *Methods Cell Biol.*, **18**, 429.
10. Harvey,W. and Atkinson,B.G. (1982) *Can. J. Biochem.*, **60**, 21.
11. Gurney,T. and Foster,D.N. (1977) *Methods Cell Biol.*, **16**, 45.
12. Marzluff,W.F. and Huang,R.C.C. (1975) *Proc. Natl. Acad. Sci. USA*, **72**, 1082.
13. Roeder,R.G. (1976) in *RNA Polymerase*, Losick,R. and Chamberlain,M. (eds.), Cold Spring Harbor Laboratory Press, New York, p. 285.
14. Wagner,F.K., Katz,L. and Penman,S. (1967) *Biochem. Biophys. Res. Commun.*, **28**, 152.
15. Chirgwin,J.M., Przbyla,A.E., MacDonald,R.J. and Rutter,W.J. (1979) *Biochemistry (Wash.)*, **18**, 5294.
16. Benecke,B.-J., Ben-Ze'ev,A. and Penman,S. (1978) *Cell*, **14**, 931.
17. McMaster,G.K. and Carmichael,G.G. (1977) *Proc. Natl. Acad. Sci. USA*, **74**, 4935.
18. Marzluff,W.F., Pan,C.J. and Cooper,D.L. (1978) *Nucleic Acids Res.*, **5**, 4177.
19. Goody,R.S. and Eckstein,F. (1971) *J. Am. Chem. Soc.*, **93**, 6252.
20. Goody,R.S., Eckstein,F. and Sehinanu,R.H. (1972) *Biochim. Biophys. Acta*, **276**, 155.
21. Reeve,A.E., Smith,M.M., Pigiet,V. and Huang,R.C.C. (1977) *Biochemistry (Wash.)*, **16**, 4464.
22. Smith,M.M., Reeve,A.E. and Huang,R.C.C. (1978) *Cell*, **15**, 615.
23. Reeve,A.E., Shatkin,A.J. and Huang,R.C.C. (1982) *J. Biol. Chem.*, **257**, 7018.
24. Benz,E.W., Wydro,R.M., Nadel-Ginard,N. and Dina,D. (1980) *Nature*, **288**, 655.
25. Cuatrecasas,P. (1970) *J. Biol. Chem.*, **245**, 3059.
26. Smith,M.M. and Huang,R.C.C. (1976) *Proc. Natl. Acad. Sci. USA*, **73**, 775.
27. Huang,R.C.C. (1983) in *Genetic Engineering Techniques*, Huang,R.C.C., Kuo,T.T. and Wu,R. (eds.), Academic Press, p. 93.
28. Ide,G. (1981) *Biochemistry (Wash.)*, **20**, 2633.
29. Hipskind,R.A. and Reeder,R.H. (1980) *J. Biol. Chem.*, **255**, 7896.
30. Sun,I.Y., Johnson,E.M. and Allfrey,V.G. (1979) *Biochemistry (Wash.)*, **18**, 4572.
31. Washington,L.D. and Stallcup,M.R. (1982) *Nucleic Acids Res.*, **10**, 8311.
32. Connolly,B.A., Rowaniuk,P.J. and Eckstein,F. (1982) *Biochemistry (Wash.)*, **21**, 1983.
33. Eckstein,F. and Goody,R.S. (1976) *Biochemistry (Wash.)*, **15**, 1685.
34. Stallcup,M.R. and Washington,L.D. (1983) *J. Biol. Chem.*, **258**, 2802.
35. Brown,A. and Marzluff,W.F. (1982) *Biochemistry (Wash.)*, **21**, 4303.
36. Contreras,R. and Fiers,W. (1981) *Nucleic Acids Res.*, **9**, 215.
37. Dale,R.M.K., Livingston,D.C. and Ward,D.C. (1973) *Proc. Natl. Acad. Sci. USA*, **70**, 2238.
38. Crouse,G.F., Fodor,E.J.B. and Doty,P. (1976) *Proc. Natl. Acad. Sci. USA*, **73**, 1564.
39. O'Malley,B.W., Tsai,M.J., Tsai,S.Y. and Towle,H.C. (1978) *Cold Spring Harbor Symp. Quant. Biol.*, **42**, 605.
40. Carrol,A.R. and Wagner,R. (1979) *J. Biol. Chem.*, **254**, 9339.
41. Reeve,A.E. and Johnson,L.P. (1979) *J. Chromatogr.*, **177**, 127.
42. Schibler,U., Marcu,K.B. and Perry,R.P. (1978) *Cell*, **15**, 1495.
43. Rave,N., Crkvenjakov,R. and Boedtker,H. (1979) *Nucleic Acids Res.*, **11**, 3559.

44. Bailey,J.M. and Davidson,N. (1976) *Anal. Biochem., 70*, 75.
45. Sittman,D.B., Graves,R.A. and Marzluff,W.F. (1983) *Proc. Natl. Acad. Sci. USA, 80*, 1849.
46. Weintraub,H., Larsen,H. and Groudine,M. (1981) *Cell, 24*, 333.
47. Hofer,E., Hofer-Warbinek,R. and Darnell,J. (1982) *Cell, 29*, 887.
48. D'Alessio,J.M. (1982) in *Gel Electrophoresis of Nucleic Acids − A Practical Approach*, Rickwood,D. and Hames,B.D. (eds.), IRL Press Ltd., Oxford and Washington, DC, p. 173.
49. Smith,M.M., Reeve,A.E. and Huang,R.C.C. (1978) *Biochemistry (Wash.), 17*, 493.
50. Cooper,D.L. and Marzluff,W.F. (1978) *J. Biol. Chem., 253*, 8375.
51. Furuichi,Y., Morgan,M., Muthukrishnan,S. and Shatkin,A. (1975) *Proc. Natl. Acad. Sci. USA, 72*, 362.
52. Blanchard,J.M., Weber,J., Jelinek,W. and Darnell,J.E. (1978) *Proc. Natl. Acad. Sci. USA, 75*, 5344.
53. Yang,Y.W., Lerner,M.R., Steitz,J.A. and Flint,S.J. (1981) *Proc. Natl. Acad. Sci. USA, 78*, 1371.
54. Stoute,J. and Marzluff,W.F. (1982) *Biochem. Biophys. Res. Commun., 107*, 1279.

CHAPTER 5

Transcription of Chromatin

R. STEWART GILMOUR

1. INTRODUCTION

Isolated eukaryotic chromatin can support the transcription of RNA *in vitro* by added RNA polymerase. Therefore this provides a cell-free approach for studying the mechanisms which regulate gene expression. Furthermore, since chromatin can be reconstituted from its basic structural components, this system should allow examination of the biological importance of these components and their interaction. A detailed review of the background to this work has been published elsewhere (1).

With the development of cDNA probes for specific genes, chromatin transcripts can be analysed for specific mRNA sequences present in extremely small amounts. Using this approach, a number of research groups have reported the synthesis of tissue-specific mRNA from isolated chromatin by exogenous RNA polymerase (2 – 6). However, the validity of these and subsequent results has been challenged by a number of criticisms which also apply to *in vitro* transcription systems in general. This chapter reviews the techniques involved in the transcription of isolated chromatin and alerts the reader to potential artifacts which must be taken into consideration when planning experiments of this kind.

2. ISOLATION OF CHROMATIN

A wide range of methods has been utilised for the isolation of chromatin. However, no single procedure appears to be optimal for all tissue types. The choice of method should take into consideration the following points:

(i)   The starting nuclei should be prepared as free from cytoplasmic material as possible. This can be done effectively by stripping off the outer nuclear membrane with a non-ionic detergent such as Triton X-100.

(ii) The soluble components of the nucleoplasm should be removed by lysis and extraction of the nuclei in low ionic strength buffers. However, this must be carried out in conditions which do not also remove proteins bound to the chromatin itself.

(iii) It is not advisable to subject chromatin to shearing or sonication. While such treatments may render chromatin easier to handle, they can also create free DNA ends as well as opening up the chromatin structure by removing protein, and hence the transcriptional properties of the chromatin may be altered.

Table 1. Isolation of Chromatin[a].

1.	Hand homogenise 1 g (wet weight) of tissue in a Dounce homogeniser (0.025 mm clearance) in 36 ml of 5 mM $MgCl_2$, 1 mM DTT, 3 mM $CaCl_2$, 0.1% Triton X-100, 2 mM Tris-HCl (pH 7.5) for 2 min. Leave the homogenate on ice for 5 min.
2.	Add 4 ml of 2 M sucrose and homogenise for another 30 sec.
3.	Layer 20 ml of the homogenate over 15 ml of 2 M sucrose, 1 mM DTT, 5 mM $MgCl_2$, 0.28 M NaCl, 2 mM Tris-HCl (pH 7.5) and then centrifuge in a swing-out rotor at 30 000 g for 1 h at 4°C. The pellet is nuclei.
4.	Resuspend the pellet in 20 ml of 2 mM EDTA, 1% Triton X-100, 0.28 M NaCl, 1 mM Tris-HCl (pH 7.9) and centrifuge in a swing-out rotor at 12 000 g for 10 min at 4°C.
5.	Wash the pellet three times in 2 mM EDTA, 1 mM Tris-HCl (pH 7.9), by resuspension and centrifugation as in step 4. The washed pellet (a viscous gel) is chromatin.

[a]If the particular tissue under investigation is rich in endogenous proteases, it is advisable to add 50 mM sodium bisulphite *or* 1 mM di-isopropyl fluorophosphate *or* 1 mM PMSF.

(iv) Some tissues are rich in proteases and nucleases. Therefore, steps should be taken to minimise chromatin degradation by performing all operations at 4°C and using inhibitors where appropriate.

Reference to previous reviews (7,8) will give an appreciation of the diverse preparative procedures used on a wide variety of cells. The method described in *Table 1* (which takes account of these considerations) has proved applicable to most tissues examined in the author's laboratory.

3. SOURCE OF RNA POLYMERASE

Eukaryotic mRNAs are synthesised *in vivo* by RNA polymerase II. A method for preparing this enzyme from HeLa cells has been described recently which yields RNA polymerase II capable of initiating RNA synthesis from the authentic cap site on a variety of isolated, cloned DNA templates (9). Unfortunately, definitive experiments using this enzyme on isolated chromatin or nuclear templates are complicated by the inability to distinguish between genuine *de novo* initiation of RNA chains by exogenous enzyme and the reactivation of previously-inactive complexes. This prevents the unambiguous study of transcription of an active gene by exogenous RNA polymerase II. Furthermore, the efficiency of transcription by RNA polymerase II preparations, even with cloned DNA templates, is extremely low (only one DNA molecule in a thousand may be copied) so that the detection of specific transcripts from chromatin templates may be difficult. For these reasons, early experiments were carried out with bacterial RNA polymerases which are more easily prepared and are much more active in transcription. This chapter describes the use of bacterial RNA polymerase for transcription of chromatin *in vitro* and discusses the validity of using a prokaryotic enzyme on a eukaryotic template (Section 6). Transcription of chromatin with endogenous RNA polymerase is described in Chapter 4, Section 2.5.

In general, most *in vitro* transcription work involving isolated chromatin has utilised RNA polymerase from *Escherichia coli*. The purification of this enzyme has been well documented in the literature. The procedure of Burgess (10) describes a basic method for purification employing classical ammonium sulphate fractionation plus ion exchange and gel filtration chromatography. More rapid isolation can be achieved by fractional precipitation with polyethylene imine (11) or binding to columns of either DNA-cellulose (12) or heparin-Sepharose (13). The method preferred by the author is described in *Tables 2* and *3*. All these methods give a final product with a specific activity of about 800 units/mg protein (1 unit of enzyme incorporates 1 nmol of UTP into RNA in 10 min at 37°C). *E. coli* RNA polymerase is also available commercially from Sigma Chemical Co. and Miles Research Products.

Table 2. Reagents Required for The Preparation of *E. coli* RNA Polymerase.

Buffer 1 (10-fold concentrated)
   0.1 M $MgCl_2$
   1 mM EDTA
   0.1 M Tris-HCl, pH 7.9.

Buffer 2
   10 mM $MgCl_2$
   0.01 mM EDTA
   0.2 M KCl
   0.1 mM DTT
   5% glycerol
   50 mM Tris-HCl, pH 7.5.

Buffer 3
   10 mM $MgCl_2$
   0.1 mM EDTA
   1 mM DTT
   5% glycerol
   25.6 g ammonium sulphate (enzyme grade) per 100 ml
   10 mM Tris-HCl, pH 7.9.

Buffer 4
   1 mM EDTA
   0.2 mM DTT
   10% glycerol
   20 mM Tris-HCl, pH 7.9.

Storage buffer
   10 mM $MgCl_2$
   0.1 mM EDTA
   0.1 mM DTT
   0.1 M KCl
   60% glycerol
   10 mM Tris-HCl, pH 7.9.

DE-52 cellulose
Wash 200 g of Whatman DE-52 cellulose (pre-swollen) in 0.5 litres of 2 M NaCl for 30 min. Decant the supernatant and wash the cellulose thoroughly with distilled water. Pour the cellulose into a column (5 cm x 15 cm) and pre-equilibrate it with 10-fold concentrated Buffer 1 and then Buffer 1.

Biogel A 1.5
Pour Biogel A 1.5 into a column (3 cm x 75 cm) and pre-equilibrate this matrix with Buffer 1 containing 1 M KCl.

Table 2 continued overleaf

Transcription of Chromatin

DNA-cellulose

1. Wash 50 g of Whatman CF-11 cellulose in 500 ml each of absolute ethanol, distilled water, 1 M NaOH and finally 1 M HCl. Wash with distilled water until the pH is neutral. Allow the cellulose to air dry.

2. Dissolve 0.8 g of calf thymus DNA (Worthington) in 400 ml of 1 mM EDTA, 10 mM Tris-HCl, pH 7.5. Mix this with the dry CF-11 cellulose. When thoroughly mixed, dry under a stream of warm air or leave at 60°C in an oven overnight.

3. Using a glass rod, break up the lumps of DNA-cellulose and resuspend the powder in 400 ml of absolute ethanol.

4. Over a period of 3 h, stir the DNA-cellulose whilst simultaneously irradiating the mixture using a u.v. lamp (Gallenkamp, LH530) placed about 10 cm away.

5. Recover the DNA-cellulose by filtration and wash it three times each with 2 litres of 1 mM EDTA, 10 mM Tris-HCl, pH 7.5.

6. Pour a suspension of the DNA-cellulose[a] into a column (2 cm x 10 cm) and pre-equilibrate this with Buffer 4.

---

[a]To estimate the amount of DNA bound, heat a suspension of the DNA-cellulose at 100°C for 10 min. Remove the cellulose by centrifugation and read the A_{260} of the supernatant (1 A_{260} ≡ 50 μg DNA).

Table 3. Preparation of *E. coli* RNA Polymerase.

Assay of E. coli RNA polymerase

The following assay is a convenient method for monitoring the purification of the *E. coli* RNA polymerase.

1. Mix:
   - 25.0 μl 50 mM MgCl$_2$, 0.75 M KCl, 0.2 M Tris-HCl (pH 7.9)
   - 3.0 μl 0.1 M sodium phosphate buffer (pH 7.5)
   - 12.5 μl 1 mM DTT
   - 12.5 μl 1 mM EDTA
   - 5.0 μl 4 mM each of ATP, CTP, GTP and [^{3}H]UTP (2.5 μCi/mmol)
   - 30.0 μl calf thymus DNA (0.7 mg/ml)
   - 12.5 μl BSA (0.6 mg/ml)
   - 27.5 μl sample (plus water if necessary).

2. Incubate at 37°C for 10 min. Then pipette 100 μl of the incubation mixture onto a strip of Whatman No.3 filter paper (1.0 cm x 2.0 cm) and drop this into ice-cold 10% TCA.

3. Wash the paper strip twice in 10% TCA and then once in absolute ethanol.

4. Dry the paper strip, add scintillation fluid and determine the precipitated radioactivity using a liquid scintillation counter.

Purification Procedure.

1. Disrupt 200 g of *E. coli* cells (strain MRE 600) in 200 ml of Buffer 2 with 500 g of glass beads[a] using a Vibromix agitator in five 2 min bursts. Maintain the temperature below 15°C during cell lysis.

2. Add 1.5 ml DNase (1 mg/ml) to the lysate and leave on ice for 30 min.

3. Pour off the liquid and keep this on ice. Wash the glass beads twice with Buffer 2 and pool these washes with the original supernatant to give a final volume of 350 ml.

4. Centrifuge the pooled lysate at 150 000 *g* for 3 h (for example, using the MSE 65 angle rotor at 45 000 r.p.m.).

5. Decant the supernatant and slowly add 23.1 g of solid ammonium sulphate per 100 ml of supernatant whilst stirring on ice. Continue to stir on ice for another 30 min.

6. Centrifuge the mixture at 10 000 *g* for 10 min.

7. Recover the supernatant and discard the pellet. To the supernatant add 10.75 g ammonium sulphate per 100 ml of supernatant with stirring on ice.

8. Centrifuge the mixture at 10 000 g for 10 min.

9. Discard the supernatant and collect the pelleted material. Resuspend this in 300 ml of Buffer 3 using hand homogenisation.

10. Centrifuge the mixture at 10 000 g for 10 min.

11. Discard the supernatant. Dissolve the pelleted material in sufficient Buffer 1 to give a conductivity less than Buffer 1 containing 0.13 M KCl.

12. Apply the sample (300 – 400 ml) to the column of DEAE-cellulose (DE-52; 5 cm x 15 cm) pre-equilibrated with Buffer 1 (*Table 2*). The sample should be applied by gravity feed over a period of about 3 h.

13. Wash the column with Buffer 1 containing 0.13 M KCl until the A_{260} of the eluate is less than 0.1.

14. Apply a linear salt gradient using 300 ml each of Buffer 1 containing 0.13 M KCl and Buffer 1 containing 0.4 M KCl. Collect 10 ml fractions.

15. Examine every second fraction for RNA polymerase activity using the assay described above. This enzyme is usually eluted in the first few fractions after application of the salt gradient.

16. Pool those fractions containing RNA polymerase activity and precipitate the enzyme by adding 35 g of solid ammonium sulphate per 100 ml of solution. Stir on ice for 30 min.

17. Centrifuge the mixture at 10 000 g for 10 min. Discard the supernatant and dissolve the precipitated enzyme in 5 ml of Buffer 1 containing 1 M KCl.

18. Load this sample onto the column (3 cm x 75 cm) of Biogel A 1.5 pre-equilibrated in Buffer 1 containing 1 M KCl (*Table 2*). Elute the column using this buffer at a flow rate of 0.5 ml/min and collect 10 ml fractions. Assay every second fraction for RNA polymerase activity; this enzyme usually elutes in the first few fractions which contain protein.

19. Pool those fractions containing RNA polymerase and precipitate the enzyme by adding 1.5 volumes of saturated ammonium sulphate dissolved in Buffer 4. Stir on ice for 30 min.

20. Centrifuge the mixture at 10 000 g for 10 min. Discard the supernatant and dissolve the precipitated enzyme in Buffer 4. Dialyse against this buffer overnight at 4°C.

21. Load the sample onto the column (2 cm x 10 cm) of DNA-cellulose (*Table 2*; 3.5mg DNA/ml packed volume) pre-equilibrated in Buffer 4. Elute the column with Buffer 4 until the A_{260} of the eluate is less than 0.05.

22. Elute the RNA polymerase with Buffer 4 containing 0.65 M KCl. Precipitate the enzyme by adding 1.5 volumes of saturated ammonium sulphate as in step 19.

23. Centrifuge the mixture at 10 000 g for 10 min. Dissolve the pellet in 2 ml of storage buffer (*Table 2*) and store at − 20°C.

---

[a]The glass beads (~ 0.1 mm diameter) should be pre-washed in concentrated HCl for 15 min then washed with distilled water until the pH is neutral.

4. OPTIMAL CONDITIONS FOR *IN VITRO* TRANSCRIPTION

4.1 Problems of Endogenous RNA

In early experiments, studies of RNA transcripts from chromatin templates were limited to the characterisation of populations of RNA sequences, defined operationally in terms of hybridisation specificity. Subsequently, the availability of cDNA probes and, later, the development of mapping techniques using nuclease S1 offered precise assays for specific mRNA sequences present in minute quantities. Unfortunately, the amount of endogenous RNA, initiated *in vivo* and bound to the isolated chromatin, is generally in the same range as the yield of RNA expected after *in vitro* synthesis by added RNA polymerase. The extent of contamination by endogenous RNA can be determined by hybridisation to a specific DNA probe. This value can then be subtracted from the total amount of RNA present after transcription *in vitro* to give a measure of the amount of the specific RNA transcribed by the exogenous RNA polymerase. However, in most cases, this procedure has proved unsatisfactory since the hybridisation reactions are RNA-driven and thus minute differences in endogenous RNA contamination between samples will produce misleading results. Because of this, earlier evidence for selective gene expression from isolated chromatin (2 – 6) must be regarded as inadequate.

Separation of endogenous and newly synthesised RNA after *in vitro* transcription is an alternative solution to the problem. One approach is to use chemically-mercurated UTP (Hg-UTP) which is incorporated into RNA by bacterial RNA polymerase using chromatin as template. The *in vitro* transcripts are separated from endogenous RNA by affinity chromatography on thiolated Sepharose; unsubstituted endogenous RNA fails to bind to this matrix while most of the mercurated RNA transcript binds tightly and can be recovered by elution with buffer containing a thiol reagent. The purified RNA transcript is then analysed by hybridisation to a gene-specific cDNA probe. This approach was used by several research groups for the analysis of *in vitro* transcripts of chromatin from a variety of sources (14 – 18). Since then a number of serious criticisms of this method have been made (19 – 21). In particular, under certain experimental conditions, *E. coli* RNA polymerase was shown to transcribe endogenous RNA as well as DNA. This produces double-stranded hybrid RNA molecules containing endogenous RNA and mercurated 'anti-sense' RNA which are retained on thiolated Sepharose. Fortunately, unlike transcription of DNA, copying of RNA by *E. coli* RNA polymerase is insensitive to actinomycin D. Thus this simple test indicates the extent to which RNA-primed transcription has taken place. By heating the sample to denature any double-stranded RNA prior to affinity chromatography, endogenous RNA can be separated from both the mercurated anti-sense RNA strands and the authentic transcripts copied from the DNA. Provided a sense-specific cDNA probe is used in the subsequent analysis, the purified anti-sense strand will not interfere since it has the same sense as the cDNA probe itself. Therefore hybridisation will occur only between cDNA and RNA transcribed from the DNA template.

Although, using the above procedure, authentic transcription can be distinguished from spurious copying of RNA, it is obviously desirable to optimise the conditions of *in vitro* transcription of isolated chromatin so that only DNA-dependent transcription takes place. To investigate this problem the author has examined the *in vitro* transcription of β-globin genes from mouse foetal liver chromatin by *E. coli* RNA polymerase. This has led to the definition of transcription conditions which exclusively favour DNA-dependent copying (22,23). Crucial in this respect is the nature of the divalent cation present. In the presence of Mg^{2+}, transcription of the β-globin genes is DNA-dependent and asymmetric. Thus high concentrations (250 $\mu g/ml$) of actinomycin D inhibit transcription by up to 95%, the mercurated transcripts hybridise to globin cDNA and cause it to be retained on thiolated Sepharose and, finally, no significant amounts of anti-sense β-globin RNA can be detected by hybridisation to anti-sense globin cDNA (i.e., DNA complementary to globin cDNA and having the same 'sense' as globin mRNA). In contrast, pronounced *RNA*-dependent transcription takes place when Mg^{2+} is replaced by Mn^{2+}. Under these conditions, only 40 – 50% of RNA synthesis is sensitive to actinomycin D and hybridisation with anti-sense globin cDNA confirms that endogenous globin mRNA has been copied. In addition, there is evidence that the newly-formed, double-stranded hybrid RNA molecules can act as templates for the further synthesis of (mercurated) globin sequences of the same polarity as the original mRNA (23). Clearly, heat denaturation prior to chromatography on thiolated Sepharose would not abolish this background.

4.2 DNA-dependent Transcription of Chromatin

This section describes the detailed methods for *in vitro* transcription of chromatin by *E. coli* RNA polymerase and isolation of the RNA transcripts by affinity chromatography on thiolated Sepharose, based on the considerations discussed in Section 4.1.

The preparation of Hg-UTP is described in *Table 4*. CTP can be mercurated in the same way and used for *in vitro* transcription instead of Hg-UTP. Both compounds are also available commercially from Sigma and Boehringer. The incubation conditions for DNA-dependent transcription of chromatin by *E. coli* RNA polymerase and subsequent extraction of the RNA are described in *Table 5*. Normally, mercury-substituted ribonucleoside triphosphates bind tightly to RNA polymerase and thus can cause severe inhibition of enzyme

Table 4. Preparation of Mercurated UTP[a].

1.	Dissolve 24 mg of UTP in 2 ml of 0.1 M sodium acetate (pH 6.0) and incubate the solution at 50°C for 1 h with an equal volume of 0.1 M mercuric acetate dissolved in the same buffer.
2.	Add 1 ml (packed volume) of Dowex chelating resin (mesh 50 – 100; Sigma) previously washed in distilled water and adjusted to pH 7.0 with 0.1 M NaOH. Mix and then leave the slurry on ice for 5 min.
3.	Pack the slurry into a disposable 5 ml plastic syringe plugged with glass wool and centrifuge (500 *g* for 10 min) the solution into a clean tube.
4.	Repeat steps 2 and 3.
5.	Precipitate the Hg-UTP with two volumes of absolute ethanol. Recover the precipitate by centrifugation.
6.	Repeat step 5.
7.	Dry the Hg-UTP under vacuum and dissolve it in 2 ml of distilled water. The concentration of Hg-UTP should be ~ 20 mM.

[a]Adapted from ref. 19.

Table 5. *In Vitro* Transcription of Chromatin.

---

1. Prepare each reaction mixture (2 ml final volume) to contain:
    0.8 mM each of ATP, GTP, CTP
    0.5 mM UTP
    0.3 mM Hg-UTP (*Table 4*)
    1 mg of chromatin DNA
    200 – 400 units of *E. coli* RNA polymerase (*Table 3*)
    14 mM 2-mercaptoethanol
    5 mM MgCl$_2$
    0.15 M KCl
    40 mM Tris-HCl (pH 7.9).

2. Incubate the mixture at 37°C for 1 h and then stop the reaction by adding 100 μl of 10% sarcosine and 50 μl of 0.1 M EDTA.

3. Sonicate the solution sufficiently to reduce the viscosity (two or three periods of sonication, each 10 sec long) then add Proteinase K (Merck) to 50 μg/ml final concentration and incubate at 37°C for 15 min.

4. Deproteinise the solution by vortexing it with an equal volume of phenol/chloroform[a] (1:1 v/v). Separate the phases by centrifugation (1000 g for 5 min). Remove the upper (aqueous) layer and add 3 volumes of absolute ethanol. Mix and then leave the solution at −20°C for 1 h to precipitate the nucleic acids.

5. Recover the precipitate by centrifugation (5000 g for 10 min) and remove the residual ethanol under vacuum.

6. Redissolve the nucleic acid precipitate in 0.5 ml of 10 mM Tris-HCl (pH 7.9), 50 mM NaCl and load this onto a column (60 cm x 1 cm) of Sephadex G50 equilibrated with the same buffer. Collect the void volume and precipitate the nucleic acids with ethanol as in steps 4 and 5.

7. Finally, dissolve the nucleic acids in 0.5 ml of 10 mM Tris-HCl (pH 7.9). Store at −20°C or proceed directly to chromatography on thiolated Sepharose as described in *Table 6*.

---

[a]Hydroxyquinoline must not be added to the phenol since this promotes removal of mercury groups from the RNA.

Table 6. Purification of Mercurated RNA Transcripts using Thiolated Sepharose.

---

1. Wash a 1 ml column of thiolated Sepharose with 5 ml of column buffer (10 mM Tris-HCl, pH 7.9, 0.2 M NaCl, 1% SDS) supplemented with 0.3 M 2-mercaptoethanol.

2. Wash the column with 15 ml of column buffer (without 2-mercaptoethanol).

3. Boil the nucleic acid sample (from *Table 5*, step 7) for 4 min and then cool it rapidly on ice.

4. Add an equal volume of ice-cold, 2-fold concentrated column buffer (without 2-mercaptoethanol) to the nucleic acid sample, mix and load the mixture onto the pre-equilibrated column of thiolated Sepharose. Leave the sample on the column, without elution, for 10 min to enable mercurated RNA to bind to the matrix. Next elute unbound RNA using column buffer lacking 2-mercaptoethanol.

5. Next elute the mercurated RNA from the column with 2 ml of column buffer containing 0.3 M 2-mercaptoethanol.

6. Precipitate the mercurated RNA with ethanol and recover it by centrifugation (*Table 5*, steps 4 and 5).

7. Repeat step 6.

8. Wash the precipitate of mercurated RNA with 5 ml of 75% ethanol (pre-cooled to 4°C) and recover the precipitate by centrifugation (*Table 5*, step 5).

9. Lyophilise the mercurated RNA, redissolve it in distilled water and store at −20°C.

activity. However, the inclusion of $10-20$ mM 2-mercaptoethanol during the transcription reaction acts as a competing ligand for the mercury groups and permits normal RNA synthesis. After extraction, the mercurated RNA transcripts are purified by affinity chromatography on thiolated Sepharose as described in *Table 6* and are then ready for quantitation by cDNA hybridisation (Section 5.1) or mapping using nuclease S1 (Section 5.2).

5. ANALYSIS OF RNA TRANSCRIPTS

5.1 **Hybridisation to Specific DNA Probes**

Accurate quantitation of the amount of specific transcript is achieved by RNA excess hybridisation to a specific [³H]cDNA or [³H]genomic DNA probe followed by incubation with nuclease S1 to digest any unhybridised DNA. Mercurated nucleic acids fail to hybridise to complementary probes, especially if the level of substitution by Hg-UTP is extensive (17,20). However, demercuration occurs on exposure to thiol reagents (24), and thus the RNA transcripts, isolated by binding to thiolated Sepharose and elution by thiol reagent (*Table 6*), are sufficiently demercurated (23) to be able to hybridise.

The methods for isolation of purified mRNA and reverse transcription into radiolabelled cDNA are described in detail elsewhere (ref. 25). Since the amount of specific RNA transcript is calculated directly from the radioactivity in hybridised, nuclease S1-resistant DNA, an incomplete cDNA probe will give rise to underestimates of the amount of RNA present. Therefore, before use it is advisable to fractionate the [³H]cDNA by sucrose gradient centrifugation to obtain full-length molecules (25).

The hybridisation conditions used in the author's laboratory are described in *Table 7*. Increasing amounts of RNA are titrated against a fixed amount of labelled cDNA and the amount of cDNA hybridised is measured by digestion with nuclease S1 followed by perchloric acid precipitation (*Table 7* and ref. 26). With increasing RNA input, the percentage of cDNA hybridised increases to a maximum (plateau) value. Since all reactions are carried out to a minimum R_0t of 10^{-1} mol.l^{-1}.sec to ensure complete hybridisation, the hybridisation curve reaches a plateau at the point when the input of specific RNA transcript is equivalent to the amount of [³H]cDNA present. It is advisable to calibrate this procedure by titration against pure mRNA before analysing the amount of specific RNA transcripts after *in vitro* transcription.

Figure 1 and *Table 8* show typical data obtained using this approach. For the discussion which follows it is important to note that *Figure 1* is a qualitative analysis only, in that it refers to the *type* of RNA transcripts produced and does not involve any consideration of yield. On the other hand, *Table 8* also includes data which relate to the yield of transcription under each set of conditions. The study examined the *in vitro* transcription of globin genes using chromatin from mouse foetal liver as template and compared this with globin gene transcription in isolated intact nuclei. The reaction conditions for *in vitro* transcription by exogenous RNA polymerase in nuclei are identical to those used for chromatin (*Table 5*). The added RNA polymerase enters nuclei readily and yields similar amounts of RNA transcript as when using chromatin

Table 7. Hybridisation of RNA Transcripts with DNA Probes.

1. Add a fixed amount of radiolabelled DNA (0.5 – 1.0 ng) in distilled water to aliquots of RNA transcript (0.1 – 2.0 μg). Lyophilise.

2. Dissolve each aliquot of nucleic acids in 5 μl of 0.5 M NaCl, 1 mM EDTA, 50% deionised formamide[a], 25 mM Hepes (pH 6.7). Seal each sample in a siliconised glass capillary. Heat at 100°C for 5 min and then incubate at 43°C for 2 days.

3. Expel each sample into 0.25 ml of 2.8 mM $ZnSO_4$, 0.14 M NaCl, 15 μg/ml denatured calf thymus DNA, 70 mM sodium acetate buffer, pH 4.5.

4. Add 100 μl of nuclease S1 (1000 units) and incubate at 37°C for 2 h. Then place on ice for 5 min.

5. Remove 100 μl of each sample, add this to 10 ml of Triton fluor[b] and determine the total radioactivity (T) using a liquid scintillation counter.

6. Remove 200 μl of each sample and add 50 μl of BSA (100 μg/ml) and then 50 μl of 6 N perchloric acid. Mix and leave on ice for 30 min. Centrifuge at 2000 g for 10 min. Mix 200 μl of the supernatant with 10 ml of Triton fluor[b] and determine the acid-soluble radioactivity (S) using a liquid scintillation counter.

The percentage of DNA probe in nuclease S1-resistant hybrids is given by:

$$\% \text{ DNA hybridised} = \frac{T - 0.75S}{T} \times 100$$

[a]Formamide can be deionised as follows. Stir 100 ml of formamide with 10 g of mixed bed resin [e.g. Bio-Rad AG 501-X8 (D)] for 2 h at room temperature. Filter off the resin and store the formamide at −20°C.
[b]To prepare Triton fluor, dissolve 0.75 g POPOP and 12.5 g PPO in 2.5 litres toluene. Mix this with one half volume of Triton X-100.

as template. *Figure 1* shows that very little RNA capable of hybridising to globin cDNA is present after incubation of nuclei or chromatin in the absence of *E. coli* RNA polymerase. The small amount of hybridisation observed with nuclear RNA under these conditions reflects transcription by endogenous RNA polymerase II which can be partially inhibited by the addition of 10 μg/ml α-amanitin. When the incubations are performed in the presence of *E. coli* RNA polymerase, there is a large increase in the relative amount of globin RNA transcribed, as judged by hybridisation, although the values are about 3-fold higher for the nuclear template as compared with the chromatin template. The increased transcription using nuclei is virtually unaffected by α-amanitin, indicating that the large increase in globin gene transcription is due to the exogenous *E. coli* RNA polymerase.

Table 8 both quantifies the data presented in *Figure 1* (see above) and extends the analysis to examine the efficiency of globin mRNA trancription. As expected, the amount of globin RNA transcripts per unit weight of template (as DNA) is 3-fold higher using nuclei compared with chromatin. Presumably, mechanical stress and structural disruption occur during the isolation of chromatin which reduce the overall specificity of transcription. Clearly, from the point of view of template specificity, it is preferable to keep the nuclear morphology intact. An important control is the effect of actinomycin D on *in*

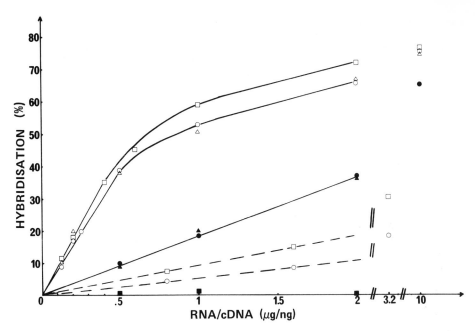

Figure 1. Saturation hybridisation analysis of *E. coli* RNA polymerase transcripts of foetal liver nuclei and chromatin. Varying amounts of purified mercurated RNA transcripts (*Table 6*) were hybridised to 0.05 ng of radiolabelled globin cDNA and the nuclease S1 resistance of the cDNA was determined as described in *Table 7*. For additional details, see the first footnote to *Table 8*. The curves shown refer to: (□——□) complete nuclear transcription; (○——○) complete nuclear transcription + 10 μg/ml α-amanitin; (△——△) complete nuclear transcription + 10 μg/ml α-amanitin + 250 μg/ml actinomycin D; (□- - -□) nuclei without added polymerase; (○- - -○) nuclei without added polymerase but + 10 μg/ml α-amanitin; (●——●) complete chromatin transcription, (▲——▲) complete chromatin transcription + 250 μg/ml actinomycin D; (■) chromatin without added polymerase.

vitro transcription. With both chromatin and nuclei, 250 μg/ml actinomycin D reduces the yield of total RNA transcripts to about 20% of its original value (*Table 8*). However, the residual RNA transcripts still produced contain the same percentage of globin RNA sequences as RNA synthesised in the uninhibited reactions (*Figure 1; Table 8*, third column). This suggests that the residual synthesis of RNA is not due to anomalous copying of endogenous mRNA sequences but represents the incomplete inhibition of conventional DNA-primed synthesis by actinomycin D. This interpretation is supported by the relatively small effect on the proportion of globin RNA transcripts produced when incubations are deliberately contaminated with relatively large amounts of purified globin mRNA (*Table 8*). The small amount of additional globin RNA 'transcripts' which is observed (*Table 8*, third column), together with the low levels of globin-specific transcription which appear to survive α-amanitin or actinomycin D inhibition (*Table 8*), probably represents background non-specific adsorption of endogenous RNA to thiolated Sepharose; this normally is equivalent to about 0.05 – 0.10% of the endogenous RNA.

Table 8. Titration of Globin RNA in Chromatin and Nuclear Transcripts[a].

Transcription conditions	Yield of transcription (μg RNA/ mg DNA)	Globin RNA in transcripts (ng/μg RNA)	Globin RNA transcripts per template (ng globin RNA/ mg DNA)
Chromatin transcripts			
– polymerase	_[b]	_[b]	_[b]
+ polymerase	1.5	0.27	0.41
+ polymerase + added globin mRNA[c]	1.5	0.36	0.54
+ polymerase + actinomycin D[d]	0.3	0.31	0.09
Nuclear transcripts			
– polymerase	_[b]	_[b]	0.17[e]
+ polymerase + α-amanitin[f]	_[b]	_[b]	0.10[e]
+ polymerase	1.31	1.16	1.52
+ polymerase + α-amanitin	1.31	1.07	1.40
+ polymerase + added globin mRNA[c]	1.14	1.94	2.21
+ polymerase + actinomycin D[d]	0.30	1.07	0.32

[a]The method of nuclei preparation is given in *Table 1*, steps 1 – 3, and the subsequent steps of the same table describe the preparation of chromatin (i.e., ruptured, salt-washed nuclei). Nuclear pellets were resuspended in 5 mM $MgCl_2$, 25% glycerol, 50 mM Tris-HCl, pH 7.9, at a DNA concentration of 2 mg/ml and 1 ml aliquots were incubated for *in vitro* transcription as described for chromatin in *Table 5*. Mercurated RNA transcripts were purified by affinity chromatography on thiolated-Sepharose (*Table 6*) and then globin RNA transcripts were quantitated using a radiolabelled globin cDNA probe as described in *Table 7*. All figures are derived from saturation values of separate titration analyses.
[b]Not measurable.
[c]Exogenous globin mRNA added (0.5 μg/mg DNA template).
[d]Actinomycin D; 250 μg/ml.
[e]In these incubations, *E. coli* tRNA was added as carrier and the isolated RNA was titrated with cDNA to give a measurement of endogenous globin RNA transcripts.
[f]α-Amanitin; 10 μg/ml.

5.2 Mapping of Transcription Initiation Sites using Nuclease S1

The preceding analysis using cDNA usefully quantitates gene-specific transcription. However, it does not reveal the fidelity of the copying process, that is, whether the transcripts have been correctly initiated. For instance, DNA sequences in chromatin that are accessible to RNA polymerase may be transcribed in a more or less random fashion rather than from the specific initiation sites used *in vivo*. Fortunately, the development of mapping with nuclease S1 (27,28) and the availability of cloned DNA containing the gene sequence of interest, permits investigation of the site or sites of initiation of transcription.

5.2.1 *Principle of Mapping with Nuclease S1*

Theoretically there are two approaches possible using mapping with nuclease S1. Either ^{32}P-labelled RNA transcripts can be hybridised to an unlabelled DNA probe or unlabelled RNA transcripts can be hybridised to a ^{32}P-labelled

DNA probe. The former approach, which is the procedure of choice in certain situations, is described in Chapter 3, Section 3.3. However, the latter approach offers higher resolution and is described here.

Location of the initiation site for transcription by this technique requires a specific genomic restriction fragment (DNA probe) which extends from within the first exon (usually) of the gene in question to some point upstream beyond the putative initiation (cap) site of the gene. This restriction fragment is end-labelled with ^{32}P, denatured and its strands separated by gel electrophoresis. The strand complementary to the RNA under investigation is recovered and hybridised in excess to the unlabelled RNA transcript preparation. RNA hybridised to the labelled DNA probe will protect it from subsequent digestion with nuclease S1 on the 3' side but not on the 5' side of the transcription in-itiation site. The exact size of the protected DNA fragment is then determined by gel electrophoresis and compared with that resulting from protection with authentic mRNA. If transcription *in vitro* has initiated at the correct site, the DNA fragments protected by the RNA transcripts and authentic mRNA should be identical in size. Failing this, the size of the DNA fragment protected by the RNA transcripts will indicate the actual site of initiation *in vitro*. It is important to realise that a meaningful answer is obtained only when the 3' end of the cDNA probe is fully protected from degradation by nuclease S1. This is usually guaranteed if the 3' end of the cDNA corresponds to a sequence in the first exon, that is, near the beginning of the coding region. An additional ad-vantage is that termination effects are minimised by this procedure.

5.2.2 *Procedure*

Protocols for preparing a ^{32}P-labelled, single-stranded DNA probe and use of this probe for mapping with nuclease S1 are described in *Tables 9* and *10*, respectively. Mapping using nuclease S1 is not an absolute diagnostic tool. If too little enzyme is used, mismatched hybrids will contaminate good hybrids and a confused (and misleading) pattern of fragments is obtained. This is over-come by using a series of nuclease S1 incubation conditions (*Table 10*) so as to range from under-digestion of hybrids to an enzyme excess where contaminant nucleases or perhaps the stringency of the conditions cause breakdown of even good hybrids. This is usually carried out simultaneously with the RNA transcribed *in vitro* and with authentic cellular RNA to compare the point at which hybrid breakdown occurs.

Figure 2 illustrates the application of nuclease S1 mapping in the analysis of *in vitro* transcription of the human epsilon (ϵ) globin gene in chromatin and nuclei from human K562 erythroleukaemia cells by *E. coli* RNA polymerase. These cells constitutively synthesise ϵ-globin. *Figure 2a* shows a region of the genomic DNA clone containing the ϵ-globin gene in plasmid pMX (ref. 31). A suitable probe for nuclease S1 mapping is provided by the 371-bp *Mbo*II frag-ment since this extends from inside the first exon of the ϵ-globin gene to a region 5' to the putative initiation or cap site. *Figure 2b* shows the use of the 5' end-labelled *Mbo*II fragment to analyse RNA transcribed *in vitro* from nuclei and chromatin isolated from K562 cells. Lanes 1 and 2 represent

Table 9. Preparation of ^{32}P-labelled DNA Probe.

1. Dissolve $0.5-1.0$ μg of the DNA probe[a] in 50 μl of 50 mM Tris-HCl, pH 8.5. Add 1 unit of calf intestine alkaline phosphatase and incubate at 37°C for 30 min to dephosphorylate the DNA. Heat at 60°C for 5 min to inactivate the enzyme.

2. Extract the sample with an equal volume of phenol pre-equilibrated with this buffer. Extract the aqueous phase with 200 μl of ether. Add sodium acetate to the aqueous phase to 0.1 M final concentration and then 3 volumes of ethanol. Keep at -20°C overnight or -70°C for 30 min to precipitate the DNA.

3. Recover the DNA by centrifugation (10 000 g for 10 min). Wash the pellet with 75% aqueous ethanol. Lyophilise the pellet.

4. Dissolve the DNA in 2.5 μl of 10 mM Tris-HCl, pH 8.0. Add 17.5 μl of 1 mM spermidine, 0.1 mM EDTA, 20 mM Tris-HCl (pH 9.5). Incubate at 90°C for 2 min. Place on ice.

5. Add 2.5 μl of 0.1 M MgCl$_2$, 50 mM DTT, 50% glycerol, 0.5 M Tris-HCl (pH 9.5). Mix. Next add the whole sample to 125 μCi of [γ-^{32}P]ATP (>5000 Ci/mmol) previously lyophilised and re-dissolved in 2.5 μl of water.

6. Add 20 units of polynucleotide kinase (PL Biochemicals) and incubate at 37°C for 30 min.

7. Add 12.5 μl of 7.5 M ammonium acetate,
    5.0 μl of yeast tRNA (2 mg/ml)
    90.0 μl of ethanol.
    Leave on dry ice for 10 min. Recover the DNA by centrifugation (10 000 g for 5 min).

8. Dissolve the DNA pellet in 25 μl of water and repeat the precipitation as in step 7.

9. Wash the DNA pellet with 75% aqueous ethanol. Dissolve the pellet in 80% deionised formamide[b] in 2.5 mM EDTA, 0.1 M Tris-borate, pH 8.3.

10. Heat at 90°C for 5 min to denature the DNA. Cool to room temperature and then load the sample onto a 6% polyacrylamide gel (40 cm long, 1 mm thick) prepared in 2.5 mM EDTA, 0.1 M Tris-borate, pH 8.3. The sample wells should be about 1 cm deep and 1 cm wide. Electrophorese in the same buffer as the gel buffer at 150 V at room temperature for 24 h.

11. Locate the separated strands[c] of the DNA by brief autoradiography at room temperature. To do this, wrap the wet gel (still on one of the glass plates after electrophoresis) in domestic 'cling-film'. Next, in a photographic dark room, place a sheet of X-ray film (e.g., Fuji RX) over the gel and leave for ~2 min. Then develop the film.

12. Cut the DNA bands from the gel using the autoradiograph as a guide. Elute the DNA from each band separately as follows. Macerate the gel slice in 0.2 ml of 1% SDS, 0.2 M sodium acetate (pH 4.8) and incubate at 37°C for 16 h.

13. Remove the gel fragments by centrifugation. Recover the eluted DNA by precipitation with ethanol (*Table 5*, steps 4 and 5). Dissolve each DNA pellet in 50 μl of water.

14. Determine which DNA strand is complementary to the RNA transcript under study. This is most easily done by a preliminary hybridisation experiment using the protocol described in *Table 10* but with radiolabelled mRNA. Alternatively one can use DNA sequencing (29).

---

[a]The size of DNA probe which is most useful for nuclease S1 mapping is $200-1000$ bp.
[b]Formamide can be deionised as described in the first footnote of *Table 7*.
[c]The two strands of DNA molecules usually separate in this gel sytem because they assume different secondary structure conformations which depend on base sequence. However, sometimes the difference in conformations may not be sufficient to allow separation to occur. In this case one either uses a different DNA fragment which will strand separate as probe or, failing this, uses a double-stranded DNA probe. Since the hybridisation step prior to mapping with nuclease S1 is carried out under conditions of DNA excess (see *Table 10*, step 1), some DNA will hybridise to the RNA transcripts and may allow mapping to be carried out, even though reannealling of the DNA probe will also occur.

Table 10. Mapping of Unlabelled RNA Transcripts using Nuclease S1[a].

---

1. Anneal each sample of unlabelled RNA transcripts to the appropriate ^{32}P-labelled, single-stranded DNA probe (prepared as described in *Table 9*) under conditions of DNA excess[b]. Carry out the hybridisation at 57°C for 16 h in 10 μl of 80% deionised formamide[c], 0.4 M NaCl, 1 mM EDTA, 40 mM Pipes (pH 6.4 with NaOH). Initially, to determine the optimal conditions for mapping with nuclease S1, it is advisable to set up a series of hybridisations (see step 3 below).

2. After hybridisation, add 200 μl of ice-cold 0.25 M NaCl, 200 μg/ml denatured calf-thymus DNA, 1 mM $ZnSO_4$, 30 mM sodium acetate (pH 4.6).

3. Incubate the mixture under a variety of digestion conditions with nuclease S1 to determine the most reliable protocol in each particular case (see Section 5.2.2). Typically, incubate a series of hybridisations at 10°, 20° or 40°C with 2000 units of nuclease S1 for 1, 2 or 5 h.

4. Precipitate the nucleic acid with ethanol as described in *Table 5*, steps 4 and 5.

5. Dissolve the DNA pellet in a small volume of 80% deionised formamide[c], 1% Bromophenol blue, 2.5 mM EDTA, 0.1 M Tris-borate (pH 8.3).

6. Size the nuclease S1-resistant hybrids by electrophoresis on a 6% polyacrylamide sequencing gel. Details of the procedures of gel preparation are given in Section 3 of reference 29 which describes the preparation of an 8% sequencing gel. For 6% gels, this author prepares a gel stock solution consisting of:

    120 ml 50% acrylamide; 2% bisacrylamide
    460 g urea
    100 ml 2.5 mM EDTA, 0.1 M Tris-borate (pH 8.3)
    Water to 1 litre.

    To prepare a 6% gel, mix:
     60 ml gel stock solution
    100 μl TEMED
    400 μl 10% ammonium persulphate.

    Introduce the gel mixture between two glass plates (20 cm x 40 cm), held 1 mm apart, using a pipette. Set a slot former in position. After polymerisation, remove the slot former. Heat the samples at 60°C for 15 min and then load 5 μl of each into separate wells using a finely drawn-out capillary. Electrophorese the gel (vertically) in 2.5 mM EDTA, 0.1 M Tris-borate (pH 8.3) at 42 watts constant power (1400 V and 30 mA) until the Bromophenol blue marker dye has moved ~35 cm. The gel should be at a temperature of ~60°C during the run.

7. After electrophoresis, transfer the gel to a sheet of filter paper (Whatman No. 3) and dry it using a gel dryer (e.g., Bio-Rad gel dryer for 1 h). In a photographic dark room, overlay the dried gel with a sheet of Kodak X-OMAT or Fuji RX film and place the 'sandwich' in an autoradiography cassette. Expose for $1-7$ days at -70°C.

    Although the size of the protected DNA fragment yields useful information about the site of initiation *in vitro* (see Section 5.2.1), the *exact* site of initiation can be determined by electrophoresis of a set of Maxam and Gilbert sequencing reaction mixtures (30) of the same labelled DNA probe on the same gel. The point in the sequence ladder which is coincident with the protected DNA fragment represents the nucleotide sequence of the initiation site used for transcription *in vitro*.

---

[a]An alternative protocol is given in Chapter 2, *Table 6*.
[b]It is important that the concentration of single-stranded DNA probe should be in excess over the hybridising RNA. To ensure this, it is advisable to carry out a series of preliminary hybridisations (see *Table 7*), each with a different amount of DNA present. Clearly, under conditions of DNA excess, further increase in the amount of DNA added has no effect upon the amount of radioactive RNA transcript hybridised. In all hybridisations the amount of total nucleic acid present should be made up to 50 μg by the addition of yeast tRNA.
[c]Formamide can be deionised as described in the first footnote of *Table 7*.

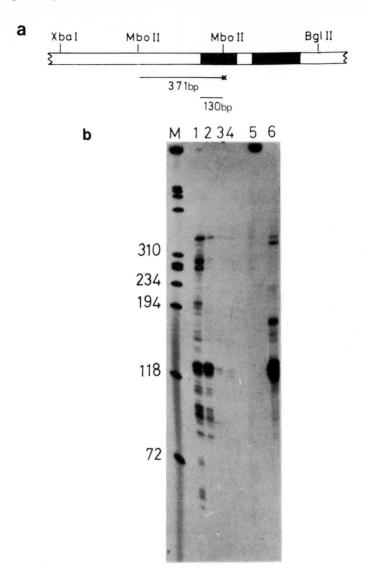

Figure 2. Identification of the initiation site of human ε-globin RNA synthesis *in vitro* by mapping with nuclease S1. (**a**) Restriction enzyme map of part of recombinant plasmid pMX (ref. 31) showing the region of human genomic DNA containing the transcriptional start point for the human ε-globin gene. The left- and right-hand black blocks represent the first and second exons, respectively. The 371-bp fragment obtained by *Mbo*II digestion of the *Xba*I-*Bgl*II fragment was 5′ end-labelled and the coding (+) strand (371 nucleotides) shown in the diagram was separated (*Table 9*) and used as a DNA probe for nuclease S1 mapping of the initiation site of RNA transcribed *in vitro*. The fragment protected (130 nucleotides) from digestion by nuclease S1 (see below) is shown. (**b**) K562 erythroleukaemia cells were induced for 5 days with 1 mM haemin and used to prepare nuclei and chromatin (*Table 1*). Mercurated RNA was synthesised *in vitro* from K562 nuclei or chromatin (*Table 5*) and then purified (*Table 6*). The DNA probe described in (**a**) was hybridised in excess to the RNA transcripts, digested with nuclease S1, and then the protected fragments of the probe were separated by electrophoresis and detected by autoradiography

as described in *Table 10*. **Lanes 1** and **3**, mercurated RNA from K562 nuclei (1 mg DNA) incubated with 10 μg/ml α-amanitin with or without 200 units *E. coli* RNA polymerase, respectively; **lanes 2** and **4**, mercurated RNA from K562 chromatin (5 mg DNA) incubated with 10 μg/ml α-amanitin with or without 200 units *E. coli* RNA polymerase, respectively; **lane 5**, RNA synthesised from 20 μg pATHεG DNA incubated with 100 units *E. coli* RNA polymerase; **lane 6**, 10 μg K562 cell total RNA; **M,** size markers.

nuclease S1-resistant DNA after hybridisation of the coding (+) strand of the *Mbo*II fragment to nuclear and chromatin RNA transcripts, respectively. Lane 6 shows an identical analysis of total cellular K562 RNA. The same two major protected fragments, each about 130 nucleotides in length, are seen in all cases. This defines the *in vitro* transcription initiation site at about 130 nucleotides upstream from the *Mbo*II site in the first exon (*Figure 2a*). Presumably two fragments are observed because of heterogeneity in either the nuclease S1 digestion or the exact site of initiation *in vitro*. In the absence of *E. coli* RNA polymerase in the reaction mixtures for *in vitro* transcription, very little protection of the DNA probe from digestion is seen (lanes 3 and 4). The presence of α-amanitin in the incubations ensures that none of the RNA analysed in lanes 1 and 2 is due to residual endogenous RNA polymerase II activity. Indeed, synthesis of the specific ε-globin transcripts can be abolished by the inclusion of 400 μg/ml rifampicin (data not shown), indicating that the transcription is due to *de novo* initiation by the exogenous *E. coli* RNA polymerase. It is worth noting that 5-fold more chromatin was required to elicit the same amount of nuclease S1-resistant hybrid as K562 nuclei when compared on a DNA basis. Therefore, although the same initiation sites for *in vitro* trancription of the ε-globin gene are observed for both templates, chromatin provides a far less efficient template for *E. coli* RNA polymerase. In all cases, transcription is asymmetric since no hybridisation occurs with the non-coding (−) strand of the *Mbo*II DNA probe (data not shown).

6. VALIDITY OF USING *E. COLI* RNA POLYMERASE

The use of *E. coli* RNA polymerase to examine accessible DNA sequences in templates has largely been discontinued since Zasloff and Felsenfeld (19) demonstrated that many experiments in which *E. coli* RNA polymerase was used in the presence of manganese ions were subject to serious artifactual distortion due to RNA-dependent transcription by this enzyme (Section 4.1). As described above, the use of mercurated nucleotides provides a reliable method for isolating RNA synthesised *in vitro* from chromatin when the correct ionic conditions of synthesis are employed. Under these conditions, *E. coli* RNA polymerase synthesises transcripts from both mammalian nuclei and chromatin which closely resemble those produced by endogenous RNA polymerase II (23,32). Initiation on both nuclear and chromatin templates by *E. coli* RNA polymerase appears to occur at exactly the same sites as *in vivo*.

Because of the large evolutionary gulf between *E. coli* and mammals, it has seemed inherently unlikely that *E. coli* RNA polymerase would recognise the same initiation signals as RNA polymerase II in mammalian cells. Moreover, while there is a strong resemblance between the Pribnow box in prokaryotic DNA and the TATA box in eukaryotic DNA, the distance between these sequences and the initiation site is quite different; about 12 bp in the former and 25 – 30 bp in the latter. It is therefore surprising that, in at least K562 nuclei and chromatin, *E. coli* RNA polymerase appears to initiate mainly at the major initiation site used by the endogenous RNA polymerase II. Presumably the chromatin conformation around the initiation site of the active ε-globin gene determines access by the bacterial enzyme. This is supported by the inability to obtain specific hybridisation of the *Mbo*II ε-globin DNA probe with RNA transcribed by *E. coli* RNA polymerase from naked plasmid DNA containing ε-globin gene sequences (*Figure 2b*, lane 5). Further study of this phenomenon may provide insight into the conformation of chromatin at the sites of mammalian promoters.

7. FRACTIONATION AND RECONSTITUTION OF CHROMATIN

Various attempts have been made in the past 10 years to fractionate chromatin and to reassemble it *in vitro* from its constituents in order to understand the interactions which govern the known structural and biological characteristics of chromatin.

7.1 Fractionation of Chromatin

Numerous methods have been described for the separation of chromatin into individual components. Nearly all of these employ high concentrations of salt, often in the presence of denaturing agents such as urea or guanidinium chloride, to dissociate chromatin proteins from the DNA. For example, chromatin proteins can be obtained by dissociating chromatin in 2 M NaCl, 5 M urea followed by ultracentrifugation to pellet the DNA (33,34). The proteins can then be fractionated further by a number of procedures. This method does not, however, remove RNA from the protein fraction. An alternative which removes both RNA and DNA from chromatin protein involves ultracentrifugation in 27% (w/w) CsCl and 4 M urea (35).

Chromatin can also be fractionated without recourse to prolonged ultracentrifugation by using hydroxylapatite chromatography (36,37). This is one of the few single-step methods for separating chromatin into its components. A suitable protocol is given in *Table 11*. In this procedure, DNA, non-histone proteins (NHP) and RNA are adsorbed onto hydroxylapatite in 2 M NaCl, 5 M urea. Histones are not retained on the column. By elution with different concentrations of sodium phosphate, DNA and NHP can be eluted separately and a crude fractionation of the proteins themselves achieved. However, the method does not separate endogenous nuclear RNA from the bulk of the NHP. Histones and NHP can also be separated on Bio-Rex 70 (ref. 38).

An alternative approach is to isolate the individual components of chromatin separately. Thus, high molecular weight DNA can be prepared by

Table 11. Protocol for Chromatin Fractionation[a].

1. Homogenise the chromatin (prepared as in *Table 1*) in chromatin buffer (2 M NaCl, 5 M urea, 1 mM sodium phosphate buffer, pH 6.8) in a Teflon-glass homogeniser at a concentration of 0.5 mg/ml DNA.
2. Centrifuge at 15 000 *g* for 15 min. Retain the supernatant.
3. Homogenise the pellet once more as in step 1.
4. Retain the supernatant and pool it with that from step 2.
5. Sonicate the pooled supernatants for two periods of 15 sec (e.g., using an MSE ultrasonic disintegrator at 1.5 A).
6. Centrifuge at 15 000 *g* for 15 min.
7. Load 50 – 100 ml of supernatant onto a column (16 cm x 25 cm) of hydroxylapatite (Bio-Rad Labs.) pre-equilibrated in chromatin buffer. Load the sample by gravity feed so as to give a flow rate of 5 – 10 ml/h.
8. (i) After loading, wash the column with chromatin buffer until the (unretained) histones are eluted.
(ii) Then elute the non-histone proteins in two fractions by washing the column successively with chromatin buffer containing 50 mM sodium phosphate (pH 6.8) and chromatin buffer containing 0.2 M sodium phosphate buffer (pH 6.8)[b]. The second fraction of non-histone proteins which elutes at the higher salt concentration contains the bulk of the chromosomal RNA. A small additional fraction of tightly-bound non-histone proteins can be recovered by elution with chromatin buffer containing 2 M guanidinium chloride.
(iii) Finally, the DNA can be eluted by washing the column with chromatin buffer containing 0.5 M sodium phosphate buffer (pH 6.8).

[a]Based on the original method of MacGillivray *et al.* (36). Virtually a complete recovery of material is obtained for chromatin containing up to 10 mg DNA.
[b]Alternatively use a concentration gradient of sodium phosphate. This resolves the non-histone proteins somewhat better than the batch elution procedure.

the method of Chambon (39), histones by salt extraction (40) and NHP can be differentially extracted from chromatin with 6 M urea, 50 mM phosphate buffer, pH 7.5 (41,46).

7.2 Reconstitution of Chromatin

Several reviews consider the various procedures that have been used for the selective reconstitution of chromatin from dissociated components (8,42,43). The basic protocol is identical irrespective of the detailed method used; samples of dissociated chromatin or of separate components in stoichiometric amounts are mixed at high ionic strength and then the salt concentration is gradually reduced to a physiological value by dialysis over several hours. This procedure may or may not be carried out in the presence of urea. A typical protocol for chromatin reconstitution is given in *Table 12*.

7.3 Fidelity of Reconstitution

It is clear from a number of structural studies including thermal denaturation (44), circular dichroism (45), X-ray diffraction (46) and nuclease digestion pat-

Table 12. Procedure for Reconstitution of Chromatin.

1. Mix high molecular weight DNA (0.5 mg/ml) with chromosomal proteins in the ratio of 1:1.5 for histones and 1:0.5 for non-histone proteins in 2 M NaCl, 5 M urea, 10 mM Tris-HCl, pH 8.0.

2. Dialyse the reconstitution mixture against 20 volumes of this buffer at 4°C for 2 h.

3. Progressively decrease the salt concentration of the mixture by sequential dialysis (2 h each) against 1 mM PMSF, 10 mM Tris-HCl, pH 6.8, containing 2.0 M, 1.0 M, 0.8 M, 0.6 M, 0.4 M and 0.2 M NaCl. Note that urea is absent during these dialyses.

4. Recover the insoluble, reconstituted nucleoprotein by centrifugation at 2000 *g* for 10 min. Wash the nucleoprotein pellet with 1 mM EDTA, 10 mM Tris-HCl, pH 8.0.

terns (47) that reconstituted chromatin may regain some of the structural properties of native chromatin. The *biological* fidelity of reconstitution can be operationally assessed by *in vitro* transcription of the reconstituted chromatin with exogenous *E. coli* RNA polymerase (Section 4.2). The RNA transcripts may be analysed by hybridisation to specific cDNA probes (Section 5.1) and by mapping with nuclease S1 (Section 5.2) and the results compared with transcripts from native chromatin. One of the major drawbacks of this approach is the inevitable re-incorporation of endogenous RNA sequences into the chromatin during the reconstitution process. However, the use of mer-curated nucleotides allows a direct isolation of transcripts arising by *in vitro* transcription from these templates (Section 4.2). Using chromatin from livers of foetal mice, mercurated transcripts from native and reconstituted templates have been compared by cDNA hybridisation (1,43). These studies indicated that the abundance of globin RNA transcripts synthesised from reconstituted chromatin was intermediate between that expected for a random transcription of naked DNA and that of native chromatin. A comparison of chromatins reconstituted by different salt/urea regimens (Section 6.2) indicated varying degrees of inaccuracy of reconstitution as judged by specific transcript analysis (1). Further examination of these templates by electron microscopy confirmed that none of the reconstitution methods tested appears to be particularly effective as judged by the irregularity of nucleosome formation (1). In addition, long stretches of naked DNA were observed and, in places, large amorphous aggregates of DNA and protein. These data can be explained by assuming that a defined temporal sequence of specific histone-histone interactions must occur for nucleosomes to be reformed correctly. The presence of urea throughout the reconstitution procedure suppresses these interactions, and its removal during the final stages of reconstitution results only in general non-specific DNA-histone interaction. In fact, other work on nucleosome reforma-tion *in vitro* (48,49) using isolated histones demonstrates the importance of the histone (H3-H4) tetramer and the histone (H2A-H2B) dimer as intermediates in the process. However, even under these carefully controlled conditions, nucleosome assembly is incomplete (48). Indeed, faithful reconstitution has only been demonstrated *in vitro* with extracts from *Xenopus* oocytes (50) and in SV40 genomes isolated from SV40-infected cells (51).

In summary, the role of nucleosome formation as a prerequisite for transcriptional specificity is not clear. The observation that eukaryotic RNA polymerase II preparations contain proteins which direct initiation by recognition of promoter DNA sequences clearly suggests that control mechanisms may not require a total reconstitution of chromatin structure for a meaningful demonstration of their effect (52).

8. REFERENCES

1. Gilmour,R.S. (1978) in *The Cell Nucleus*, Vol. **6**, Busch,H. (ed.), Academic Press, New York and London, p. 329.
2. Axel,R., Cedar,H. and Felsenfeld,G. (1973) *Proc. Natl. Acad. Sci. USA*, **70**, 2029.
3. Gilmour,R.S. and Paul,J. (1973) *Proc. Natl. Acad. Si. USA*, **70**, 3440.
4. Steggles,A.W., Wilson,G.W., Kantor,J.A., Picciano,D.J., Falvey,A.K. and Anderson,F.W. (1974) *Proc. Natl. Acad. Sci. USA,* **71**, 1219.
5. Barret,T., Maryanka,D., Hamlyn,P.H. and Gould,H.J. (1974) *Proc. Natl. Acad. Sci. USA,* **71**, 5057.
6. Chiu,J.F., Tsai,Y.H., Sakuma,K. and Hnilica,L.S. (1975) *J. Biol. Chem.,* **250**, 9431.
7. MacGillivray,A.J. (1976) in *Subnuclear Components*, Birnie,G.D. (ed.), Butterworths, London, p.209.
8. Stein,G.S and Stein,J.L. (1978) in *Methods in Cell Biology,* Vol. **19**, Stein,G., Stein,J.S. and Kleinsmith,L.J. (eds.), Academic Press, Inc., New York and London, p. 379.
9. Manley,J.L., Fire,A., Cano,A., Sharp,P.A. and Gefter,M.L. (1980) *Proc. Natl. Acad. Sci. USA,* **77**, 3855.
10. Burgess,R.R. (1969) *J. Biol. Chem.,* **244**, 6160.
11. Hondo,H.G. and Blatti,S.P. (1977) *Biochemistry (Wash.),* **16**, 2334.
12. Bautz,E.I.C.R. and Dunn,J.J. (1969) *Biochem. Biophys. Res. Commun.,* **34**, 230.
13. Smith,S.S. and Brown,R. (1978) *Eur. J. Biochem.,* **82**, 309.
14. Smith,M.M. and Huang,R.C.C. (1976) *Proc. Natl. Acad. Sci. USA,* **73**, 755.
15. Crouse,G.F., Fodor,E.J. and Doty,P. (1976) *Proc. Natl. Acad. Sci. USA,* **70**, 1564.
16. Biesmann,H., Gjerset,R.A., Levy,W.B. and McCarthy,B.J. (1976) *Biochemistry (Wash.),* **15**, 4356.
17. Beebee,T.J.C. and Butterworth,P.H.W. (1964) *Eur. J. Biochem.,* **66**, 543.
18. Towle,H.C., Tsai,M.J., Tsai,S.Y. and O'Malley,B.W. (1976) *J. Biol. Chem.,* **252**, 2396.
19. Zasloff,M. and Felsenfeld,G. (1977) *Biochem. Biophys. Res. Commun.,* **75**, 598.
20. Zasloff,M. and Felsenfeld,G. (1977) *Biochemistry (Wash.),* **16**, 5135.
21. Giesecke,K., Sippel,A.E., Nguyen-Hu,M.C., Groner,B., Hynes,N.E., Wurtz,T. and Schutz,G. (1977) *Nucleic Acids Res.,* **4**, 3943.
22. Pays,E., Donaldson,D. and Gilmour,R.S. (1979) *Biochim. Biophys. Acta,* **562**, 112.
23. Pays,E. and Gilmour,R.S. (1981) *Biochim. Biophys. Acta,* **653**, 356.
24. Van Brockhoven,C and De Wachter,R. (1978) *Nucleic Acids Res.,* **5**, 2133.
25. Harrison,P.R., Birnie,G.D., Hell,A., Humphries,S., Young,B.D. and Paul,J. (1974) *J. Mol. Biol.,* **84**, 539.
26. Young,B.D., Harrison,P.R., Gilmour,R.S., Birnie,G.D., Hell,A., Humphries,S. and Paul,J. (1974) *J. Mol. Biol.,* **84**, 555.
27. Berk,A.J. and Sharp,P.A. (1977) *Cell,* **12**, 421.
28. Weaver,R.F. and Weissman,C. (1979) *Nucleic Acids Res.,* **6**, 1175.
29. Davies,R.W. (1982) in *Gel Electrophoresis of Nucleic Acids − A Practical Approach*, Rickwood,D. and Hames,B.D. (eds.), IRL Press Ltd., Oxford and Washington, DC, p. 117.
30. Maxam,A.M. and Gilbert,W. (1980) in *Methods in Enzymology,* Vol. **65**, Grossman,L. and Moldave,K. (eds.), Academic Press Inc., New York and London, p. 39.
31. Alan,M., Grindlay,G.J., Stefani,L. and Paul,J. (1982) *Nucleic Acids Res.,* **10**, 5133.
32. Brown,D.K.T., Pragnell,I.B. and Paul,J. (1980) *Eur. J. Biochem.,* **104**, 459.
33. Graziano,S.L. and Huang,R.C. (1971) *Biochemistry (Wash.),* **10**, 4470.
34. Gadski,R.A. and Chae,C.B. (1976) *Biochemistry (Wash.),* **15**, 3812.
35. Gilmour,R.S. and Paul,J. (1975) in *Chromosomal Proteins and their Role in the Regulation of Gene Expression,* Stein,G.S. and Kleinsmith,L.J. (eds.), Academic Press Inc., New York and London, p. 19.

36. MacGillivray,A.J., Cameron,A., Krause,R., Rickwood,D. and Paul,J. (1972) *Biochim. Biophys. Acta,* **277**, 384.
37. Rickwood,D. and MacGillivray,A.J. (1975) *Eur. J. Biochem.,* **51**, 593.
38. Morris,M. and Gould,H.J. (1971) *Proc. Natl. Acad. Sci. USA,* **68**, 481.
39. Chambon,P. (1973) *Eur. J. Biochem.,* **36**, 32.
40. Noll,M., Zimmer,S., Engel,A. and Dubochett,J. (1980) *Nucleic Acids Res.,* **8**, 21.
41. Barrett,T., Maryanka,D., Hamlyn,P.H. and Gould,H.J. (1974) *Proc. Natl. Acad. Sci. USA,* **71**, 5057.
42. Gilmour,R.S. (1978) in *Methods in Cell Biology,* Vol. **19**, Stein,G., Stein,J.S. and Kleinsmith,L.J. (eds.), Academic Press Inc., New York and London, p. 373.
43. Gould,H.J., Maryanka,D., Fey,S.J., Cowling,G.J. and Allan,J. (1978) in *Methods in Cell Biology,* Vol. **19**, Stein,G., Stein,J.S. and Kleinsmith,L.J. (eds.), Academic Press Inc., New York and London, p. 387.
44. Kleiman,L. and Huang,R.C. (1972) *J. Mol. Biol.,* **64**, 1.
45. Boseley,P.G., Bradbury,E.M., Butler-Browne,G.S., Carpenter,B.G. and Stpehens,R.M. (1976) *Eur. J. Biochem.,* **62**, 21.
46. Richards,M. and Pardon,J.F. (1970) *Exp. Cell Res.,* **62**, 184.
47. Axel,R., Melchior,W., Solner-Webb,B. and Felsenfeld,G. (1974) *Proc. Natl. Acad. Sci. USA,* **71**, 4101.
48. Oudet,P., Gross-Bellard,M. and Chambon,P. (1975) *Cell,* **4**, 181.
49. Ruiz-Carrillo,A. and Jorcano,J.L. (1979) *Biochemistry (Wash.),* **18**, 760.
50. Laskey,R.A., Mills,A.D. and Morris,N.R (1977) *Cell,* **10**, 237.
51. Griffith,J. (1975) *Science (Wash.),* **187**, 1202.
52. Davison,B.L., Egly,J.M., Mulvihill,E.R. and Chambon,P. (1983) *Nature,* **301**, 680.

CHAPTER 6

In Vivo Gene Expression Systems In Prokaryotes

NEIL G. STOKER, JULIE M. PRATT AND I. BARRY HOLLAND

1. INTRODUCTION

This chapter describes the three major *in vivo* expression systems in pro-karyotes; the u.v.-irradiated host system, 'minicells' and 'maxicells'. These are all coupled transcription-translation systems, and will be discussed with reference to *Escherichia coli* and bacteriophage λ. Systems using other bacterial species or bacteriophages will be referred to only briefly.

(i) *The u.v.-irradiated host system. E. coli* is heavily u.v. irradiated to minimise host protein synthesis and then infected with phage λ carrying cloned genes. [35S]methionine is included in a subsequent incubation to label phage-encoded proteins and the products are analysed by SDS-PAGE. As an extra sophistication, synthesis of most phage λ proteins may be blocked by the use of a host lysogenic for λ, or carrying a plasmid coding for the phage λ trans-criptional repressor, the *c*I gene product.

(ii) *The minicell system.* A mutant strain of *E. coli (minA, minB)* produces large numbers of small anucleate cells (minicells) into which plasmids present in the bacterial host will segregate. Since no chromosomal DNA is present, in-cubation of the purified minicells with [35S]methionine allows specific labelling of plasmid-encoded proteins.

(iii) *The maxicell system.* A u.v.-sensitive strain of *E. coli* is given a low dose of u.v. irradiation and damaged DNA is extensively degraded during a subse-quent incubation. Due to their small target size compared with that of the bacterial chromosome, a proportion of any plasmid molecules present will sur-vive. Thus, incubation with [35S]methionine allows preferential labelling of plasmid-encoded proteins. This is known as the maxicell system to distinguish it from the minicell system.

Each of these three systems has its advantages and disadvantages. The choice of which *in vivo* gene expression system to use will depend mainly on the vector used for cloning and the gene which has been cloned. The relative merits of each system are listed in *Table 1*. The properties of the Zubay *in vitro* system described in Chapter 7 are also included in *Table 1* for comparison. The characteristics of the plasmids used as examples in this chapter are shown in

Table 1. Advantages and Disadvantages of *In Vivo* and *In Vitro* Prokaryotic Gene Expression Systems.

System	Advantages	Disadvantages
Phage λ-infection of u.v.-irradiated host	1. Phage λ is frequently used for the initial cloning of genes. This can be crucial for genes whose presence in more than a single copy is lethal, since the use of a standard multicopy plasmid vector would result either in no clones being obtained or, inpossibly, in the cloning of a gene which has acquired a mutation reducing its deleterious effects.	1. Clearly limited to phage λ vectors. 2. Residual background synthesis of host polypeptides below 20 000 mol.wt. can obscure identification of cloned gene products in this size range. 3. The system makes relatively inefficient use of the labelled amino acids supplied, and therefore fluorography and long exposure times may be needed to detect the labelled polypeptide products.
Minicells	1. Very efficient labelling system. 2. Purified minicells can be stored for at least a year. 3. In principle, minicells can be used to examine genes cloned into either plasmid or phage λ vectors.	1. High copies of certain genes when present in plasmids can disturb minicell formation and/or efficient purification of minicells. 2. The introduction of additional mutations (e.g., *recA*) into the minicell-producing strain can have the same effect. 3. Quantitative fractionation of minicells is difficult.
Maxicells	1. Simple procedure. 2. Easier than minicell system when screening large numbers of plasmids.	
In vitro (Zubay)	1. Can be used with any kind of cloned DNA including small linear fragments. This allows unambiguous assignment of cloned DNA products (rather than those coded by the vector) to relatively small regions of DNA. 2. Extremely efficient labelling system allowing rapid analysis. 3. Can be used to study gene expression. 4. Has considerable advantages when uncharacterised gene products must be identified.	1. Relatively complex system to set up initially. 2. Interpretation can sometimes be obscured by the presence of some abortive translation products and/or proteolytic breakdown products especially in the case of large proteins ($>60 000$ mol. wt.).

Table 2 and *Figure 1* while details of the bacteriophages used are given in *Figure 2*. Media and buffers common to the three *in vivo* expression systems are described in *Table 3*.

Table 2. Characteristics of Selected Plasmids[a].

Plasmid	Size (kb)	Copy number (per genome)	Genes	Gene products
pACYC184	4.0	18	Chloramphenicol resistance (Cm^R)	Chloramphenicol acetyl transferase (CAT); 25 000 mol. wt.
			Tetracycline resistance (Tc^R)	Tetracycline-resistance protein (Tet); 34 000 mol. wt.
pLG517	10.4	18	*dacC*	Penicillin-binding protein 6 (PBP6); 40 000 mol. wt.
			Tc^R	Tetracycline-resistance protein (Tet); 34 000 mol. wt.
pBS42	12.2	6	*dacA*	Penicillin-binding protein 5 (PBP5); 42 000 mol. wt.
			Kanamycin resistance (Km^R)	Kanamycin phosphotransfer-ase I (Kan); 27 000 mol. wt.

[a]Diagrammatic representations of these plasmids are shown in *Figure 1*.

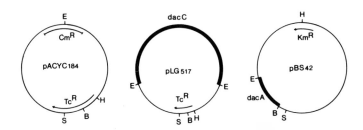

Figure 1. Diagrams of plasmids used as examples in this chapter. Thick black lines represent cloned DNA. Restriction enzyme sites are indicated as follows: B, *Bam*HI; E, *Eco*RI; H, *Hind*III; S, *Sal*I. Further details are given in *Table 2*.

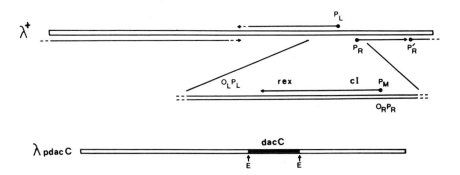

Figure 2. Diagram of bacteriophages used as examples in this chapter. Bacteriophage λ$^+$; the promoters P_L and P_R are indicated together with the direction of transcription during λ infection. These promoters are repressed in lysogens by the *cI* product which is transcribed together with *rex*, a gene of unknown function, from the P_M promoter. For other details of bacteriophage λ biology, see reference 3. Bacteriophage λ*pdacC* is a λ-cloning vector carying an *Eco*RI fragment, *dacC*, encoding the *E. coli* penicillin-binding protein 6 (PBP6).

Table 3. Composition of Media and Buffers.

Luria agar[a]

     1% tryptone (Oxoid)
     0.5% yeast extract (Oxoid)
     0.5% NaCl
     1.5% Agar (Oxoid No. 3)
     The final pH should be 7.4.

M9-minimal medium[b]

     40 mM Na_2HPO_4
     20 mM KH_2PO_4
     8 mM NaCl
     20 mM NH_4Cl
     1 mM $CaCl_2$
     10 mM $MgSO_4$
     Autoclave the $CaCl_2$ and $MgSO_4$ separately from the other reagents and then mix after they have cooled. Store at room temperature.

Nutrient broth

     2.5% (w/v) nutrient broth powder (Oxoid No. 2) in water. Autoclave at 15 p.s.i. for 15 min and store at room temperature.

Nutrient agar

     Nutrient broth containing 1.5% agar (Oxoid No. 3)

Trypticase agar[c]

     1% Trypticase peptone
     0.5% NaCl
     1.5% agar (Oxoid No. 3)

Bacterial buffer

     50 mM Na_2HPO_4
     20 mM KH_2PO_4
     70 mM NaCl
     0.4 mM $MgSO_4$
     Autoclave and then store at room temperature.

Lambda buffer

     10 mM $MgSO_4$
     0.05% (w/v) gelatin
     6 mM Tris-HCl, pH 7.2
     Autoclave and then store at room temperature.

SDS-sample buffer

     0.125 M Tris-HCl, pH 6.8
     20% glycerol
     2% SDS
     5% 2-mercaptoethanol
     0.001% (w/v) Bromophenol blue
     Store at room temperature.

STET buffer

     8% (w/v) sucrose
     5% Triton X-100
     50 mM EDTA
     50 mM Tris-HCl, pH 8.0
     Autoclave and then store at 4°C.

---

[a]'Soft' Luria agar contains only 0.7% agar.
[b]From ref. 6
[c]'Soft' Trypticase agar contains only 0.7% agar.

2. THE U.V.-IRRADIATED HOST SYSTEM

2.1 Introduction

In the u.v.-irradiated host system (1,2), transcription of the bacterial chromosome is prevented by administering a large dose of u.v. irradiation. In a subsequent infection with phage λ, only the phage DNA should be a suitable template for transcription. The addition of [^{35}S]methionine then allows phage-encoded proteins to be specifically labelled. The biology of phage λ, and its use as a cloning vector have been recently reviewed (3).

If the bacterial strain which is irradiated is lysogenic for phage λ, it will contain the repressor protein (the *c*I gene product) which will bind to any superinfecting phage DNA, preventing all transcription from the major promoters, P_L and P_R. Only the immunity region (encoding the *c*I and *rex* gene products) which has its own promoter, should be transcribed under these conditions (ref. 3; see also *Figure 3*). Any bacterial genes with their own promoters, which have been cloned into the superinfecting phages, will also be transcribed, thus allowing the preferential labelling of these gene products. Clearly this approach is not appropriate for genes cloned into a phage vector without a bacterial promoter or where the bacterial genes have very weak promoters, since in these cases the expression of the bacterial genes is dependent upon transcription from the strong phage promoters P_L or P_R.

Although the infection of *E. coli* with phage λ will be discussed exclusively here, similar systems have been used to identify proteins encoded by P22 specialised transducing phages in *Salmonella typhimurium* (4), and proteins encoded by phage ϕ29 in *Bacillus subtilis* (5).

2.2 Requirements

The requirements for use of the u.v.-irradiated host system are listed in *Table 4*.

Table 4. Requirements for the U.V.-irradiated Host System.

Bacterial strains[a]
  E. coli 159 *uvrA*
  or *E. coli* 159 *uvrA* (λ*ind*)

Bacteriophages
  For example, phage λ$^+$ (at 10^{10} p.f.u./ml)

Solutions
  M9-maltose medium [M9-minimal medium[b] containing 0.2% maltose as carbon source
  (added after being autoclaved separately)]
  M9-maltose medium containing 20 mM MgCl$_2$
  Lambda buffer[b]
  [^{35}S]methionine (10 μCi/μl; 1000 Ci/mmol; Amersham International)
  L-methionine (8 mg/ml)
  Bacterial buffer[b]
  SDS-sample buffer[b]

---

[a]See Section 2.3 for choice of bacterial host strain.
[b]Prepared as described in *Table 3*

2.3. **Bacterial Host Strain**

E. coli 159 (*uvrA gal rpsL*) is widely used as the host strain (1). The *uvrA* mutation inactivates one of the cell's major DNA repair pathways, making the cell more sensitive to u.v. irradiation. The strain is phage λ-sensitive, prototrophic and free of suppressor-tRNA mutations. The particular lysogen used is *E. coli* 159(λ*ind*). Irradiation with u.v. normally induces excision of an integrated phage (prophage) but phage λ*ind* carries a mutation which prevents this. However, the dose of u.v. irradiation used is so high that this is probably an unnecessary precaution since the prophage itself would normally be severely damaged.

Table 5. Preparation of Phage[a].

---

1. Grow an overnight culture of the host bacterial strain in nutrient broth supplemented with 10 mM $MgCl_2$.
2. Prepare serial phage dilutions (100 μl final volume) in lambda buffer. Add 50 μl of bacterial culture to each dilution and leave at room temperature for 10 min to allow phage adsorption to occur.
3. To each dilution, add 3 ml of molten, 'soft' Luria agar[b] already at 50°C. Mix briefly and pour the mixture as an overlay onto freshly-prepared Luria agar[a] plates.
4. Incubate the plates at 37°C overnight. If the phage carry the *c*I857 mutation, incubate the plates at 42°C for 20 min before transferring to 37°C overnight.
5. Select one of the phage dilution plates on which the plaques are just confluent. Pipette 3 ml of lambda buffer onto the surface and leave for 30 min at room temperature.
6. Pipette off the buffer (now containing the phage) into a McCartney bottle. Remove any debris by centrifugation at 3000 g for 10 min (e.g. using a Sorvall SS34 rotor at 5000 r.p.m.). Add a few drops of chloroform to kill any surviving bacteria and store the phage suspension at 4°C[c].
7. Titre the phage suspension as described in *Table 6*.

---

[a]The compositions of media and buffers are given in *Table 3*.
[b]Some phage λ derivatives, particularly those carrying non-λ DNA, do not grow as vigorously as the wild-type phage. Consequently, they form much smaller plaques on rich media. If this is the case, better results may be obtained by using less-rich media such as Trypticase agar *(Table 3)*.
[c]To prepare larger quantities of phage, more plates should be prepared using the phage dilution just sufficient for confluent lysis.

Table 6. Titration of Phage[a].

---

1. Grow an overnight culture of the bacterial host strain in nutrient broth supplemented with 10 mM $MgCl_2$.
2. Prepare serial dilutions (100 μl final volume) of the phage in lambda buffer. Add 50 μl of bacterial culture to each dilution and leave at room temperature to allow phage adsorption to occur.
3. To each dilution, add 3 ml of molten, 'soft' Luria agar[b] already at 50°C. Mix briefly and pour the mixture as an overlay onto freshly-prepared Luria agar[b] plates.
4. Incubate the plates at 37°C overnight.
5. Count the number of plaques formed at each dilution and calculate the phage titre as plaque forming units (p.f.u.) per ml.

---

[a]The compositions of media and buffers are given in *Table 3*.
[b]As described in the second footnote to *Table 5*, a less-rich medium such as Trypticase agar *(Table 3)* might be more suitable for phages which grow poorly.

2.4 **Preparation of Phage**

Methods for the preparation and titration of phage are described in *Tables 5* and *6* and elsewhere (6,7). Ideally, the phage stocks should have a titre of at least 10^{10} plaque-forming units (p.f.u.)/ml, so that the cells can be infected at high multiplicity [5 − 10 p.f.u./cell; see Section 2.5 (iv)]. If necessary, the phage preparation can be concentrated by polyethylene glycol precipitation and/or CsCl density gradient centrifugation (*Table 7* and refs. 6−8). The phage preparation must be extensively dialysed before use in order to remove

Table 7. Concentration of Phage Preparations.

Phages can readily be concentrated either by precipitation with polyethylene glycol or by centrifugation in CsCl density gradients.

Polyethylene glycol precipitation
1. To the phage suspension, add approximately two-thirds the volume of 25% (w/v) polyethylene-glycol 6000, 1.25 M NaCl.
2. Incubate at 4°C for at least 1 h; if more convenient, leave overnight.
3. Centrifuge at 3600 *g* for 10 min at 4°C (e.g., Sorvall GS3 rotor at 5000 r.p.m.).
4. Discard the supernatant and drain the phage pellet.
5. Resuspend the phage pellet in 1 − 2 ml of lambda buffer[a].

CsCl gradient centrifugation; step gradient method
1. Prepare solutions of CsCl with densities of 1.7, 1.5 and 1.3 g/ml (refractive indices of 1.3952, 1.3800 and 1.3687, respectively) by mixing a saturated-CsCl solution with lambda buffer.
2. Pipette 5 ml of the most dense solution (1.7 g/ml) into a clear ultracentrifuge tube (2.5 x 8.75 cm). On top of this, carefully layer 5 ml of the solution with density 1.5 g/ml followed by 5 ml of the solution with density 1.3 g/ml. Finally, layer on the phage suspension.
3. Centrifuge at 60 000 *g* for 2−6 h at 15°C (e.g., Beckman SW27 rotor at 22 000 r.p.m.).
4. The phage particles form a sharp white band in the centre of the layer with a density of 1.5 g/ml. This is best observed by illuminating the ultracentrifuge tube from above. Having located the phage band visually, smear a little grease onto the side of the tube and pierce the tube through the grease with a syringe. Using the syringe, recover the phages and transfer the preparation to a glass McCartney bottle.
5. Add a few drops of chloroform and store the preparation at 4°C. The phage particles are quite stable in CsCl solution. However, if necessary, the CsCl may be removed by dialysis against lambda buffer.

CsCl gradient centrifugation; self-forming gradient method
1. Adjust the phage suspension to a density of 1.5 g/ml (refractive index = 1.3800) with a saturated solution of CsCl.
2. Transfer the mixture to a clear ultracentrifuge tube (e.g., 1.5 x 7.5 cm). If necessary, fill the rest of the tube using a solution of CsCl in lambda buffer with the same density. Mix well.
3. Centrifuge at 100 000 *g* for 40 h at 15°C (e.g., Beckman 50 Ti rotor at 35 000 r.p.m.).
4. Remove the phage band and treat it as in steps 4 and 5 of the step gradient method described above.

[a]The composition of lambda buffer is given in *Table 3*.

endogenous methionine, otherwise incorporation of label may be prohibitively low. Phage stocks are usually stored in glass bottles previously cleaned with chromic acid or concentrated sulphuric acid, in order to remove any traces of detergent to which phages are extremely sensitive. A few drops of chloroform are routinely added to the stocks to kill any bacteria present. Before use, the chloroform should be removed by bubbling the phage suspension with sterile air or nitrogen.

2.5 Protocol for Labelling Phage-encoded Proteins

The steps involved in using the u.v.-irradiated host system are outlined in *Table 8*. However, certain aspects of the procedure are worth discussing in more detail.

(i)   There are good reasons why the host strain is grown in M9-minimal medium containing maltose as the carbon source. A λ phage can only infect *E. coli* if the *lamB* gene product is present on the cell surface. This protein, which promotes maltose transport, also acts as a receptor for the phage. It is normally present at only a few molecules per cell, but the

Table 8. Labelling Phage-encoded Proteins in the U.V.-irradiated Host System[a].

1.	Grow the host strain in M9-maltose medium overnight at 37°C.
2.	Dilute to $A_{450} = 0.1$ and grow at 37°C to $A_{450} = 0.64$ (~ 2×10^8 cells/ml). The doubling time of *E. coli* 159 is approximately 120 min under these conditions[b].
3.	Transfer 10 ml of this culture to a Petri dish (9 cm diameter). The depth of liquid should not be greater than ~0.3 cm. Irradiate with u.v. (254 nm) to a dose of 1200 Jm^{-2}, shaking gently to ensure no cells escape the irradiation.
4.	For each sample, transfer 500 μl of the culture to a 1.5 ml microcentrifuge tube. Sediment the cells (microcentrifuge, 2 min) and discard the supernatant.
5.	Resuspend the cells in 100 μl of prewarmed M9-maltose medium containing 20 mM MgCl$_2$. Add the λ phage in $10 - 100$ μl of lambda buffer to a multiplicity of infection of $5 - 10$ p.f.u./cell. To the control sample (no phage) add an equivalent volume of lambda buffer alone.
6.	Incubate at 37°C for 10 min.
7.	Add 400 μl of pre-warmed M9-maltose medium and incubate at 37°C for a further 20 min.
8.	Add $10 - 100$ μCi of [^{35}S]methionine (10 μCi/μl; 1000 Ci/mmol) and incubate at 37°C for 10 min.
9.	Add 200 μl of unlabelled L-methionine (8 mg/ml) and incubate at 37°C for 5 min.
10.	Chill the samples on ice for 2 min. Sediment the cells (microcentrifuge, 2 min) and discard the supernatant.
11.	Resuspend the cell pellet in bacterial buffer and centrifuge the cells again. Discard the supernatant.
12.	Resuspend the cell pellet in 25 μl of bacterial buffer and add 25 μl of SDS-sample buffer. Incubate for 3 min in a boiling water bath.
13.	Apply 25 μl of each sample to an SDS-polyacrylamide slab gel[c]. Store the remainder of the sample at -20°C.
14.	After electrophoresis, fluorograph the gel to detect the labelled proteins[c].

[a]The compositions of media and buffers are given in *Table 3;* details of host strains and phages are given in *Table 4*.
[b]Glycerol may be added to a final concentration of 0.4% which reduces the generation time to 90 min.
[c]Suitable protocols are given in ref. 10.

THE PRACTICAL APPROACH SERIES

Series Editors: D. Rickwood, University of Essex and B. D. Hames, University of Leeds

BOOKS

As laboratory-bench manuals, books in the Practical Approach series provide comprehensive coverage of current experimental techniques in a variety of specific areas within the life sciences, medicine and biotechnology.

The books' unique fusion of thorough step-by-step instructions with clear explanations of theoretical principles, results in a valuable guide to laboratory procedures. In addition, each book contains a wealth of expert advice on frequently encountered problems. Together with an easily referenced format, the Practical Approach series provides a readable source of information for researchers, from the senior undergraduate to the post-doctorate level.

Gel Electrophoresis of Proteins
Edited by B. D. Hames, University of Leeds, and D. Rickwood, University of Essex
1981, 305 pp
Soft 0 904147 22 3; £16.00/US$29.00

Gel Electrophoresis of Nucleic Acids
Edited by D. Rickwood, University of Essex, and B. D. Hames, University of Leeds
1982, 260 pp
Soft 0 904147 24 X; £16.00/US$29.00

Iodinated Density Gradient Media
Edited by D. Rickwood, University of Essex
1983, 254 pp
Soft 0 904147 51 7; £16.00/US$29.00

Transcription and Translation
Edited by B. D. Hames and S. J. Higgins, University of Leeds
1984, 348 pp
Soft 0 904147 52 5; £16.50/US$30.00

Centrifugation (2nd Edition)
Edited by D. Rickwood, University of Essex
1984, 364 pp
Soft 0 904147 55 X; £16.50/US$30.00

Oligonucleotide Synthesis
Edited by M. J. Gait, MRC Laboratory of Molecular Biology, Cambridge
1984, 232 pp
Soft 0 904147 74 6; £16.50/US$30.00

Microcomputers in Biology
Edited by C. R. Ireland and S. P. Long, University of Essex
1985, 338 pp
Soft 0 904147 57 6; £18.00/US$33.00

Companion software
0 904147 97 5; £50.00(+£7.50 VAT in UK)/US$90.00
Combined pack (book and software)
0 904147 98 3; £60.00(+ £6.75 VAT in UK)/US$110.00

Mutagenicity Testing
Edited by S. Venitt, Royal Cancer Hospital, Sutton, and J. M. Parry, University College of Swansea
1985, 368 pp
Soft 0 904147 72 X; £18.00/US$33.00

Affinity Chromatography
Edited by P. D. G. Dean, Agricultural Genetics Co. Ltd., W. S. Johnson, University of Liverpool, and F. A. Middle, P&S Biochemicals
1985, 232 pp
Soft 0 904147 71 1; £16.50/US$30.00

Virology
Edited by B. W. J. Mahy, Animal Virus Research Institute
1985, 280 pp
Soft 0 904147 78 9; £16.50/US$30.00

DNA Cloning Volumes I and II
Edited by D. M. Glover, Imperial College of Science and Technology, London
Volume I:
1985, 204 pp
Soft 0 947946 18 7; £17.00/US$32.00
Volume II:
1985, 260 pp
Soft 0 947946 19 5; £17.00/US$32.00

Plant Cell Culture
Edited by R. A. Dixon, Royal Holloway and Bedford New College, London
1985, 252 pp
Soft 0 947946 22 5; £16.50/US$30.00

Immobilised Cells and Enzymes
Edited by J. Woodward, Oak Ridge National Laboratory
1985, 192 pp
Soft 0 947946 21 7; £16.00/US$29.00
Hard 0 947946 60 9 8; £25.00/US$45.00

Nucleic Acid Hybridisation
Edited by B. D. Hames and S. J. Higgins, University of Leeds
1985, 264 pp
Soft 0 947946 23 3; £16.50/US$30.00
Hard 0 947946 61 6; £25.00/US$45.00

Animal Cell Culture
Edited by R. I. Freshney, University of Glasgow
February 1986, 264 pp
Soft 0 947946 33 0; £16.50/US$30.00
Hard 0 947946 62 4; £25.00/US$45.00

Human Cytogenetics
Edited by D. E. Rooney, St. Mary's Hospital Medical School, London and B. H. Czepulkowski, St. Bartholomew's Hospital, London
July 1986, 250 pp
Soft 0 947946 70 5; £16.50/US$30.00
Hard 0 947946 71 3; £25.00/US$45.00

Human Genetic Diseases
Edited by K. E. Davies, University of Oxford
June 1986, 152 pp
Soft 0 947946 75 6; £15.00/US$27.00
Hard 0 947946 76 4; £22.00/US$40.00

Photosynthesis: Energy Transduction
Edited by M. F. Hipkins, University of Glasgow, and N. R. Baker, University of Essex
June 1986, 208 pp
Soft 0 947946 51 9; £16.50/US$30.00
Hard 0 947946 63 2; £25.00/US$45.00

Drosophila
Edited by D. B. Roberts, University of Oxford
August 1986, 310 pp
Soft 0 947946 45 4; £17.00/US$32.00
Hard 0 947946 66 7; £26.00/US$47.00

HPLC of Small Molecules
Edited by C. K. Lim, Clinical Research Centre, Medical Research Council, London
October 1986, 350 pp
Soft 0 947946 77 2; £18.00/US$34.00
Hard 0 947946 78 0; £27.00/US$49.00

◇ IRL PRESS

ORDER FORM

Please fill out this form carefully, noting the following points:

- All orders direct to the publisher should be prepaid. There is no charge for postage or packing on pre-paid orders, but this will be added to orders sent without pre-payment. Without prior agreement books will only be despatched after receipt of correct payment. US customers should give a street address below (delivery is by UPS).
- US$ prices are valid in North and South America and Japan; £ prices are valid elsewhere. Prices are subject to change without notice.

PLEASE SUPPLY

Quantity	Title	Price US$/£	Amount
	TOTAL		

PAYMENT

☐ **Payment enclosed**

Cheque/bank draft/postal or money order/UNESCO coupons made payable to IRL Press for

£/US$ _____

☐ **Charge my credit card**

☐ American Express   ☐ VISA (except France)   ☐ MasterCard/Access

Card no ☐☐☐☐☐☐☐☐☐☐☐☐☐☐☐☐

Expiry date _____

Signature _____   Date _____

☐ **Payment sent separately**

☐ National or international giro to account no 597 8351

☐ Bank transfer to Midland Bank plc, 220 High Holborn, London WC1V 7BZ; bank sorting code 40-03-27; account no 9108414 3 (£) or 68587170 (US$). Sufficient funds must be remitted to pay all bank handling charges, including those of any forwarding bank in the UK.

☐ **Please send pro forma invoice**

Name
(BLOCK CAPITALS)

Address _____

_____

_____ code _____

_____

Return to

IRL Press Ltd, PO Box 1, Eynsham, Oxford OX8 1JJ, UK
IRL Press Inc, PO Box Q, McLean, VA 22101-0850, USA

LATEST TITLES

Mammalian Development
Edited by M. Monk, MRC Mammalian Development Unit, London
August 1987, 288 pp

Soft 1 85221 029 X: £18.00/US$34.00
Hard 1 85221 030 3: £27.00/US$49.00

Mitochondria
Edited by V. M. Darley-Usmar, The Wellcome Research Laboratories, D. Rickwood and M. T. Wilson, University of Essex
August 1987, 340 pp

Soft 1 85221 033 8: £18.00/US$34.00
Hard 1 85221 034 6: £29.00/US$52.00

Lymphokines and Interferons
Edited by M. J. Clemens, St George's Hospital Medical School, A. G. Morris, University of Warwick, and A. J. H. Gearing, UK National Institute for Biological Standards and Control
September 1987, 392 pp

Soft 1 85221 035 4: £22.00/US$42.00
Hard 1 85221 036 2: £33.00/US$63.00

DNA Cloning Volume III
Edited by D. M. Glover, Imperial College of Science and Technology, London
September 1987, 272 pp

Soft 1 85221 048 6: £17.00/US$32.00
Hard 1 85221 049 4: £26.00/US$47.00

Volumes I, II and III:
1 85221 069 9: £43.00/US$82.00

Lymphocytes
Edited by G. G. B. Klaus, National Institute for Medical Research, London
September 1987, 256 pp

Soft 1 85221 019 2: £17.00/US$32.00
Hard 1 85221 018 4: £26.00/US$47.00

1987 TITLES

Carbohydrate Analysis
Edited by M. F. Chaplin, Polytechnic of the South Bank, and J F Kennedy, University of Birmingham
January 1987, 244 pp
Soft 0 94746 44 6; £17.00/US$32.00
Hard 0 94746 68 3; £26.00/US$47.00

Plasmids
Edited by K. G. Hardy, Biogen SA
February 1987, 192 pp
Soft 0 94746 81 0; £16.00/US$30.00
Hard 0 94746 82 9; £25.00/US$45.00

Biochemical Toxicology
Edited by K. Snell, University of Surrey, and B. Mullock, Robens Institute of Industrial and Environmental Health and Safety
February 1987, 272 pp
Soft 0 94746 52 7; £17.00/US$32.00
Hard 0 94746 67 5; £26.00/US$47.00

Electron Microscopy in Molecular Biology
Edited by J. Sommerville, University of St Andrews, and U. Scheer, German Cancer Research Center, Heidelberg
March 1987, 256 pp
Soft 0 94746 54 3; £17.00/US$32.00
Hard 0 94746 64 0; £26.00/US$47.00

Spectrophotometry & Spectrofluorimetry
Edited by D. A. Harris, University of Oxford, and C. L. Bashford, St George's Hospital Medical School, London
March 1987, 192 pp
Soft 0 94746 46 2; £16.00/US$30.00
Hard 0 94746 69 1; £25.00/US$45.00

Teratocarcinomas and Embryonic Stem Cells
Edited by E. J. Robertson, University of Cambridge
March 1987, 268 pp
Soft 1 85221 004 4; £17.00/US$32.00
Hard 1 85221 005 2; £26.00/US$47.00

Nucleic Acid and Protein Sequence Analysis
Edited by M. Bishop, University of Cambridge, and C. Rawlings, ICRF, London
March 1987, 400 pp
Soft 1 85221 006 0; £20.00/US$36.00
Hard 1 85221 007 9; £30.00/US$54.00

Biological Membranes
Edited by J. B. C. Findlay, University of Leeds, and W. H. Evans, NIMR, London
May 1987, 312 pp
Soft 0 94746 83 7; £18.00/US$34.00
Hard 0 94746 84 5; £27.00/US$49.00

Steroid Hormones
Edited by B. Green, University of Strathclyde, and R. E. Leake, University of Glasgow
May 1987, 264 pp
Soft 0 94746 53 5; £17.00/US$32.00
Hard 0 94746 65 9; £26.00/US$47.00

Neurochemistry
Edited by A. J. Turner, University of Leeds, and H. S. Bachelard, St Thomas's Hospital Medical School, London
July 1987, 256 pp
Soft 1 85221 027 3; £17.00/US$32.00
Hard 1 85221 028 1; £26.00/US$47.00

Prostaglandins and Related Substances
Edited by C. Benedetto, Università di Torino, R. G. McDonald-Gibson, Brunel University, S. Nigam, Freie Universität Berlin, and T. F. Slater, Brunel University
July 1987, 330 pp
Soft 1 85221 031 1; £18.00/US$34.00
Hard 1 85221 032 X; £29.00/US$52.00

lamB gene can be induced to produce much greater amounts of the receptor protein by growth of the bacteria in maltose (9). The induction of the *lamB* gene can therefore be used to increase the efficiency of phage infection. No glucose is added to the medium since this causes catabolite repression of the maltose operon.

(ii) *E. coli* strains 159 and 159(λ*ind*) grow slowly in M9-minimal medium containing maltose, with a doubling time of approximately 120 min. If desired, glycerol may be added to a final concentration of 0.4% (v/v), which reduces the doubling time to about 90 min. If the A_{450} of the culture either exceeds or is less than the required value of 0.64 after the incubation at 37°C (*Table 8*, step 2), but the cells are growing exponentially, then the culture can be diluted or concentrated (by centrifugation) as

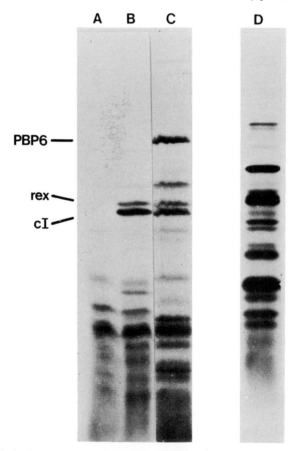

Figure 3. Analysis of phage-encoded proteins synthesised in u.v.-irradiated cells. Phage-specified proteins were labelled as described in *Table 8*. Samples were analysed by SDS-PAGE (15% slab gel) and fluorography. **Lanes A, B** and **C** (host strain *E. coli* 159 λ*ind*) were exposed for 3 days, while **lane D** (host strain *E. coli* 159) was exposed for 2 h. **(A)** control; **(B)** λ^+; **(C)** λp*dacC*; **(D)** λ^+. **Lane D** shows the range of phage λ-encoded proteins which are synthesised in a non-lysogenic host whereas in a host lysogenic for λ the only major proteins synthesised are rex and cI **(lane B)**, the products of the *rex* and *cI* genes of the immunity region of phage λ. **Lane C** shows the synthesis of the extra protein, PBP6, encoded by λp*dacC* (see *Table 2*).

necessary. It is important to use a known number of cells per experiment to ensure that the correct multiplicity of infection is achieved.

(iii) After irradiation (*Table 8*, step 3), the cells are concentrated 5-fold in medium containing 20 mM $MgCl_2$ (steps 4 and 5). The presence of some magnesium ions is essential for both the stability of the phage particles and for phage adsorption, but the high magnesium ion concentration is also important to prevent a depression of protein synthesis which is otherwise observed on infection of cells with a high multiplicity of phage λ (1).

(iv) A multiplicity of infection of 5 – 10 is usually optimal; a value of less than 5 results in lower incorporation of labelled methionine while too high a multiplicity of infection might result (in the case of the infection of a lysogen) in the repressor being titrated out, leading to the expression of phage genes. When using a host lysogenic for λ phage, the correct multiplicity of infection is indicated by the expression of only two major phage proteins, the *cI* and *rex* gene products, after infection by wild-type phage ($λ^+$) (see *Figure 3*, lane B).

(v) Non-infected host cells should also be incubated with [^{35}S]methionine in order to monitor the level of residual expression of bacterial chromosomal genes. For these control incubations, an equivalent volume of lambda buffer is added instead of the phage (*Table 8*, step 5).

(vi) After the addition of phage to the test incubations, the samples are incubated at 37°C for 10 min to allow phage adsorption (*Table 8*, step 6). Then, after dilution of the samples, [^{35}S]methionine is added to label the phage-encoded proteins and incubation is continued for an additional 10 min. This is followed by a brief incubation with unlabelled methionine in order to allow the synthesis of labelled proteins to be completed.

(vii) The cells are washed once in buffer before being resuspended in a small volume of buffer and prepared for analysis by SDS-PAGE using a slab gel. As incorporation is often quite low, it is usually necessary to fluorograph the gel after electrophoresis in order to detect the labelled protein products. Both SDS-PAGE and fluorography have been described in detail in an earlier volume of this series (10).

Examples of phage-encoded proteins labelled after infection of u.v.-irradiated cells are shown in *Figure 3*.

2.6 Potential Problems

2.6.1 *High Background*

A high background of [^{35}S]methionine incorporation in the control (without added phage) indicates that the u.v. dose during irradiation (*Table 8*, step 3) was too low. With the phage-infected samples, a high u.v. dose gives a low background, but this is accompanied by a general reduction in protein synthesis (11). How high a background can be tolerated will depend on the efficiency of expression of the genes being studied, so the optimal u.v. dose may have to be determined empirically. Generally, doses between 400 and 1200 Jm^{-2} are used. It is advisable to use a dosimeter for reproducible results.

2.6.2 *Low Incorporation*

The efficiency of incorporation of [^{35}S]methionine, especially when using a lysogen as host, is often low in comparison with minicell and maxicell systems. Certain simple procedures should help to maximise incorporation.

(i)   The phage lysate should be of a sufficiently high titre (10^{10} p.f.u./ml) and should be dialysed thoroughly against lambda buffer *(Table 3)* before use.
(ii) The specific activity of the [^{35}S]methionine should be the highest available (≥ 1000 Ci/mmol).
(iii) The u.v. dose for host cell irradiation should be the lowest compatible with acceptable background incorporation (see Section 2.6.1).
(iv) After SDS-PAGE, the polyacrylamide gel should be fluorographed for maximum sensitivity of detection of the labelled products. It may also help to preflash the X-ray film before use to hypersensitise it (10,12).

2.7 Use of a Non-lysogenic Host

Proteins synthesised from λ transducing phages can also be detected by infection of a u.v.-irradiated non-lysogen of *E. coli*. In this case, the absence of repressor in the cells results in the transcription of phage genes and cloned genes from P_L and P_R. This can be useful if the chromosomal genes have very weak promoters, or have been separated from their own promoters.

2.8 Use of Phages Carrying Other Immunities

2.8.1 *Repression of 434 and 21 Immunities*

Although only phages carrying the λ immunity region are discussed above, many cloning vectors carry the phage 434 or 21 immunity region. In order to repress the major transcriptional units in these phages, *E. coli* 159 (λ*imm*434) or 159 (λ*imm*21) should be used as the host strain, respectively. However, these prophages do not repress the superinfecting phages as efficiently as the *imm*λ prophage. Therefore, when working with phages carrying phage 434 immunity, the authors use *E. coli* 159 carrying the plasmid pGY101 as the host strain. Plasmid pGY101 carries the 434 repressor gene, and produces approximately 70 times more repressor protein than the λ*imm*434 prophage (13). A plasmid carrying the phage 21 repressor gene has also been constructed (14) which could be used with *imm*21 phages. An alternative method of avoiding the expression of phage genes of an *imm*21 phage is to reduce all incubation times to a minimum (15) since the number of infecting phages which escape immunity seems to increase significantly with longer incubations.

2.8.2 *Hybrid-immunity phages*

A sophistication of the technique is the use of hybrid-immunity phages (16) to determine the direction in which cloned genes are transcribed. These phages have been constructed with, for example, the left half of the *imm*λ region and the right half of the *imm*434 region, allowing P_L and P_R to be controlled independently. Additionally, the presence of amber mutations in the *N* gene allows transcription from P_L to be controlled by the use of suppressing and

Table 9. Requirements for the Minicell System.

Bacterial strains

    E. coli DS410 (see Section 3.3) carrying appropriate plasmids.

Solutions

    Nutrient broth[a]

    Antibiotics for plasmid selection

    M9-minimal medium

    M9-minimal medium containing 20% (w/v) sucrose

    M9-minimal medium containing 30% (v/v) glycerol

    Methionine assay medium (10.5% w/v) (Difco)

    [^{35}S]methionine (10 μCi/μl; 1000 Ci/mmol)

    L-methionine (8 mg/ml)

    Bacterial buffer[a]

    SDS-sample buffer[a]

---

[a]Prepared as described in *Table 3*.

non-suppressing hosts. Expression of genes cloned downstream from P_L may be examined as usual by infection of an irradiated host. Due to the anti-termination properties of the N protein, this transcription in a suppressing host will proceed through all the cloned genes, unless a very strong terminator has been cloned. Since their expression will be increased or diminished by the powerful phage λ transcription (17), comparison of the relative expression of cloned genes in suppressing and non-suppressing hosts should allow the direction of transcription of each gene to be determined. The phages are often difficult to work with, but this approach has been successfully used in several studies (15,18,19).

3. THE MINICELL SYSTEM

3.1 Introduction

Strains of *E. coli* which carry both *minA* and *minB* mutations divide asymmetrically, so that approximately one in two divisions results in the formation of a small anucleate cell (20). These minicells, which are capable of supporting DNA, RNA and protein synthesis, may be separated from the larger, viable cells by their differential sedimentation through sucrose gradients. When minicell-producing strains carry small multicopy plasmids, most minicells will contain, by chance, at least one plasmid molecule although no bacterial chromosome will be present. These minicells will synthesise plasmid-encoded proteins under essentially *in vivo* conditions for considerable periods of time. The detailed properties of minicells as well as the properties of minicell-producing mutants isolated from other bacterial species have been reviewed elsewhere (21).

3.2 Requirements

The requirements for the minicell system are given in *Table 9*.

3.3 The Escherichia Coli Minicell-producing Strain, DS410

The original minicell-producing strain isolated (20) was *E. coli* P678-54 (also called χ925) with the genotype F$^-$ *minA minB thr leu thi ara lacY gal malA xyl mtl tonA rpsL supE*. Subsequently, Dougan and Sherratt (22) isolated a *thr$^+$ leu$^+$ mal$^+$ lac$^+$ sup^o* derivative, DS410, which is now widely used for minicell production.

3.4 Suitable Plasmids

All plasmids examined appear to segregate into *E. coli* minicells. Low-copy-number plasmids (e.g., pSC101) segregate quite efficiently (23). Even large conjugative plasmids such as F are found in minicells, although segregation is not always particularly efficient in these cases (21). Preparation of plasmid and transformation of the minicell host is carried out as described in *Tables 10* and *11*, respectively.

3.5 Purification of Minicells

Before minicells can be purified, continuous sucrose gradients (10 − 30% w/v) should be prepared. These may be formed in the usual way using a gradient maker. Alternatively, they may be prepared by a simple freeze-thaw method (see *Table 12*). This procedure results in the formation of perfectly adequate, reproducible continuous sucrose gradients, and allows large numbers of gradients to be prepared at once. Minicell purification is then carried out as described in *Table 13*. Certain steps are worth discussing in more detail here.

(i)   The bacteria are grown in rich medium to stationary phase (usually over-

Table 10. Preparation of Plasmid DNA[a,b].

1.	Grow an overnight culture of the plasmid-containing strain in nutrient broth.
2.	Transfer 1.5 ml of bacterial culture to a microcentrifuge tube and pellet the cells by centrifugation in a microcentrifuge for 1 min.
3.	Discard the supernatant and resuspend the pellet in 75 μl of STET buffer.
4.	Add 6 μl of 10 mg/ml lysozyme and mix briefly.
5.	Immediately place the tube in a boiling water bath for 1 min.
6.	Centrifuge (microcentrifuge, 10 min) to pellet the chromosomal DNA-protein complex.
7.	Transfer the supernatant to a fresh tube and add an equal volume of propan-2-ol. Leave at −70°C for 10 min or −20°C for 1 h to precipitate the nucleic acid.
8.	Centrifuge (microcentrifuge, 10 min) to recover the nucleic acid.
9.	Discard the supernatant and resuspend the pellet in 100 μl of 0.3M sodium acetate, pH 5.6. Add 0.3 ml of absolute ethanol and then precipitate and recover the nucleic acid as in steps 7 and 8.
10.	Discard the supernatant and dry the pellet under vacuum.
11.	Resuspend the pellet in 30 μl of 10 mM Tris-HCl (pH 7.5), 1 mM EDTA.
12.	Store at 4°C until use; 10 μl of the preparation should be sufficient for good transformation (see *Table 11*).

[a]A small-scale plasmid preparation is adequate for transformation purposes. The protocol described here is derived from the method of Holmes and Quigley (24) but other protocols (e.g. ref. 25 or Chapter 7, *Table 8*) are equally suitable.
[b]The compositions of media and buffers are given in *Table 3*.

Table 11. Preparation of Competent Cells and Their Transformation[a].

1.	Grow an overnight culture of the bacterial strain in nutrient broth at 37°C.
2.	Dilute the culture 50-fold into fresh nutrient broth and grow to $A_{600} = 0.2$. Chill on ice.
3.	Transfer 30 ml of the culture to a centrifuge tube and pellet the cells by centrifugation at 3000 g per 10 min at 4°C (e.g. Sorvall SS34 rotor at 5000 r.p.m.).
4.	Discard the supernatant and resuspend the pellet in 15 ml of ice-cold 0.1 M $CaCl_2$. Leave on ice for 20 min.
5.	Pellet the cells as in step 3; at this stage they will form a 'halo' on the bottom of the tube.
6.	Discard the supernatant and resuspend the pellet in 1 ml of ice-cold 0.1 M $CaCl_2$. Store at 4°C until use[b]. These are now 'competent' cells, i.e., competent to undergo transformation.
7.	Transfer 0.2 ml of the competent cells to a microcentrifuge tube and add 0.1 μg of plasmid DNA. Also set up a control consisting of competent cells with no DNA added. Leave both tubes on ice for 1 h.
8.	Heat-shock the cells by transferring the tubes to a 42°C water bath for 5 min.
9.	Dilute each set of cells into 2 ml of nutrient broth and incubate at 37°C with shaking for at least 1 h in order to allow the plasmid-encoded drug resistance to be expressed.
10.	Plate 100 μl of serial dilutions of the cells onto nutrient agar plates containing the relevant antibiotics and incubate at 37°C overnight.
11.	Pick a suitable colony. Ensure genetic homogeneity by streaking at least once to single colonies before use.

[a]The compositions of media and buffers are given in *Table 3*.
[b]The competence of the cells increases for ~24 h at 4°C (26) before viability decreases drastically. Alternatively, competent cells may be resuspended in 0.1 M $CaCl_2$, 30% glycerol and frozen at −70°C (27). These frozen cells are stable for several months.

Table 12. Preparation of Sucrose Gradients.

1.	Pipette 35 ml of M9-minimal medium[a] containing 20% (w/v) sucrose into as many tubes as will be required. The centrifuge tubes should be transparent to allow bands to be visualised. Polysulfone tubes (50 ml capacity) are ideal.
2.	Place the tubes at −70°C for at least 1 h, or until the gradients are completely frozen.
3.	When required for use, place the tubes at 4°C overnight to allow the gradients to thaw.

[a]The composition of M9-minimal medium is given in *Table 3*.

night). Minicells continue to be produced for several hours in stationary phase (21).

(ii) The initial low speed centrifugation (*Table 13*, step 3) is included to remove some of the normal (nucleated) cells. This helps to prevent overloading of the sucrose gradients.

(iii) Two sequential sucrose gradient centrifugation steps are involved in minicell preparation (*Table 13*, steps 5 and 9). *Figure 4B* shows the profile expected after the first of these. The viable cells form a wide band towards the bottom of the tube and some may have pelleted. The minicell band is usually visible above the viable cells, although the two bands may not be completely separate. Only the top two-thirds of the minicell band should be recovered (*Table 13,* step 7). This further reduces the risk of contamination with viable cells. After this minicell preparation has been fractionated by centrifugation in the second sucrose gradient, the minicell band is clearly separated from viable cells (*Figure 4C*). Again only the top

Table 13. Purification of Minicells[a].

1.	Grow a 400 ml culture of *E. coli* DS410 or its plasmid-containing derivative in nutrient broth containing the relevant antibiotics. Continue the incubation until the cells are well into the stationary phase of growth [see Section 3.5(i)].
2.	Chill the cells at 4°C for 10 min.
3.	Centrifuge the cells at 700 *g* for 5 min (e.g., Sorvall GS3 rotor at 2000 r.p.m.) and then transfer the supernatant to fresh centrifuge bottles.
4.	Harvest the minicells and the remaining whole cells by centrifugation at 11 000 *g* for 15 min at 4°C (e.g., Sorvall GS3 rotor at 8000 r.p.m.). Discard the supernatant.
5.	Resuspend the cell pellet in 6 ml of M9-minimal medium and layer this onto two 10–30% (w/v) sucrose gradients[b].
6.	Centrifuge the gradients in a swing-out rotor at 4000 *g* for 18 min at 4°C (e.g., Sorvall HB4 rotor at 5000 r.p.m.).
7.	Recover the top two-thirds of the minicell band (see *Figure 4B*) from each of the two gradients using a Pasteur pipette and pool.
8.	Add an equal volume of M9-minimal medium and harvest the minicells by centrifugation at 16 000 *g* for 10 min (e.g., Sorvall HB4 rotor at 10 000 r.p.m.). Discard the supernatant.
9.	Resuspend the minicells in 3 ml of M9-minimal medium and layer onto one 10–30% (w/v) sucrose gradient.
10.	Centrifuge the gradient as in step 6. Remove the top two-thirds of the minicell band (see *Figure 4C*) into a test tube on ice.
11.	Add an equal volume of M9-minimal medium, and measure the final volume. Read the optical density at 600 nm.
12.	Transfer the minicells to a centrifuge tube and harvest as in step 8.
13.	Resuspend the minicells in M9-minimal medium containing 30% (v/v) glycerol to $A_{600} = 2.0$ (~2 x 10^{10}/ml).
14.	Transfer the cell suspension to a sterile microcentrifuge tube and store at −70°C. The minicells are stable for at least a year under these conditions.

[a]The compositions of media and buffer and details of strains are given in *Tables 3* and *9*.
[b]Prepared as described in *Table 12*.

two-thirds of the minicell band should be recovered (*Table 13*, step 10).

(iii) It is advisable to check the purified minicells (*Table 13*, step 12) under the microscope for contamination by viable cells (see *Figure 5* and Section 3.7) before storage in minimal medium containing 30% (v/v) glycerol (step 13). They may be stored in this way for at least a year without apparent loss of activity (see *Figure 6*).

3.6 Protocol for Labelling Plasmid-encoded Proteins

The purified minicells are labelled as described in *Table 14*. Levy (28) reported that minicells contain some very stable mRNA species, with a half-life of 40–80 min. Most of this mRNA encodes outer membrane proteins. This finding was confirmed by Reeve (29), who showed significant translation of stable mRNA during periods of up to 90 min incubation. Therefore, it is advisable to pre-incubate minicells for a period of 60–90 min at 37°C before adding [^{35}S]methionine. The authors find that 1 h of pre-incubation appears to be sufficient.

While not essential, the inclusion of Difco methionine assay medium (which contains all the common amino acids except methionine) in the incubation with [^{35}S]methionine improves the efficiency of isotope incorporation.

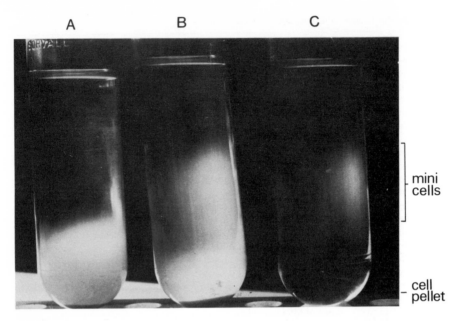

Figure 4. Purification of minicells on sucrose gradients. Minicells were purified by centrifugation through two sucrose gradients prepared as described in *Table 13*. **(A)** a non-minicell-producing strain, first gradient; **(B)** *E. coli* strain DS410, first gradient *(Table 13)*; **(C)** DS410, second gradient *(Table 13)*. The apparent unevenness in the bands is an artefact of the lighting not the separation method.

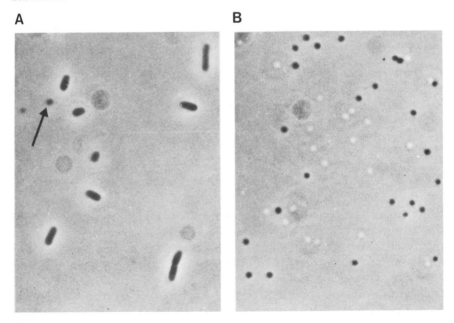

Figure 5. Phase-contrast microscopy showing minicell purification. A starting culture of DS410 **(A)** and the purified minicells **(B)** were examined by phase-contrast microscopy (1250x magnification). A minicell visible in **(A)** is marked with an arrow.

Table 14. Labelling Plasmid-encoded Proteins in Minicells[a].

1.	Thaw the frozen minicell suspension[b] completely and transfer 100 μl to a sterile micro-centrifuge tube. Refreeze the rest of the minicells.
2.	Pellet the minicells (microcentrifuge, 3 min) and discard the supernatant. Resuspend the cell pellet in 200 μl of M9-minimal medium and add 3 μl of 10.5% (w/v) Difco methionine assay medium.
3.	Incubate the minicells at 37°C for 90 min.
4.	Add 25 μCi of [^{35}S]methionine (10 μCi/μl; 1000 Ci/mmol), and incubate at 37°C for 60 min.
5.	Add 10 μl of unlabelled L-methionine (8 mg/ml), and incubate for a further 5 min.
6.	Harvest the minicells (microcentrifuge, 3 min) and discard the supernatant.
7.	Resuspend the minicells in 50 μl of bacterial buffer and add 50 μl of SDS-sample buffer. Heat for 3 min in a boiling water bath.
8.	Apply 25 μl of the sample to an SDS-polyacrylamide slab gel[c] and store the remainder of the sample at −20°C.
9.	After electrophoresis, autoradiograph or fluorograph the gel to detect labelled proteins[c].

[a]The compositions of media and buffers are given in *Table 3*.
[b]Minicell preparations (prepared as described in *Table 13*) can be thawed and re-frozen many times without appreciable loss of activity. The cells can be thawed at 37°C.
[c]Suitable protocols are given in ref. 10.

Examples of plasmid-encoded proteins synthesised in minicells are shown in *Figure 6*.

3.7 Potential Problems

The major problem likely to be encountered with minicells is that of contamination of the minicell preparation with viable cells. The degree of contamination is best estimated prior to use in one of two ways. The first is to dilute the final minicell preparation (A_{600} = 2.0; *Table 13*, step 13) with bacterial buffer *(Table 3)* and then plate aliquots onto nutrient agar. After incubation at 37°C overnight, the number of colonies are counted. Minicells fail to grow since they are anucleate. Contamination of a minicell preparation with 10^5 viable cells/ml or less should be acceptable. Alternatively, the degree of contamination can be checked by phase-contrast microscopy (see *Figure 5*). To do this, the minicells (A_{600} = 2.0) are examined under the microscope before being frozen (*Table 13*, step 14). Using a magnification of 400x, one parental cell per field indicates approximately 10^5 cells/ml which is acceptable (29). Unfortunately, unless there is gross contamination, neither method described is entirely reliable. Furthermore, the degree of contamination which can be tolerated may depend upon the relative strength of the promoter of the genes of interest, and will only be revealed when the minicells are incubated with [^{35}S]methionine and the labelled proteins are analysed by SDS-PAGE and autoradiography.

If there are continued problems with contamination leading to an unacceptably high background, sodium ampicillin (100 μg/ml) may be added to the incubation to kill viable cells. If the minicells carry a plasmid which encodes a β-lactamase, then D-cycloserine (100 μg/ml) may be used instead.

The authors have found that a relatively poor separation of minicells from

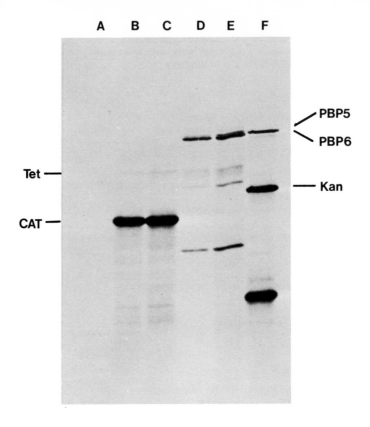

Figure 6. Labelling of plasmid-encoded proteins in minicells. Minicells were purified and labelled with [^{35}S]methionine (*Tables 13* and *14*). Samples were analysed by SDS-PAGE (15% slab gel) and autoradiographed for 12 h. **(A)** DS410; **(B)** and **(C)** DS410 carrying pACYC184; **(D)** and **(E)** DS410 carrying pLG517; **(F)** DS410 carrying pBS42. Minicell preparations **(B)** and **(D)** had been stored at −70°C for 1 year, while minicells **(A)** and **(C)** were freshly prepared. The Tet, CAT, Kan, PBP5 and PBP6 proteins are defined in *Table 2*.

whole cells occurs if *E. coli* DS410 carries a *recA* mutation. Also, the cloning of certain genes coding for envelope proteins in *E. coli* into multicopy plasmids might result in disturbed division patterns in DS410, with consequent poor yields of purified minicells. Therefore, if working with a plasmid which must be maintained in a *recA* host, or which carries a gene which disturbs minicell formation in this way, it may be necessary to use an alternative gene expression system, such as the maxicell stystem (see Section 4).

3.8 Fractionation of Minicells

In comparison with normal bacterial cells, the fractionation of minicells into various membrane, cytoplasmic and periplasmic fractions is difficult (21,30). In particular, quantitative recovery of membrane fractions may not be possible so that it may be desirable to use one of the other gene expression systems for this purpose.

3.9 **Appearance of Precursor Polypeptides**

It is not uncommon for the precursor forms of processed proteins to be seen in minicells (e.g., *β*-lactamase; see ref. 30). While this is not a handicap, it is useful to be aware of it. The same phenomenon is true of maxicells (see Section 4).

3.10 **Infection of Minicells with Bacteriophages**

Purified minicells may be infected by bacteriophages, and so used as a coupled transcription-translation system for genes encoded by the bacteriophages, as in the u.v.-irradiated host system (Section 2). In the case of phage λ, expression of phage-encoded proteins may be largely prevented by infection of minicells containing a plasmid carrying the phage repressor gene. This use has been reviewed by Reeve (29).

4. THE MAXICELL SYSTEM

4.1 **Introduction**

E. coli contains several systems for repairing DNA damaged by u.v. irradiation and other mutagens (31). Mutations in the *recA* and *uvrA* genes inactivate the major repair systems, making cells which carry these mutations extremely sensitive to u.v. irradiation. Direct inhibition of transcription by the presence of u.v.-induced pyrimidine dimers is accompanied by extensive degradation of the chromosomal DNA by intracellular nucleases under these conditions. Subsequent incubation of these 'maxicells' therefore results in very little protein synthesis due to the lack of transcribable DNA. However, if such strains carry a multicopy plasmid, many plasmid molecules will survive the u.v. treatment due to their small size compared with that of the bacterial chromosome. These plasmid molecules will consequently serve as templates for transcription in subsequent incubations, and the addition of labelled amino acids therefore leads to preferential labelling of plasmid-encoded proteins.

4.2 **Choice of Bacterial Strain for Maxicell Production**

E. coli CSR603 is a strain often used to produce maxicells. CSR603 is deficient in all the major DNA-damage repair pathways, since it carries *recA, uvrA* and *phr* mutations. This strain is therefore extremely sensitive to u.v. irradiation, a single u.v.-induced lesion in the chromosome being lethal. The cells are easily transformed with small plasmids, and large plasmids may be conjugated into them. However, CSR603 is sometimes difficult to maintain; survival on plates is poor, and in our hands the background incorporation due to transcription of residual host templates is rather variable.

Sancar *et al*. (32) reported that strains carrying only a *recA* mutation could be used to generate maxicells. Therefore, the authors have investigated the ability of strain CSH26 (Δ*recA*) as an alternative to CSR603. Although CSH26 *recA* is less sensitive to u.v. irradiation than CSR603, the background expres-

171

sion from chromosomal DNA can be reduced to that obtained with CSR603 under optimal conditions (see *Figure 7*). In addition, CSH26 *recA* has the advantage of being generally easier to grow and to maintain, and is more reliable than CSR603. It should be emphasised that it is essential to inactivate the *recA* gene in order to produce maxicells. A *recA*$^+$ *uvrA* strain, for example, would not be suitable. This is because the recA protein not only functions directly in recombinational repair, but it also, in some way, regulates the activity of a major cellular nuclease, exonuclease V. This enzyme is the product of the *recB* and *recC* genes, and is responsible for virtually all the degradation of chromosomal DNA in u.v.-irradiated cells.

Many cloning experiments involve transformation into a *recA* strain in order to minimise any rearrangements of the cloned DNA sequences. It is therefore convenient that the same strain should be capable of serving as a maxicell strain, facilitating the identification of the cloned gene products without the need to transfer the plasmid to another host for maxicell production.

4.3 Suitable Plasmids

The maxicell system, as with minicells, is more reliable if the plasmid used is relatively small, and is present at a high copy-number. Sancar *et al.* (33) found that ColE1-like multicopy plasmids not damaged by irradiation continued to replicate, with plasmid DNA levels increasing 10-fold after 6 h incubation in cells where 80% of the chromosomal DNA was degraded. They also showed that a tetramer of pBR322 (17.4 kb) could be used. Finally, the proteins encoded by a 40 kb plasmid which is present at only 1 or 2 copies per cell have recently been identified in maxicells (N.Mackman, personal communication), showing that a wide variety of plasmids may be used with this technique.

4.4 Procedure

The requirements for the production and use of maxicells are given in *Table 15*. Preparation of plasmid and transformation of the maxicell strain is carried out as described previously for the minicell system (*Tables 10* and *11*, respectively). The protocol for labelling plasmid-encoded proteins in maxicells is outlined in *Table 16*, but it is worth emphasising certain points.

(i)   The cells are grown to mid-exponential phase in M9-minimal medium supplemented with casamino acids. This medium is used instead of a rich medium such as nutrient broth because the latter contains u.v.-absorbing material which would interfere with the irradiation of the cells (*Table 16*, step 3).

(ii) After irradiation (step 3), the cells are incubated at 37°C for 1 h to allow those cells which have been damaged to cease growth. Any viable, growing cells, which would otherwise overgrow the u.v.-damaged cells, will lyse during the subsequent overnight incubation with ampicillin (30 μg/ml) or D-cycloserine (100 μg/ml). This long incubation also allows the damaged chromosomal DNA to be degraded.

Table 15. Requirements for the Maxicell System.

Bacterial strains[a]

      E. coli CSR603; *phr1 recA1 uvrA6 thr1 leuB6 argE3 thi1 ara14 lacY1 galK2 xyl5 mtl1 supE44 tsx33 gyrA96 rpsL31*

      or E. coli CSH26 *recA; (recA)Δ (pro lac)Δ thi ara rpsL*

      or derivatives carrying appropriate plasmids.

Solutions

      M9-minimal medium[b] containing 0.2% casamino acids

      Antibiotics for plasmid selection

      M9-minimal medium

      Sodium ampicillin (3 mg/ml)

      Methionine assay medium (10.5% w/v) (Difco)

      [^{35}S]methionine (10 μCi/μl; 1000 Ci/mmol)

      L-methionine (8 mg/ml)

      Bacterial buffer[b]

      SDS-sample buffer[b]

---

[a]For choice of bacterial strain, see Section 4.2.
[b]Prepared as described in *Table 3*.

Table 16. Labelling Plasmid-encoded Proteins in the Maxicell System[a].

---

1. Grow a culture of the maxicell strain[b] (previously transformed with the desired plasmid) at 37°C overnight in minimal medium containing casamino acids.
2. Dilute the culture 20-fold into minimal medium containing casamino acids. Grow at 37°C to $A_{450} = 0.5$.
3. Transfer 3 ml of the culture to a sterile Petri dish (5 cm diameter). Irradiate with u.v. (254 nm) to a dose of 1.5 Jm^{-2}, shaking gently to ensure no cells escape irradiation.
4. Transfer 2 ml of the irradiated culture to a sterile test tube and incubate, with shaking, at 37°C for 1 h.
5. Add sodium ampicillin[c] to a final concentration of 30 μg/ml and incubate the culture, with shaking, for 16−24 h at 37°C.
6. Transfer 1 ml of the culture to a sterile 1.5 ml microcentrifuge tube and harvest the cells by centrifugation (microcentrifuge, 2 min). Discard the supernatant.
7. Resuspend the cell pellet in M9-minimal medium and harvest the cells again by centrifugation as in the step 6.
8. Repeat step 7.
9. Resuspend the cells in 0.5 ml of M9-minimal containing 30 μg/ml of sodium ampicillin. Add 3 μl of 10.5% (w/v) Difco methionine assay medium.
10. Incubate at 37°C for 1 h.
11. Add 25 μCi of [^{35}S]methionine and incubate at 37°C for 1 h.
12. Add 10 μl of unlabelled L-methionine (8 mg/ml) and incubate for a further 5 min.
13. Harvest the cells by centrifugation (microcentrifuge, 2 min). Discard the supernatant.
14. Resuspend the cells in 50 μl of bacterial buffer and add 50 μl of SDS-sample buffer. Heat for 3 min in a boiling water bath.
15. Apply 25 μl of the sample to an SDS-polyacrylamide slab gel[d] and store the remainder of the sample at −20°C.
16. After electrophoresis, autoradiograph or fluorograph the gel to detect the labelled proteins[d].

---

[a]The compositions of media and buffers are given in *Tables 3* and *15*.
[b]Suitable strains are described in *Table 15*.
[c]If the plasmid encodes β-lactamase, other antibiotics should be used instead (see Section 4.5).
[d]Suitable protocols are given in ref. 10.

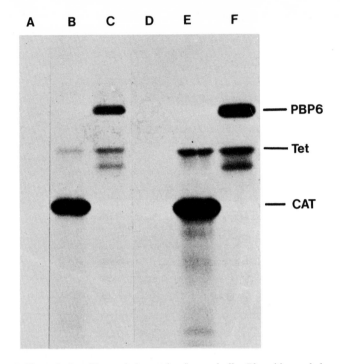

Figure 7. Labelling of plasmid-encoded proteins in maxicells. Plasmid-encoded proteins were labelled with [^{35}S]methionine in maxicells *(Table 16)*. Samples were analysed by SDS-PAGE (15% slab gel) and fluorographed for 24 h. **(A)** CSR603, **(B)** CSR603 carrying pACYC184, **(C)** CSR603 carrying pLG517, **(D)** CSH26*recA*, **(E)** CSH26*recA* carrying pACYC184, **(F)** CSH26*recA* carrying pLG517. For details of the Tet, CAT and PBP6 proteins, see *Table 2*.

(iii) If the strain used is *phr*$^+$, and can thus remove u.v.-induced damage by photoreactivation, the tubes should be covered with aluminium foil immediately after irradiation and kept covered through steps 4 and 5.

(iv) After the lysis of viable growing cells, the remaining cells are resuspended in M9-minimal medium. Difco methionine assay medium may be added (step 9) to replenish the amino acid pools and thus ensure efficient protein synthesis.

(v) After labelling the plasmid-specific proteins with [^{35}S]methionine, the cells are chased with unlabelled methionine to complete the synthesis of labelled proteins and are then prepared for analysis by SDS-PAGE.

Examples of plasmid-encoded proteins labelled in maxicells are shown in *Figure 7*.

4.5 Potential Problems

If there is substantial incorporation of [^{35}S]methionine into non-plasmid-encoded proteins, it may be necessary to increase the u.v. dose in order to increase the damage to host chromosomal DNA. However, in practice, the dose can be varied considerably without altering the effectiveness of the system. Plating for viable cells (which should be < 20 viable cells/ml with the extremely u.v.-sensitive strain CSR603) is an indication of the efficiency of irradiation. When using a strain which lacks only the *recA*-dependent repair pathways, doses of up to 40 Jm^{-2} may be used (32,34). If the contamination of the irradiated culture with growing cells persists, ampicillin (*Table 15*) may be added to a concentration of 30 μg/ml, provided the plasmid carried does not encode β-lactamase (the *bla* gene product). If the plasmid does carry this gene, D-cycloserine (100 μg/ml) may be used instead. Alternatively, a β-lactam antibiotic which is relatively resistant to the β-lactamase, such as cefuroxime (35) or a combination of ampicillin and clavulanic acid (36) may be used.

5. CORRELATION OF GENES AND GENE PRODUCTS

Once the proteins encoded by a plasmid or phage have been identified using one of the *in vivo* systems described above, various methods may be used to correlate particular cloned genes with specific protein bands.

(i)   Smaller fragments carrying the gene of interest may be subcloned from the original plasmid or phage and examined in an expression system. Purified fragments may also be used as templates in the *in vitro* expression systems described in Chapter 7.

(ii) A mutant allele may be cloned and compared with the wild-type allele. Expression of an amber mutant could be examined in a suppressing and a non-suppressing host, while a missense mutation may alter the charge of the protein, altering its mobility in two-dimensional gel electrophoresis.

(iii) A plasmid may be mutagenised with transposons. Insertion of a transposon into the gene of interest will cause the truncation or disappearance of the gene product, although it must be borne in mind that the transposon may have polar effects.

(iv) When working with a gene cloned into phage λ, an alternative to subcloning fragments is to isolate a family of deletion derivatives of the phage. This is easily accomplished by plating the phage on agar containing EDTA (37).

(v)   The most conclusive method is to compare the DNA sequence of the gene with the amino acid sequence of the labelled protein. Partial amino acid sequences may be obtained from the labelled protein excised from a polyacrylamide gel (38).

If, however, no gene product can be found for a gene which is carried by a plasmid or phage, there are a number of possible explanations, a few of which are summarised here.

(i)   The protein may lack methionine. In this case, [^{35}S]cysteine could be used and is commercially available, although it is not an amino acid which occurs frequently in proteins. Alternatively, [^{14}C]leucine or a high specific activity [^{3}H]amino acid mixture could be used.

(ii) The protein may be very unstable, being degraded before it can be detected by SDS-PAGE. To reduce this possibility, a protease inhibitor such as PMSF can be added before the cells are lysed, to minimise degradation. It may also help to label for only a short period of time (e.g., 5 min or less) so that detection of the gene product is not obscured by the continued synthesis of more stable proteins. Alternatively, the Zubay *in vitro* system described in Chapter 7 may be used, since peptides which are unstable *in vivo* appear to be more stable in this system.

(iii) The gene may have a weak promoter or may have been separated from its own promoter. If cloned in phage λ, it might be possible to use the powerful P_L promoter (see Sections 2.7 and 2.8). If cloned in a plasmid, a strong promoter may be introduced by standard recombinant DNA techniques.

(iv) The protein may aggregate under the electrophoresis conditions used, so that it fails to enter the separating gel. This can occur with very hydrophobic proteins. Solubilisation of the proteins in the SDS-sample buffer at 37°C rather than 100°C can sometimes prevent this (39,40).

6. ACKNOWLEDGEMENTS

Part of this work was supported by MRC Project Grant Award No. G82037 14CB.

7. REFERENCES

1. Ptashne,M. (1966) *Proc. Natl. Acad. Sci. USA,* **57**, 306.
2. Jaskunas,S.R., Lindahl,L. and Nomura,M. (1975) *Proc. Natl. Acad. Sci. USA,* **72**, 6.
3. Brammar,W.J. (1982) in *Genetic Engineering,* Vol. 3, Williamson,R. (ed.), Academic Press Inc., London, p. 53.
4. Koshland,D. and Botstein,D. (1980) *Cell,* **20**, 749.
5. Pène,J.J., Murr,P.C. and Barrow-Carraway,J. (1973) *J. Virol.,* **12**, 61.
6. Miller,J.H. (1972) *Experiments in Molecular Genetics,* published by Cold Spring Harbor Laboratory Press, New York.
7. Davis,R.W., Botstein,D. and Roth,J.R. (1980) *Advanced Bacterial Genetics,* published by Cold Spring Harbor Laboratory Press, New York.
8. Yamamoto,K.R., Alberts,B.M., Benzinger,R., Lawhorne,L. and Treiber,G. (1970) *J. Virol.,* **40**, 734.
9. Englesberg,E. and Wilcox,G. (1974) *Annu. Rev. Genet.,* **8**, 219.
10. Hames,B.D. (1981) in *Gel Electrophoresis of Proteins: A Practical Approach,* Hames,B.D. and Rickwood,D. (eds.), IRL Press Ltd., Oxford, p. 1.
11. Murialdo,H. and Siminovitch,L. (1971) in *The Bacteriophage Lambda,* Hershey,A.D. (ed.), Cold Spring Harbor Laboratory Press, New York, p. 711.
12. Laskey,R.A. and Mills,A.D. (1975) *Eur. J. Biochem.,* **56**, 335.
13. Levine,A., Bailone,A. and Devoret,R. (1979) *J. Mol. Biol.,* **131**, 655.
14. Hedgpeth,J., Ballivet,M. and Eisen,H. (1978) *Mol. Gen. Genet.,* **163**, 197.
15. Lutkenhaus,J.F. and Wu,H.C. (1980) *J. Bacteriol.,* **143**, 1281.
16. Brammar,W.J. (1977) *Biochem. Soc. Trans.,* **5**, 1633.
17. Ward,D.F. and Murray,N.E. (1979) *J. Mol. Biol.,* **133**, 249.
18. Murray,N.E. and Kelley,W.S. (1979) *Mol. Gen. Genet.,* **175**, 77.
19. Wilson,G.G. and Murray,N.E. (1979) *J. Mol. Biol.,* **132**, 471.
20. Adler,H.I., Fisher,W.D., Cohen,A. and Hardigree,A.A. (1967) *Proc. Natl. Acad. Sci. USA,* **57**, 321.
21. Frazer,A.C. and Curtiss,R. (1975) *Curr. Top. Microbiol. Immunol.,* **69**, 1.
22. Dougan,G. and Sherratt,D. (1977) *Mol. Gen. Genet.,* **151**, 151.
23. Meagher,R.B., Tait,R.C., Betlach,M. and Boyer,H.W. (1974) *Cell,* **10**, 521.
24. Holmes,D.S. and Quigley,M. (1981) *Anal. Biochem.,* **114**, 193.
25. Birnboim,H.C. and Doly,J. (1979) *Nucleic Acids Res.,* **7**, 1513.

26. Dagert,M. and Ehrlich,S.D. (1979) *Gene, 6*, 23.
27. Morrison,D.A. (1979) in *Methods in Enzymology,* Vol. **68**, Wu,R. (ed.), Academic Press Inc., London and New York, p. 326.
28. Levy,S.B. (1975) *Proc. Natl. Acad. Sci. USA,* **72**, 2900.
29. Reeve,J. (1979) in *Methods in Enzymology,* Vol. **68**, Wu,R. (ed.), Academic Press Inc., London and New York, p. 493.
30. Clement,J.M., Perrin,D. and Hedgpeth,J. (1982) *Mol. Gen. Genet.,* **185**, 302.
31. Hanawalt,P.C., Cooper,P.K., Ganesan,A.K. and Smith,C.A. (1979) *Annu. Rev. Biochem.,* **48**, 783.
32. Sancar,A., Wharton,R.P., Seltzer,S., Kacinski,B.M., Clarke,N.D. and Rupp,W.D. (1981) *J. Mol. Biol.,* **148**, 45.
33. Sancar,A., Hack,A.M. and Rupp,W.D. (1979) *J. Bacteriol.,* **173**, 692.
34. Isberg,R.R., Lazaar,A.L. and Syvanen,M. (1982) *Cell,* **30**, 883.
35. Neu,H.C. and Fu,K.P. (1978) *Antimicrob. Agents Chemother.,* **13**, 657.
36. Reading,C. and Cole,M. (1977) *Antimicrob. Agents Chemother.,* **11**, 852.
37. Stoker,N.G., Broome-Smith,J.K., Edelman,A. and Spratt,B.G. (1983) *J. Bacteriol.,* **155**, 847.
38. Hedgpeth,J., Clement,J.M., Marchal,C., Perrin,D. and Hofnung,M. (1980) *Proc. Natl. Acad. Sci. USA,* **77**, 2621.
39. Teather,R.M., Müller-Hill,B., Abrutsch,U., Aichele,G. and Overath,P. (1978) *Mol. Gen. Genet.,* **159**, 239.
40. Stoker,N.G., Pratt,J.M. and Spratt,B.G. (1983) *J. Bacteriol.,* **155**, 854.

CHAPTER 7

Coupled Transcription-Translation in Prokaryotic Cell-Free Systems

JULIE M. PRATT

1. INTRODUCTION

Although in eukaryotes the two processes of transcription and translation are often studied separately, in prokaryotes translation is rarely studied in the absence of coupled transcription. The lack of success with bacterial mRNA-primed translation systems arises largely for two reasons. First, it is difficult to prepare intact mRNA from *Escherichia coli*, the bacterial species usually studied, due to high levels of endogenous nuclease which cause extensive degradation of mRNA during its isolation. Secondly, the translation of bacterial mRNAs usually begins while transcription is still occurring (i.e., transcription and translation are closely coupled), whereas in eukaryotes the mRNA is not normally translated until synthesis is complete. Completed bacterial mRNA molecules may be unable to bind to ribosomes because of secondary structure near the initiation site. DNA-directed, cell-free protein synthesis in a coupled transcription-translation system overcomes both problems. Two such systems have found the widest application; the pre-incubated *E. coli* S30 extract system devised by Zubay (1,2) and the fractionated system of Gold and Schweiger (3). Both systems are described in this chapter.

2. PREPARATION OF THE CELL-FREE SYSTEM OF ZUBAY

2.1 Introduction

The cell-free system devised by Zubay (1,2) involves the preparation of a crude extract from *E. coli* which contains all the enzymes and factors necessary for transcription and translation, although the extract must be supplemented with amino acids, an energy regenerating system and certain cofactors. The original method has since received minor modifications (4,5) and is becoming increasingly more widely used. The following sections describe in detail the modified method which, if followed precisely, will give active extracts every time.

2.2 Apparatus and Reagents

A total of 16 litres of water is required for the preparation of solutions. Treat the water with diethylpyrocarbonate (DEPC) to inactivate RNase activity by adding 1 ml of DEPC per litre of distilled water slowly, with stirring, in a fume cupboard. Continue stirring for 1 h, then autoclave the water and store it at 4°C.

Table 1. Apparatus to be DEPC-treated and Autoclaved.

---

4 x 5 litre flasks with cotton-wool bungs
6 x 1 litre flasks with screw top
16 x 50 ml polypropylene centrifuge tubes with caps
6 x 25 ml beakers
4 x 100 ml measuring cylinders
4 x 50 ml measuring cylinders
6 – 10 McCartney bottles
3 – 4 stirring bars (large)
20 – 30 Pasteur pipettes
3 x 100 ml Büchner flasks
1 rubber bung for Büchner flask
3 spatulas (various sizes)
1 homogenising pestle.

Table 2. Reagents required for the Zubay System.

Reagent	Commercial source[a]	Maximum requirement
Ammonium acetate (Analar)	BDH Chemical Co.	5 g
Glucose (Analar)	BDH Chemical Co.	500 g
KH_2PO_4 (Analar)	BDH Chemical Co.	200 g
K_2HPO_4 (Analar)	BDH Chemical Co.	1 kg
Calcium acetate	Fisons Scientific Apparatus	5 g
Glacial acetic acid	Fisons Scientific Apparatus	100 ml
Magnesium acetate.$4H_2O$ (Analar)	Fisons Scientific Apparatus	500 g
Polyethylene glycol-6000	Fisons Scientific Apparatus	5 g
Potassium acetate (SLR)	Fisons Scientific Apparatus	5 g
Amino acids (all 20 commonly occurring; see *Table 4*)	Sigma Chemical Co.	1 g (total wt.)
ATP (Na_2) (from equine muscle; vanadium free)	Sigma Chemical Co.	2 g
3′,5′-Cyclic AMP (Na)	Sigma Chemical Co.	100 mg
CTP (Na)	Sigma Chemical Co.	100 mg
Diethyl pyrocarbonate	Sigma Chemical Co.	25 ml
DL-DTT	Sigma Chemical Co.	5 g
Folinic acid (calcium leucovorin)	Sigma Chemical Co.	100 mg
GTP (Na)	Sigma Chemical Co.	100 mg
2-Mercaptoethanol	Sigma Chemical Co.	20 ml
Phosphoenol pyruvate (K)	Sigma Chemical Co.	2 g
Pyruvate kinase (2000 U/ml) (from rabbit muscle)	Sigma Chemical Co.	0.5 ml
Thiamine-HCl	Sigma Chemical Co.	100 mg
Tris	Sigma Chemical Co.	500 g
Transfer RNA (*E. coli* deacylated tRNA)	Sigma Chemical Co.	100 mg
UTP (Na) from yeast	Sigma Chemical Co.	100 mg
[^{35}S]Methionine (1000 Ci/mmol; ~10 μCi/μl)	Amersham International	1 mCi
Yeast extract	Oxoid	100 g

[a]Other commercial sources of these reagents may be equally satisfactory.

The glassware and other apparatus required for a standard preparation of the Zubay system are listed in *Table 1*. Treat all glassware, centrifuge tubes and plastic tubing with DEPC by filling (or soaking) the apparatus for 1 h in

Table 3. Stock Solutions for the Zubay System.

Stock solution	Volume (ml)	Autoclave?	Storage temperature
Amino acid mixture A (see *Table 4*)	5	No	$-20°C$
Amino acid mixture B (see *Table 4*)	5	No	$-20°C$
38 mM ATP, pH 7.0	20	No	$-20°C$
88 mM CTP⎫ 88 mM GTP⎬ pH 7.0 88 mM UTP⎭	1	No	$-20°C$
50 mM 3′,5′-Cyclic AMP, pH 7.0	1	No	$-20°C$
0.10 M DTT	100	No	$-20°C$
0.55 M DTT	5	No	$-20°C$
Folinic acid (2.7 mg/ml)	1	No	$-20°C$
0.1 M Magnesium acetate	10	Yes	$4°C$
1.4 M Magnesium acetate	500	Yes	$4°C$
3.0 M Magnesium acetate	100	Yes	$4°C$
Methionine (8 mg/ml)	1	No	$-20°C$
0.42 M Phosphoenol pyruvate (PEP), pH 7.0	10	No	$-20°C$
40% Polyethylene glycol-6000	10	Yes	$-20°C$
6.0 M Potassium acetate	500	Yes	$4°C$
Salts stock solution: 1.4 M ammonium acetate⎫ 2.8 M potassium acetate⎬ 0.38 M calcium acetate⎭	10	Yes	$-20°C$
Transfer RNA (17.4 mg/ml)	1	No	$-20°C$
TA buffer (10 mM Tris-acetate, pH 7.0)	500	Yes	$4°C$
1.0 M Tris-acetate, pH 8.2	500	Yes	$4°C$
2.2 M Tris-acetate, pH 8.2	100	Yes	$4°C$
TE buffer (10 mM Tris-HCl, 1 mM EDTA, pH 7.5)	500	Yes	$4°C$

water to which DEPC (1 ml/litre of water) has been added just before use. Drain the apparatus and autoclave immediately. Also cut $8-12$ pieces of dialysis tubing (2.5 cm diameter, $30-40$ cm long) and boil these for 20 min in 0.1 M $NaHCO_3$, 10 mM EDTA. Discard the solution and repeat the procedure once or twice more until the solution is no longer brown. Rinse the dialysis tubing extensively in DEPC-treated water and store it in sterile DEPC-treated water at $4°C$. The tubing must not be stored longer than $1-2$ days.

A standard preparation of the Zubay system requires the reagents described in *Table 2*. The commercial sources listed are those routinely used by the author but other suppliers may be equally satisfactory. Prepare the stock solutions given in *Tables 3* and *4* using DEPC-treated water. Adjust the pH of alkaline solutions with acetic acid (glacial or a 10-fold dilution of glacial acetic acid in DEPC-treated water) and the pH of acidic solutions with 2 M Tris, having autoclaved the diluted acetic acid and 2 M Tris before use. Place the stock solutions into DEPC-treated glass bottles and then autoclave and store as indicated *(Table 3)*. Finally, using these stock solutions, prepare the low molecular weight mixture (LM mixture) as described in *Table 5*.

Table 4. Preparation of the Amino Acid Mixtures.

Amino acid	Weight (mg)	
	Mixture A	*Mixture B*
Alanine	22	25
Arginine HCl	53	58
Asparagine	33	36
Aspartic acid	33	37
Cysteine	30	33
Glutamic acid	37	41
Glutamine	37	40
Glycine	19	21
Histidine HCl	48	53
Isoleucine	33	36
Leucine	33	36
Lysine	46	50
Methionine	37	–
Phenylalanine	31	34
Proline	29	32
Serine	26	29
Threonine	30	33
Tryptophan	51	56
Tyrosine	45	50
Valine	29	32

For each amino acid mixture, resuspend the amino acids listed in 5 ml final volume of water. The final total concentration of amino acids in mixture A is 50 mM whilst in mixture B it is 55 mM.

Table 5. Preparation of the Low Molecular Weight Mixture (LM Mixture).

Stock solution	Volume (μl)	Final concentration in reaction mixture
2.2 M Tris-acetate, pH 8.2	40	56.40 mM
0.55 M DTT	5	1.76 mM
38 mM ATP, pH 7.0	50	1.22 mM
88 mM CTP ⎫ 88 mM GTP ⎬ pH 7.0 88 mM UTP ⎭	15	0.85 mM
0.42 M Phosphoenol pyruvate, pH 7.0	100	27.0 mM
Amino acid mixture B (i.e. *minus* methionine)	10	0.35 mM
40% Polyethylene glycol-6000	75	1.9%
Folinic acid (2.7 mg/ml)	20	34.6 μg/ml
50 mM 3′,5′-Cyclic AMP, pH 7.0	20	0.64 mM
Transfer RNA (17.4 mg/ml)	15	0.17 mg/ml
Salts stock solution:		
1.4 M ammonium acetate ⎫		36.0 mM
2.8 M potassium acetate ⎬	40	72.0 mM
0.38 M calcium acetate ⎭		9.7 mM

Mix the components in the order given, then divide the solution into aliquots of 100 μl volume and store these at −20°C. The LM mixture remains active for at least 2−3 months at this temperature.

Table 6. Preparation of Incomplete Rich Medium.

KH$_2$PO$_4$	56 g
K$_2$HPO$_4$	289 g
Yeast extract	10 g
Thiamine	15 mg
Each required amino acid[a]	0.5 g
Distilled water to 10 litres.	

Autoclave the medium for 45 min at 15 p.s.i.

[a]For *E. coli* MRE600, there are no required amino acids.

2.3 Preparation of Cells

E. coli strain MRE600 is favoured as the source of S30 extract for routine iden-tification of gene products using plasmids or phage λ as the DNA template since it lacks the major RNase activity of *E. coli* and is simple to culture. For the analysis of linear DNA, a *recB*[ts] strain is advisable (see Section 5.1.1). For more specific studies, for example investigation of the control of gene expres-sion (Section 4.4), an extract may be prepared from any mutant strain.

(i)   Prepare 10 litres of incomplete rich medium as described in *Table 6*.

(ii)   Add 800 ml of 25% (w/v) glucose and 100 ml of 0.1 M magnesium acetate, each of which has been autoclaved.

(iii)   Pre-incubate the medium to the required temperature (usually 37°C but 30°C for temperature-sensitive strains; see Section 5.1.1).

(iv)   Inoculate the medium with an overnight culture of the required strain to give an A_{450} of 0.07.

(v)   Aerate the culture vigorously (sparge or use a fermenter with air flow) during incubation at the appropriate temperature until an A_{450} of 1.5 − 2.0 is reached (usually overnight).

(vi)   On the day required, prepare the buffer needed to wash the cells (S30 buf-fer) by mixing 10 ml each of the following stock solutions (*Table 3*); 1.4 M magnesium acetate, 6.0 M potassium acetate, 0.1 M DTT, 1.0 M Tris-acetate (pH 8.2). Then add DEPC-treated water to 1 litre final volume. Store the buffer at 4°C.

(vii) Harvest the cells by centrifugation at 10 000 *g* for 15 min at 4°C.

(viii) Wash the cells with a total of 500 ml of ice-cold S30 buffer containing 0.5 ml of 2-mercaptoethanol per litre. Repeat this wash twice more (a total of three washes) and then estimate the wet weight of the cells.

(ix)   Store the cells as pellets at − 70°C for a maximum of 1 − 2 days.

2.4 Preparation of the S30 Extract

(i)   Preferably the day before use, clean the pressure cell of a French press thoroughly. Rinse the components of the pressure cell in DEPC-treated water and fix a piece of flexible tubing, about 60 cm long, to the side arm. Assemble the cell and flush it with DEPC-treated water. Cover the cell with aluminium foil, being careful to cover the tubing also. Store the apparatus at 4°C.

(ii) On the day chosen for preparation of the S30 extract, prepare fresh S30 buffer as described in step (vi) of Section 2.3.

(iii) Remove the cell pellets [step (ix), Section 2.3] from the −70°C freezer and allow them to thaw at 4°C for 30−60 min.

(iv) Resuspend the cells slowly in S30 buffer containing 2-mercaptoethanol (0.05 ml per litre) using 100 ml of buffer per 10 g of cells.

(v) Centrifuge the suspension at 16 000 *g* for 30 min at 4°C. Weigh the cell pellet.

(vi) Resuspend the cells carefully as follows. Spoon the cells into a Büchner flask and add 63.5 ml of S30 buffer per 50 g of cells. Connect the flask to a vacuum pump using rubber tubing and place a rubber bung in the neck of the flask. Evacuate the flask. Remove the bung and resuspend the cells by mixing with a Teflon homogeniser pestle. Evacuate the flask at frequent intervals so that the cells are resuspended fairly anaerobically. It is important that all cell clumps are dispersed. Usually this can be achieved using the pestle simply to mix the cell suspension and to squash any clumps. However, occasionally one experiences great difficulty in resuspending certain cells, for example *recB*ts. In these cases, it may be necessary to use very gentle homogenisation with a loose-fitting tube. However, homogenisation should be avoided if at all possible.

(vii) Measure the volume of the resuspended cells and place them on ice. Pipette the cells into the pressure cell of the French press using a 10 ml pipette to prevent the introduction of large lumps of cells. Avoid introducing air bubbles.

(viii) Press the cells at 8400 p.s.i. and collect the lysate on ice. Add 100 μl of 0.1 M DTT per 10 ml of lysate collected. The preparation may still be viscous and turbid at this stage. Any whole cells remaining are removed in the subsequent spin; it is not advisable to attempt to improve the degree of breakage by passage through the French press a second time since this leads to loss of activity.

(ix) Centrifuge the preparation *immediately* at 30 000 *g* for 30 min at 4°C. Meanwhile, equilibrate a rotating shaker water bath or orbital incubator to 37°C. Also thaw the solutions necessary to prepare the pre-incubation mixture *(Table 7)*.

(x) At the end of the centrifugation [step (ix)], transfer the upper four-fifths of the supernatant to a fresh tube. Estimate the volume of the supernatant and then centrifuge it again at 30 000 *g* for 30 min at 4°C. Meanwhile, prepare 7.5 ml of pre-incubation mixture *(Table 7)* per 25 ml of supernatant.

(xi) At the end of the centrifugation [step (x)], transfer the upper four-fifths of the supernatant to a small conical flask. Measure the volume and add 7.5 ml of pre-incubation mixture per 25 ml of supernatant. Cover the flask with aluminium foil and incubate with gentle swirling in a water bath (or in an orbital incubator) at 37°C for 80 min (*or* 160 min at 30°C

Table 7. Preparation of the Pre-incubation Mixture[a].

Mix the following stock solutions[b]

Pyruvate kinase	25 μl
2.2 M Tris-acetate, pH 8.2	1.0 ml
3.0 M Magnesium acetate	23 μl
38 mM ATP, pH 7.0	2.6 ml
0.42 M Phosphoenol pyruvate, pH 7.0	1.5 ml
0.55 M DTT	60 μl
Amino acid mixture A	6 μl
Distilled water (DEPC-treated) to 7.5 ml	

[a]This preincubation mixture must be prepared fresh just before use.
[b]The compositions of stock solutions are given in *Tables 2 – 4*.

for temperature-sensitive strains; see Section 5.1.1).

(xii) Pour the mixture into the prepared dialysis tubing (Section 2.2) and dialyse it against 50 volumes of S30 buffer at 4°C. Change the S30 buffer three times allowing 45 min between each change

(xiii) Centrifuge the extract at 4000 *g* for 10 min at 4°C. Subdivide the supernatant into 0.5 – 2.0 ml aliquots and store these in liquid nitrogen. If the liquid nitrogen container is properly maintained, these extracts will give reproducible activities for several years.

2.5 Optimisation of the System

2.5.1 Preparation of Plasmid DNA

Approximately 50 μg of plasmid DNA is needed for optimisation of the system. The author routinely uses the multicopy plasmid pBR325 to check the activity of newly-prepared extracts. Plasmid pBR325 has the advantage of having two easily-identifiable gene products, β-lactamase and chloramphenicol acetyltransferase (CAT). However, any multicopy plasmid would be suitable, for example, pBR322, pAT153, pACYC184. A reliable procedure for the preparation of plasmid DNA is described in *Table 8*.

Based on the author's experience using plasmid DNA prepared by a number of different methods, it is essential that the plasmid DNA is not contaminated with RNase, caesium chloride, ethanol, agarose or large amounts of salt. These points are perhaps obvious but, in the author's experience, they are the usual causes of the inability of DNAs to act as templates in *in vitro* incubations. Usually, such problems can be overcome by repeated phenol extractions, ethanol precipitation and dialysis.

2.5.2 Optimisation of the Magnesium Ion and S30 Extract Concentrations

In the author's experience, the optimum magnesium concentration for protein synthesis in the Zubay system is in the range 10 – 15 mM. The exact value depends on the particular extract preparation and is determined by varying the magnesium ion concentration and assaying aliquots of the incubation mixture for TCA-precipitable radioactivity. It may be necessary to optimise the S30

Table 8. Preparation of Plasmid DNA.

---

1. Grow a 400 ml culture of the bacterial host in nutrient broth (2.5% nutrient broth powder in water), containing the relevant antibiotics, to late log phase (e.g., A_{450} of 0.8) at 37°C.

2. Amplify the plasmid by adding chloramphenicol (to 170 μg/ml) or spectinomycin (to 300 μg/ml) and continuing the incubation at 37°C overnight[a].

3. Chill the cells at 4°C for 10 min. Then centrifuge at 4000 g for 10 min at 4°C (e.g. Sorvall GS3 rotor at 5000 r.p.m.).

4. Wash the cells by centrifugation with 40 ml of bacterial buffer (20 mM KH_2PO_4, 50 mM Na_2HPO_4, 70 mM NaCl, 40 μM $MgSO_4$, pH 7.0).

5. Resuspend the cells in 3 ml of 25% sucrose, 50 mM Tris-HCl (pH 8.0) and then add 0.5 ml of a mixture of lysozyme (10 mg/ml) and RNase (300 μg/ml).

6. Incubate at room temperature for 5 min. Then add 4 ml of 2% (v/v) Triton X-100, 50 mM Tris-HCl (pH 8.0) and invert the tube containing the cells several times until lysis is complete.

7. Centrifuge at 39 000 g for 20 min at 4°C (e.g., Sorvall SS34 rotor at 18 000 r.p.m.).

8. Decant the supernatant into a polycarbonate tube and add two-thirds this volume of 1.25 M NaCl, 25% (w/v) polyethylene glycol (mol. wt. 6000). Leave on ice for at least 1 h.

9. Centrifuge at 4000 g for 10 min at 4°C (e.g., Sorvall HB4 rotor at 5000 r.p.m.).

10. Resuspend the pellet in exactly 1.1 ml of 5 mM EDTA, 50 mM NaCl, 50 mM Tris-HCl (pH 8.0) and transfer it to a tube for the Beckman VTi65 vertical rotor[b]. Underlay the sample with 4 ml of CsCl-ethidium bromide solution[c] and seal the tube. Centrifuge at 300 000 g for 3 h or 230 000 g overnight at 15°C (VTi65 rotor at 55 000 r.p.m. or 50 000 r.p.m., respectively).

11. After centrifugation, visualise the DNA band profile using u.v. illumination. Collect the plasmid band through the tube wall using a syringe. Remove the ethidium bromide by extraction with CsCl-saturated propan-2-ol.

12. Dialyse the plasmid DNA preparation against three changes of TE buffer (10 mM Tris-HCl, pH 7.5, 1 mM EDTA).

13. Transfer to a siliconised Corex tube. Add 3 M sodium acetate, pH 5.6, to 0.3 M and then 2.5 volumes of ethanol. Leave at -20°C for $3-4$ h.

14. Recover the plasmid DNA by centrifugation at 16 000 g for 10 min at -10°C (e.g., Sorvall HB4 rotor at 10 000 r.p.m.).

15. Dry the DNA pellet under vacuum. Resuspend the pellet in 1 ml of TE buffer and extract the solution twice with an equal volume of phenol mixture[d].

16. Extract the sample twice with diethyl ether to remove contaminating phenol and then precipitate the plasmid DNA from the aqueous phase as described in steps 13 and 14.

17. Resuspend the plasmid DNA in TE buffer to $200-500$ μg/ml.

---

[a]If the plasmid cannot be amplified, prepare the plasmid from a 400 ml culture of host cells grown at 37°C overnight.
[b]Provided the relative proportions of DNA sample to CsCl-ethidium bromide mixture are maintained, the volumes may be scaled up to suit any available tube size. For example, the centrifugation may be carried out at 82 000 g for 40 h using a Beckman 50Ti rotor (at 30 000 r.p.m.).
[c]This is prepared by mixing 80 g CsCl, 8 ml of 5 mg/ml ethidium bromide and 52 ml of 5 mM EDTA, 50 mM NaCl, 50 mM Tris-HCl (pH 8.0). Adjust the refractive index to $1.3990-1.4000$ by addition of more CsCl or dilution as necessary.
[d]This is prepared by mixing 100 g phenol, 4 ml isoamyl alcohol, 0.1 g 8-hydroxyquinoline and 100 ml chloroform. This phenol mixture is stored under 10 mM Tris-HCl, pH 7.5 at 4°C.

extract concentration also. A typical optimisation experiment for both magnesium ion and S30 concentrations is shown in *Table 9*. Tubes $1-6$ are required for determination of the optimum magnesium concentration (tube 6 is a

Table 9. Optimisation of Magnesium Ion Concentration[a].

				Volume (μl)			
Tube	*DNA*[b]	*[35S]methionine*	*LM mixture*	*0.1M Magnesium acetate*	*TA buffer*	*TE buffer*	*S30 extract*[c]
1	5.0	2.0	7.5	2.0	8.5	–	5.0
2	5.0	2.0	7.5	2.5	8.0	–	5.0
3	5.0	2.0	7.5	3.0	7.5	–	5.0
4	5.0	2.0	7.5	3.5	7.0	–	5.0
5	5.0	2.0	7.5	4.0	6.5	–	5.0
6	–	2.0	7.5	3.5	2.0	5.0	5.0
7	5.0	2.0	7.5	2.5	5.0	–	8.0
8	5.0	2.0	7.5	3.5	4.0	–	8.0
9	–	2.0	7.5	3.5	–	4.0	8.0

[a]The composition of reagents other than the DNA and S30 extract is given in *Tables 2 – 5*.
[b]Plasmid DNA (e.g., pBR325) at 300 – 500 μg/ml in TE buffer.
[c]See Section 2.4 for preparation.

control without DNA) whilst tubes 7 – 9 test the effect of increasing S30 extract concentration (tube 9 is the corresponding control without DNA). The procedure is as follows:

(i)   Add the specified volumes *(Table 9)* of [35S]methionine, LM mixture, TA buffer, TE buffer, magnesium acetate and DNA to numbered, sterile microcentrifuge tubes on ice.

(ii) Centrifuge the tubes for 15 sec in a microcentrifuge to ensure that all the components collect at the bottom of the tubes.

(iii) Pre-incubate the tubes for 2 – 4 min in a shaking water bath set at 37°C.

(iv) Take one microcentrifuge tube at a time and add the S30 extract directly into the contents, mixing each gently with a separate micropipette tip. Replace the tube in the water bath.

(v)   When the S30 extract has been added to all the tubes, continue the incubation at 37°C for 30 – 60 min, shaking the tubes as vigorously as possible to keep the contents well mixed. This is important for good incorporation (1,6).

(vi) Add 10 μl of pre-warmed (37°C) unlabelled L-methionine (8 mg/ml in DEPC-treated water) to each tube, mix and incubate for a further 5 min to complete the synthesis of labelled polypeptides.

(vii) Transfer the tubes to ice. After 10 min, mix the contents of each tube by vortexing briefly. Then remove an aliquot (2 – 5 μl) to a numbered Whatman 3MM filter paper disc for estimation of the incorporation of [35S]-methionine into protein. This procedure is described in *Table 10*.
      The results of a typical experiment are shown in *Figure 1*, in which the addition of 3.5 μl of 0.1 M magnesium acetate per incubation gave optimum incorporation of [35S]methionine into protein, that is, a magnesium optimum of 11.7 mM.

(viii)To the remainder of each incubation mixture, add 30 μl of TA buffer and 30 μl of 20% glycerol, 2% SDS, 5% 2-mercaptoethanol, 0.001% (w/v) Bromophenol blue, 0.125 M Tris-HCl (pH 6.8).

Table 10. Estimation of Radioactivity Incorporated into Protein.

The following procedure allows many samples to be processed together under identical conditions and gives very reproducible results.

1. Place 2−5 μl of each radioactive sample onto a separate numbered (use pencil) Whatman 3MM filter paper disc. Allow to dry at room temperature.
2. Drop the filters, individually, into 200 ml of ice-cold 10% TCA, 0.1% methionine. Leave on ice (swirl occasionally) for 1 h.
3. Transfer the filters to 200 ml of 5% TCA, 0.1% methionine. Incubate at 90°C for 10 min.
4. Pour off the TCA and discard this. Add 200 ml of fresh, ice-cold 5% TCA, 0.1% methionine.
5. Change the TCA-methionine mixture every 15 min (3−4 changes).
6. Pour off the TCA-methionine and add just enough acetone to cover the filters. Swirl and then pour off the acetone and discard it.
7. Add fresh acetone and leave at room temperature for 10 min.
8. Remove the filters individually using forceps and allow them to dry at room temperature.
9. Place each filter into a separate scintillation vial, add scintillation fluid and count to determine the radioactivity present.

(ix) Boil the samples for 3 min and analyse the labelled polypeptide content of each by SDS-PAGE, loading a maximum of 50 μl of each sample per well.

Figure 2 (lanes 1−6) shows an SDS-PAGE analysis of the corresponding samples from the magnesium optimisation experiment of *Figure 1*, confirming that 3.5 μl of 0.1 M magnesium acetate per incubation gave the maximum [^{35}S]methionine incorporation into polypeptide products in this experiment. However, it should be noted that the magnesium optimum values deduced by TCA precipitation and by SDS-PAGE do not always agree. Thus the highest value for incorporation based on TCA precipitation may be reached at the expense of fidelity of transcription since high magnesium ion concentrations facilitate initiation of transcription at sites not normally used, in addition to authentic sites. The resulting transcripts may be translated to give aberrant polypeptides and hence increased incorporation, as judged by TCA precipitation, but an anomalous SDS-PAGE profile. High magnesium ion concentrations also reduce the fidelity of translation by causing initiation at sites other than authentic sites, again yielding high values of incorporation but abnormal gel profiles. An example of disagreement between TCA precipitation data and SDS-PAGE is shown in *Figures 1* and *2*. Although the incorporation based on TCA precipitation *(Figure 1)* is the same for the samples analysed in lanes 1 and 2 of *Figure 2*, the radioactivity in the former sample must be in low molecular weight peptides too short to be detected by the SDS-PAGE analysis but large enough to be precipitated by TCA. In all cases of disagreement, SDS-PAGE is the criterion by which the magnesium ion concentration should be decided.

188

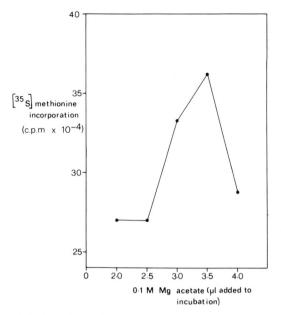

Figure 1. TCA precipitation of samples 1 − 5 from the magnesium optimisation experiment described in *Table 9*. The DNA template used was pBR325 at a concentration of 300 μg/ml.

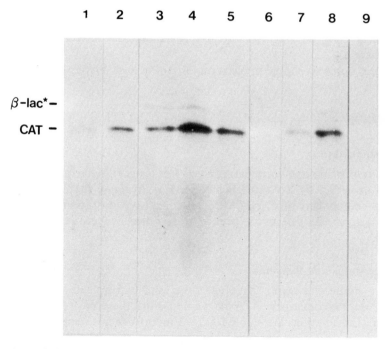

Figure 2. Analysis of magnesium optimisation incubations of *Figure 1* by SDS-PAGE (15% slab gel) and autoradiography. **Lanes 1 − 9** correspond to incubations 1 − 9 described in *Table 9*. The polypeptide products of the DNA template used (pBR325) are the β-lactamase precursor (βlac*) and chloramphenicol acetyltransferase (CAT).

The background [^{35}S]methionine incorporation of the S30 extract, determined by TCA precipitation of a control incubation lacking plasmid DNA (*Table 9, tube 6*), will vary with the magnesium ion concentration. However, at saturating DNA concentrations, a 40- to 60-fold stimulation of incorporation over the control should be seen. An extract which exhibits greater than a 10-fold stimulation usually gives perfectly acceptable results. Clearly, the control incubation should not contain plasmid-encoded polypeptide products when examined by SDS-PAGE (*Figure 2, lane 6*).

Figure 2, lanes 7−9, show SDS-PAGE analyses of incubation mixtures 7−9, respectively, of *Table 9*. When compared with *Figure 2,* lanes 2 and 4, these data indicate that 5 μl of S30 is sufficient extract to achieve maximum synthesis of polypeptides from the available template. The author often finds that the addition of more S30 extract than this causes a decrease in incorporation.

2.5.3 *Time Dependence of Incorporation*

Having optimised the magnesium ion and S30 concentrations for the system, the time dependence of incorporation should be followed.

(i)   Set up an incubation mixture of twice the usual volume (i.e., 60 μl) and remove 2 μl to numbered Whatman 3MM filter paper discs after various times of incubation.

(ii) Allow the filters to dry, TCA precipitate the labelled polypeptide products, wash the filters and determine the radioactivity all as described in *Table 10*.

Protein synthesis should proceed rapidly for at least 40 min. A typical result is shown in *Figure 3*.

3. PREPARATION OF THE CELL-FREE SYSTEM OF GOLD AND SCHWEIGER

3.1 **Introduction**

An alternative method of preparing a cell-free transcription-translation system is that described by Gold and Schweiger (3), involving the fractionation of various components of *E. coli* with subsequent reconstitution to produce an active system. This system has been as widely used as the Zubay system (Section 2).

3.2 **Apparatus and Reagents**

Prepare sterile glassware, centrifuge tubes and dialysis tubing. Prepare the stock solutions described in *Table 11* and store these at −20°C where indicated. The solutions are stable for 2−3 months at this temperature.

3.3 **Preparation of Cells**

The choice of *E. coli* strain is dictated by the same considerations as apply to the preparation of the S30 extract for the Zubay system and hence the reader is

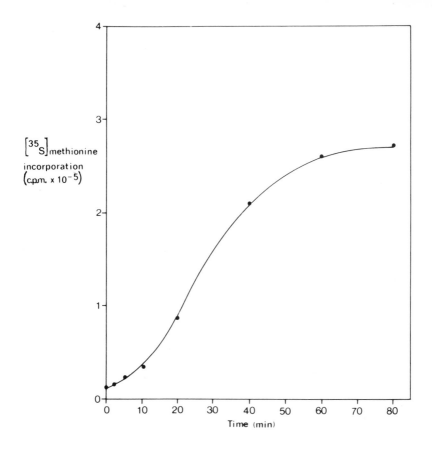

Figure 3. Time course of incorporation of [^{35}S]methionine into polypeptides synthesised *in vitro*. Plasmid pBR325 DNA (2 μg) was used as a template with an S30 extract prepared from *E. coli* MRE600.

referred to Section 2.3. For the analysis of linear DNA a *recB* strain may be used (see Section 5.1.2).

(i)   Prepare 10 litres of rich medium as described in *Table 12*.

(ii)   Pre-incubate the medium to the required temperature and inoculate with an overnight culture of the relevant strain of *E. coli* to give an A_{450} of 0.07.

(iii) Incubate the cells at 37°C with vigorous aeration (sparge), until an A_{450} of 0.6−0.7 is reached (early logarithmic growth; ~2 x 10^8 cells/ml).

(iv)   Harvest the cells by centrifugation at 10 000 *g* for 15 min at 4°C.

(v)   Wash the cells with a total of 250 ml of buffer A (*Table 11*) prepared beforehand and stored at 4°C. Repeat this wash twice more and then estimate the wet weight of the cells.

(vi)   Store the cells as pellets in liquid nitrogen until required.

Table 11. Stock Solutions for the Gold and Schweiger System[a].

Stock solution	Volume (ml)	Autoclave?	Storage temperature
Buffer A:			
10 mM Tris-HCl, pH 7.5 ⎫			
10 mM magnesium acetate ⎬ 2 litres	Yes	4°C	
22 mM ammonium acetate ⎪			
1 mM DTT ⎭			
Amino acids (each at 5 mM)[b]	5 ml	No	−20°C
0.2 M ATP[c]	2 ml	No	−20°C
0.2 M CTP[c]	0.5 ml	No	−20°C
0.2 M GTP[c]	0.5 ml	No	−20°C
0.2 M UTP[c]	0.5 ml	No	−20°C
1.0 M DTT	2 ml	No	−20°C
6 mM Folinic acid	0.5 ml	No	−20°C
0.12 M Magnesium acetate	10 ml	Yes	−20°C
1.0 M Phosphoenol pyruvate[c]	5 ml	No	−20°C
2.0 M Potassium acetate	10 ml	Yes	−20°C
Transfer RNA (*E. coli* deacylated tRNA, 10 mg/ml)	1 ml	No	−20°C
1.0 M Tris-acetate, pH 8.0	20 ml	Yes	−20°C
Radioactive amino acid[d]	−	−	−

[a]The sources of reagents given in *Table 2* may be used for this method.
[b]Each of the 20 common amino acids are included except that corresponding to the radioactive amino acid used.
[c]The pH of these solutions is not adjusted prior to use (see *Table 13*).
[d]Usually [^{3}H]leucine (5 μCi/μl; 150 Ci/mmol) or [^{35}S]methionine (10 μCi/μl; 1000 Ci/mmol).

Table 12. Preparation of Rich Medium.

Prepare and autoclave solutions I, II and III separately:

Solution I

Bacteriological peptone	100 g
Glycerol	238 ml
CaCl$_2$	0.33 g
Gelatin	0.10 g
NH$_4$Cl	10.0 g
MgSO$_4$.4H$_2$O	3.0 g
Distilled water to 500 ml	

Solution II

Glucose (50 g) dissolved in 500 ml of distilled water

Solution III

K$_2$HPO$_4$	90.0 g
KH$_2$PO$_4$	34.5 g
Distilled water to 300 ml	

To prepare the rich medium, mix the autoclaved solutions I, II and III and bring the volume to 10 litres with sterile distilled water.

3.4 Preparation of Cell-free Extract

Unless stated otherwise, all manipulations are carried out at 4°C.

(i)   Preferably the day before, wash a Sorvall Omnimix and sufficient glass beads (40 g per 10 g of cells) with sterile water and store these at 4°C. The glass beads used by Gold and Schweiger were Superbrite 100 (Minnesota Mining Co.). Also prepare a column (1 cm x 10 cm) containing 5 g of DEAE-cellulose (Serva), wash this with 50 ml of buffer A *(Table 11)* and store it at 4°C.

(ii)   Take 10 g of cells and mix with 10 ml of cold buffer A and 40 g of washed glass beads.

(iii)   Break the cells in the Sorvall Omnimix set at maximum speed for 3 min.

(iv)   Centrifuge the preparation at 16 000 *g* for 5 min to remove the glass beads.

(v)   Centrifuge the supernatant at 145 000 *g* for 20 min. Recover the supernatant.

(vi)   Centrifuge this supernatant at 198 000 *g* for 90 min to pellet the ribosomes. *Retain the supernatant on ice* for step (x) and proceed with the ribosome pellet as follows.

(vii)   Resuspend the ribosomes in approximately 10 ml of buffer A and incubate at 37°C for 90 min.

(viii) Centrifuge the suspension at 12 000 *g* for 10 min, discard the pellet and re-centrifuge the supernatant at 198 000 *g* for 90 min to pellet the ribosomes as before.

(ix)   Resuspend the ribosomes in buffer A to a concentration of 100 mg/ml (based on A_{260} of 1480 ≡ 100 mg/ml) and store the suspension in aliquots in liquid nitrogen until required.

(x)   Load the supernatant from step (vi) onto the DEAE-cellulose column already prepared [step (i)] and wash the column with 50 ml of buffer A.

(xi)   Elute the protein from the DEAE-cellulose column with buffer A containing 0.25 M ammonium chloride, monitoring fractions by A_{280}. Pool those fractions containing the protein peak.

(xii)   Divide the eluted protein into aliquots and store these in liquid nitrogen.

The ribosomes are the source of initiation factors. They should be stable for at least 3 months if stored in liquid nitrogen. The protein fraction contains the amino acid activating enzymes, elongation factors T and G and most of the DNA-dependent RNA polymerase. The protein concentration of this fraction should be about 8 – 10 mg/ml and, stored in liquid nitrogen, it should be stable for at least 3 months.

3.5 Optimisation of the System

As with the *in vitro* transcription-translation system of Zubay (Section 2.5.2), the magnesium ion concentration for the system of Gold and Schweiger should be optimised. This is usually in the range of 11 – 15 mM. The volume of the ribosome and protein fractions added should also be varied until optimum synthesis is achieved. For each of these optimisations, prepare the incubation

Table 13. Incubation Mixture for the Gold and Schweiger System[a].

1. Prepare the following mixtures:

   Mixture I:

1.0 M Tris-acetate, pH 8.0	0.50 ml
2.0 M potassium acetate	0.25 ml
Amino acids (each at 5 mM)	0.40 ml
1.0 M DTT	20 μl
Water to 2.0 ml final volume	

   Mixture II

0.2 M ATP	0.20 ml
0.2 M CTP	50 μl
0.2 M GTP	50 μl
0.2 M UTP	50 μl
1.0 M phosphoenol pyruvate	0.40 ml

   Add KOH to adjust the pH to pH 6.5 and then add water to 1.0 ml final volume.

2. For each incubation mixture, mix the stock solutions in the following proportions:

Stock solution	Volume (μl)
Mixture I	20[b]
Mixture II	5[b]
Transfer RNA (10 mg/ml)	5
6 mM Folinic acid	5
0.12 M Magnesium acetate	5
Protein fraction[c]	40
Ribosomes[c]	100
DNA (100 – 600 μg/ml)[d]	10
Radioactive amino acid[e]	100

[a]The compositions of stock solutions are given in *Table 11*.
[b]The remainder of mixtures I and II may be stored in aliquots at − 70°C for future use.
[c]See Section 3.4 for preparation of these components. The volume of these may be varied ±50% to optimise the system.
[d]Plasmid DNA (e.g., pBR325).
[e]Usually [³H]leucine (5 μCi/μl, 150 Ci/mmol) or [³⁵S]methionine (10 μCi/μl; 1000 Ci/mmol).

mixtures by mixing the reagents described in *Table 13*. As with the Zubay system, the test DNA is most conveniently a multicopy plasmid (Section 2.5.1). After incubation at 37°C for up to 40 min, estimate the incorporation of radioactive amino acid into protein by TCA precipitation as described previously *(Table 10)* and analyse the products of the incubation by SDS-PAGE (Section 2.5.2).

Having optimised the composition of the incubation mixture, the time dependence of the coupled transcription-translation system should be examined as described earlier (Section 2.5.3). Protein synthesis should proceed rapidly for at least 40 min.

3.6 Modifications of the Basic Method

A number of researchers have slightly modified the procedure of Gold and Schweiger. For further details and examples see references 6 and 7.

Table 14. Preparation of Phage DNA.

---

1. Prepare the phage as described in Chapter 6, *Table 5*, and purify the phage by CsCl density gradient centrifugation using a step gradient as described in Chapter 6, *Table 7*.

2. After centrifugation, recover the phage band using a syringe and transfer it to a siliconised Corex tube.

3. Add RNase to 2 μg/ml and leave at room temperature for 30 min.

4. Add an equal volume of TE buffer (10 mM Tris-HCl, pH 7.5, 1 mM EDTA) to dilute the CsCl. Then add an equal volume of phenol mixture[a] and mix thoroughly.

5. Centrifuge at 16 000 g for 20 min at 8°C (e.g. Sorvall HB4 rotor at 10 000 r.p.m.).

6. Transfer the aqueous layer to a fresh Corex tube and repeat the phenol extraction.

7. Dialyse the aqueous layer against several changes of TE buffer.

8. After dialysis, add 3.0 M sodium acetate, pH 5.6, to 0.3 M and then two volumes of ethanol. Mix well and leave at -70°C for 1 h.

9. Centrifuge at 16 000 g for 30 min at -10°C (e.g. Sorvall HB4 rotor at 10 000 r.p.m.).

10. Dry the phage DNA pellet under vacuum. Resuspend the pellet in a small volume of TE buffer to give a concentration of $200-500$ μg/ml. Store at -20°C.

---

[a]For the composition of the phenol mixture, see the final footnote to *Table 8*.

4. USE OF PLASMID OR PHAGE λ DNA TEMPLATES

4.1 **Experimental Approach**

Any gene cloned into a plasmid vector or phage λ can be examined by using the native recombinant DNA as template in a reaction mixture prepared according to either the Zubay (Section 2.5) or Gold and Schweiger procedures (Section 3.5). The preparation of plasmid DNA has been described previously *(Table 8)* and a suitable protocol for the preparation of phage λ DNA is given in *Table 14*. The choice of strain of *E. coli* for preparation of the cell-free extract is dictated by the aims of the experiment; MRE600 is routinely used for the identification of cloned gene products (Section 4.2) and protein precursors (Section 4.3) whereas other strains may be more appropriate for studies of the control of gene expression (Section 4.4). The incubation is carried out under optimal conditions of magnesium ion and extract concentrations and then the polypeptide products are characterised by SDS-PAGE (Section 2.5.2).

4.2 **Identification of Cloned Gene Products**

By separately analysing the labelled polypeptide products obtained when the system is supplied with cloning vector (plasmid or phage λ) or with the recombinant DNA, the gene products of any genes cloned with their own promoter (or using a promoter in the vector) can be identified. *Figure 4* shows such an example. Plasmid pBR325 and pLG281 DNAs were used as templates in a typical series of incubations (Section 2.5.2). Plasmid pLG281 is pBR325 carrying an *Eco*RI fragment from the plasmid CollIb specifying colicin I production. The polypeptides encoded by the vector pBR325 and expressed in the Zubay system are CAT and pre-β-lactamase (β-lac*) These are analysed in *Figure 4,* lane 4, but note that pre-β-lactamase is not easily visible in this

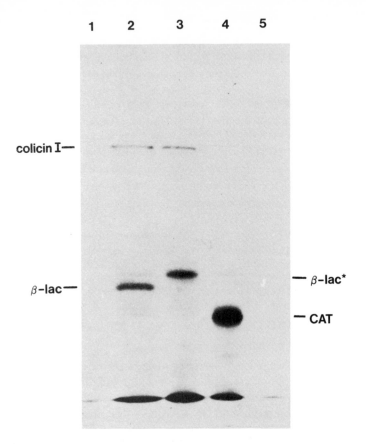

Figure 4. Identification of cloned gene products. Plasmid-encoded gene products were labelled in minicells of *E. coli* DS410 (**lanes 1** and **2**; see Chapter 6) or *in vitro* (Zubay system) (**lanes 3 – 5**; this chapter Section 2) and analysed by SDS-PAGE (11% gel) and autoradiography. **Lane 1**, DS410 minicells alone; **lane 2**, DS410 minicells carrying pLG281. **Lanes 3** and **4**, pLG281 and pBR325, respectively, in the Zubay system. **Lane 5**, control Zubay system (no DNA). The abbreviations used are: β-lac, β-lactamase; β-lac*, pre-β-lactamase; CAT, chloramphenicol acetyltransferase. These proteins, as well as colicin I, are described in Section 4.2. Pre-β-lactamase is present in **lane 4** but is not clearly visible in this exposure.

exposure. In lane 3 (pLG281 in the Zubay system), CAT has been inactivated during cloning; pre-β-lactamase is still present as expected and a new polypeptide of approximately 68 000 molecular weight (the colicin I protein) is also present. Lane 2 shows pLG281 analysed in a minicell expression system, *E. coli* DS410 (see Chapter 6, Section 3). The molecular weight of the colicin I protein synthesised *in vitro* (lane 3) is apparently identical to that synthesised under essentially *in vivo* conditions (minicells, lane 2), suggesting that colicin I is not synthesised as a precursor (see Section 4.3).

4.3 **Identification of Protein Precursors**

Many membrane proteins are initially synthesised as precursors with hydro-

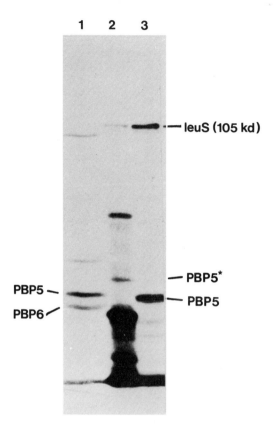

Figure 5. Identification of protein precursors by comparison of *in vivo* and *in vitro* products. SDS-PAGE (11% gel) and fluorography of [^{35}S]methionine-labelled polypeptides. **Lane 1**; authentic penicillin-binding proteins (PBPs) labelled with [^{14}C]benzyl penicillin; **lane 2**, λpBS10 (λ carrying the genes for *pbpA, rodA, leuS* and *dacA*; ref. 8) expressed in the Zubay system *in vitro*; **lane 3**, λpBS10 expressed in a u.v.-irradiated host system (Chapter 6). In the *in vitro* incubation not only the cloned genes but also some of the phage λ genes are expressed since there is no repressor protein present to prevent transcription from the λ promoters. In the u.v.-irradiated host system, repressor is present and the expression of phage genes is prevented (see Chapter 6, Section 2). The abbreviations used are: leuS, leucyl-tRNA synthetase; PBP5 and PBP6, penicillin-binding proteins 5 and 6, respectively; PBP5*, the precursor of PBP5.

phobic N-terminal extensions (signal sequences) which are subsequently removed during their integration into or passage through the cytoplasmic membrane (8). The *in vitro* system of Zubay contains membrane fragments but these are inactive as acceptors for newly-synthesised membrane proteins. The system of Gold and Schweiger is completely devoid of membrane. Either system can therefore be used to identify protein precursors. Two approaches are possible. Firstly, the same plasmid or phage λ DNA may be analysed *in vitro* and in one of the *in vivo* gene expression systems (see Chapter 6), and the profiles compared. If a precursor exists, it should be larger than the mature polypeptide and should be present after synthesis *in vitro* but absent *in vivo*, having been processed to the mature polypeptide. Two examples (9,10) are the precursor of β-lactamase (β-lac*) (*Figure 4*; compare lanes 2 and 3) and the

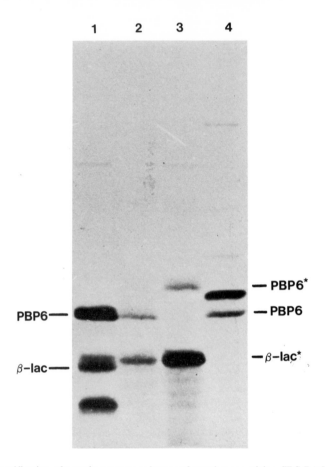

Figure 6. Identification of protein precursors by use of membrane vesicles. SDS-PAGE (11% gel) and autoradiography of [^{35}S]methionine-labelled polypeptides. The construction of pLG310 is described in the legend to *Figure 9*. **Lane 1**, DS410 minicells carrying pLG310; **lane 3**, pLG310 expressed *in vitro*; **lane 2**, pLG310 expressed *in vitro* in the presence of inverted inner membrane vesicles of *E. coli*; **lane 4**, [^{14}C]benzyl penicillin-labelled PBPs; PBP6 is indicated . The abbreviations β-lac and PBP6 refer to β-lactamase and penicillin-binding protein 6, respectively, while β-lac* and PBP6* are the corresponding precursor molecules. The precursor to PBP6 is synthesised *in vitro* (**lane 3**) and processed to the mature protein in the presence of membrane vesicles (**lane 2**).

precursor of penicillin-binding protein 5 (PBP5*), the *dacA* gene product (*Figure 5*; compare lanes 2 and 3). Alternatively, exogenous membrane vesicles may be added to the *in vitro* system to act as acceptors for the precursor polypeptides and process these to mature products. Therefore, a comparison of the polypeptides synthesised *in vitro* in the presence or absence of membrane vesicles should identify those proteins made as precursors. The precursor of penicillin-binding protein 6 (PBP6*), the *dacC* gene product, was identified in this way (9); see *Figure 6*. *Table 15* describes the preparation and use of membrane vesicles for this purpose.

Table 15. Identification of Protein Precursors using Membrane Vesicles.

Preparation of Membrane Vesicles[a]

1. Grow 4 – 6 litres of cells in nutrient broth [2.5% (w/v) nutrient broth powder in water] to an A_{600} of 0.5.

2. Harvest the cells by centrifugation at 3000 g for 10 min at 4°C. Wash the cells by centrifugation in ice-cold buffer A (1 mM DTT, 5 mM magnesium acetate, 50 mM Tris-HCl, pH 7.5).

3. Resuspend the washed cells in 40 ml of buffer A and break them using a French press at 3000 p.s.i.

4. Centrifuge the broken cells at 10 000 g for 10 min at 4°C.

5. Retain the supernatant and layer it in 8 ml aliquots over 2 ml of 20% (w/v) sucrose in buffer A. Centrifuge at 100 000 g for 1 h at 4°C.

6. Discard the supernatant and resuspend the pellet in a small volume of buffer A. Load this sample on top of two linear [20 – 50% (w/v)] sucrose gradients prepared in buffer A. Centrifuge in a swing-out rotor at 180 000 g for 20 h at 4°C.

7. Recover the zone containing turbid material located at 35 – 45% sucrose concentration and dilute it with an equal volume of buffer A.

8. Centrifuge at 100 000 g for 1 h at 4°C.

9. Resuspend the pellet in buffer B (1 mM DTT, 10 mM EDTA, 50 mM Tris-HCl, pH 7.5) to a concentration of 10 A_{280} units/ml[b].

10. Layer 8 ml portions of the sample over 2 ml of 20% (w/v) sucrose in buffer A. Centrifuge at 100 000 g for 2 h at 4°C.

11. Resuspend the pellet in buffer A at a concentration of 75 A_{280} units/ml[b] and store in aliquots at $-70°C$.

Use of Membrane Vesicles

The addition of membrane vesicles to a Zubay incubation mixture results in significant inhibition of protein synthesis (3- to 15-fold decrease). For this reason, a balance must be reached empirically between adding enough membrane vesicles to achieve processing whilst maintaining adequate protein synthesis. Use a range of volumes of vesicles, from 0.5 μl to 5.0 μl per 30 μl incubation mixture. Analyse the products by SDS-PAGE to determine the optimal conditions.

[a]This procedure is that described by Chang *et al.* (11) which uses *E. coli* ML30 as the source of vesicles.
[b]To determine the A_{280} value, dilute a small aliquot of the sample in 3% SDS at 25°C and then mix before reading the optical density.

4.4 Studies of the Control of Gene Expression

An *in vitro* system obviously lends itself to experiments in which the effects of changes in specific components may be easily studied. Indeed, invaluable information of this kind has been obtained by a number of workers (1). However, since the efficiency of protein synthesis in the coupled system relies on the production of mRNA as an intermediate, an extract high in RNase activity would be expected to seriously affect studies on the control of gene expression. MRE600, a strain which lacks the major RNase activity of *E. coli*, is routinely used as the source of the S30 extract for this reason. Nevertheless, active extracts have been prepared from several RNase[+] strains, including KN126 (ref. 5), N138*recB*[ts] (ref. 12) and LE316 (ref. 5), and so it appears that the endogenous RNase activity in *E. coli* must be relatively low, at least under the reaction conditions used. It should, therefore, be possible to prepare an active extract from any mutant strain of *E. coli*. The mutant extract could then be used as an assay for purifying specific components from a fractionated wild-type strain which re-establish transcriptional or translational activity. The use of coupled *in vitro* transcription-translation systems in studying the control of gene expression has been reviewed in detail elsewhere (1).

5. USE OF LINEAR DNA TEMPLATES

5.1 **Experimental Approach**

The examples of *in vitro* transcription and translation described above have all involved using plasmid or phage λ DNA as template. The plasmid DNA is added to the incubation in its native conformation as covalently-closed circular molecules with the predominant species being the negatively-supercoiled form. On the other hand, phage λ DNA is linear but is still a good substrate for transcription. It would seem possible, therefore, that DNA fragments generated by restriction endonuclease digestion might also function efficiently as templates. However, attempts at using small linear DNAs in *in vitro* transcription-translation systems have met with only limited success. The smaller the DNA fragment involved, the more difficult it has been to obtain meaningful results. An examination of the stability of the DNA template under normal incubation conditions, based on TCA precipitation, showed that circular plasmid DNA was very stable (<5% degraded), phage λ DNA was about 30% degraded and linearised plasmid DNA was extensively degraded (13). This degradation of linear templates results from the exonucleolytic activity present in *E. coli* extracts. The unexpectedly high stability of phage λ DNA results from its ability to circularise via long cohesive ends during the incubation, thus rendering the DNA resistant to exonucleolytic digestion. Linear DNA fragments (even linearised plasmids with short cohesive ends) do not recircularise under these incubation conditions and are therefore rapidly degraded. Even so, it is possible to obtain complete polypeptide products using small linear DNA fragments as templates if large quantities of DNA (at least 5 μg per incubation) are used (5). This allows a few DNA molecules to be transcribed prior to degradation. However, the products of genes with weak promoters would be very poorly transcribed in the time available and therefore may not be identified even with this protocol. Fortunately, the two methods of preparing *in vitro* transcription-translation systems described in this chapter can be modified so that linear DNA of any size can be efficiently used as a template.

5.1.1 *The System of Zubay*

The major source of exonucleolytic activity in *E. coli* is the enzyme exonuclease V, the product of the *recB* and *recC* genes. A mutation in just one of these loci results in inactive exonuclease V (14). Linear DNA is stable in an S30 extract prepared from a *recB* strain for at least 2 h, but the extract also supports high levels of protein synthesis in the absence of added DNA due to the

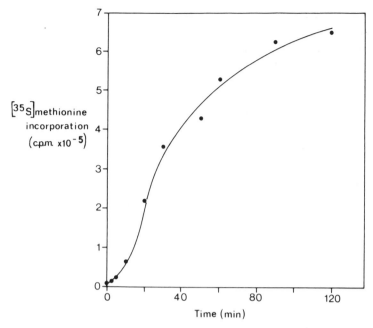

Figure 7. Protein synthesis from a linear DNA template using an exonuclease V-free extract. pBR325 digested with *Hind*III was used as a template in an *in vitro* incubation with an S30 extract derived from N138 *recB*ts. Samples (2 µl) were removed at various times during the incubation and the radioactivity incorporated into protein was estimated by TCA precipitation *(Table 10)*.

large number of contaminating chromosomal fragments (12,13). In the Zubay method there are two stages at which chromosomal DNA is normally removed. Firstly, two 30 000 *g* centrifugations [Section 2.4, steps (ix) and (x)] sediment large DNA fragments, and secondly an 80 min pre-incubation at 37°C [Section 2.4, step (xi)] in the presence of ATP allows degradation of any remaining chromosomal fragments. This latter step requires a functional exonuclease V. Therefore, a temperature-sensitive *recB* strain (N138 *recB*ts) can be used and the extract prepared at the permissive temperature (30°C). It is necessary to extend the pre-incubation period to 160 min to allow complete degradation of chromosomal fragments [Section 2.4, step (xi)]. Under the conditions of incubation used for *in vitro* transcription-translation (37°C; the non-permissive temperature) exonuclease V is inactive and exogenous linear DNA survives for at least 2 h with this extract, compared with 5 – 10 min in a *recB*+ extract. The extract will support synthesis from linear templates for approximately 100 min (*Figure 7*). The background is low (*Figure 8,* lane 2), and as little as 50 ng of DNA is needed to obtain usable results (*Figure 8,* lane 3) in comparison with 2 – 5 µg for a *recB*+ S30 extract (12).

5.1.2 *The System of Gold and Schweiger*

The problem of contaminating chromosomal DNA fragments due to the use of a *recB* strain does not arise in the method of Gold and Schweiger (Section 3)

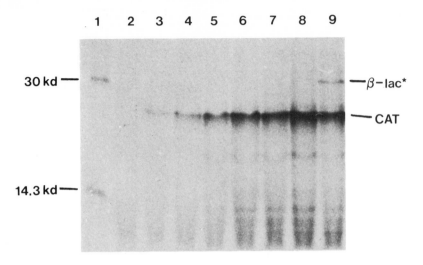

Figure 8. DNA concentration dependence of protein synthesis using linear DNA as a template. SDS-PAGE (15% gel) and autoradiography of [^{35}S]methionine-labelled polypeptides synthesised from varying amounts of pBR325 linearised with *Pst*I as a template, using an S30 extract from N138 *recB*^{ts}. **Lane 1**, protein mol. wt. standards; **lane 2**, no DNA; **lanes 3 – 8**, 50 ng, 375 ng, 0.75 μg, 1.5 μg, 3 μg and 5 μg of *Pst*I-digested pBR325, respectively. **Lane 9**, 2.5 μg of supercoiled pBR325. Pre-β-lactamase (β-lac*) is only present in **lane 9** since *Pst*I digestion inactivates this gene in the other incubations. CAT is unaffected by restriction by *Pst*I.

since any DNA fragments are removed in the DEAE-cellulose chromatography step. Extracts prepared from *recB* strains fractionated in this way have been used successfully by a number of workers to study proteins encoded by specific restriction endonuclease-generated fragments (13).

5.2 Importance of Supercoiling

One consideration to be aware of when using linear DNA as a template is the contribution, if any, of supercoiling to the efficiency of expression of certain genes. This is particularly important if the factors controlling gene expression are being investigated. A number of studies using purified transcription systems have indicated that, as a template for transcription, negatively-supercoiled DNA is more efficient than the relaxed form of the same molecule (15). It is clear, from structural considerations, that unwinding of the double helix is favoured if negative supercoils are present. Since the formation of an initiation complex between RNA polymerase and the promoter is thought to require at least partial unwinding of the double helix, the observed stimulation of supercoiling on transcription may be the result of a decrease in the activation energy needed to form the initiation complex. Supercoiling could increase the amount of transcription in two ways; by increasing the frequency of initiation at existing sites and by making new transcription sites available. This stimulation has been shown to be promoter-specific and may vary greatly in magnitude. The degree of stimulation is also apparently dependent on a number of conditions including temperature, ionic strength and the polymerase:DNA ratio

(16). The enzyme responsible for introducing negative supercoils into circular DNA molecules is DNA gyrase (17). This enzyme consists of two subunits, the *gyrA* and *gyrB* gene products. S30 extracts have considerable gyrase activity and relaxed covalently-closed DNAs are rapidly supercoiled when incubated with the extract. Unwinding of DNA to remove supercoils is carried out by the enzyme topoisomerase I (17), but this activity is low in comparison with super-coiling activity. During incubations with S30 extracts, circular DNA molecules are present in their supercoiled form. Linear DNAs, on the other hand, should not be supercoiled. If supercoiling is necessary for the efficient initiation of transcription at certain promoters then this may constitute a limitation to the use of linear DNA fragments as templates in the *in vitro* systems. Genes whose promoters require supercoiling for expression may not be identified, and *in vitro* studies on the control of gene expression may be completely unrelated to the *in vivo* situation.

One way to investigate the contribution of supercoiling to transcriptional efficiency is to compare expression from supercoiled and relaxed covalently-closed forms of the same molecule. Relaxed plasmid DNAs are rapidly super-coiled during incubation with an S30 extract but this supercoiling may be prevented by using one of the gyrase-specific inhibitors (15). Using this method, Yang *et al.* (15) reported significant reduction in the expression of a number of genes, including those encoding colicin E1 and β-galactosidase, in the absence of supercoiling. Similar experiments have been carried out using an S30 extract prepared from a DNA gyrase mutant (5). Relaxed and super-coiled forms of the plasmid ColE1 were incubated independently with this ex-tract and the template DNA checked for changes in topology. The DNAs re-mained completely relaxed or fully supercoiled, respectively. Analysis of the gene products showed clearly that, in contrast to the findings of Yang *et al.* (15), there was no apparent differential effect on the expression of any of the genes carried by ColE1. Analysis of the plasmid pBR325 (relaxed and super-coiled) with a gyrase mutant S30 extract showed stimulation of expression of all plasmid-encoded proteins in the absence of supercoiling, and β-lactamase appeared to be expressed proportionally better from the relaxed template (5).

The contribution of supercoiling to gene expression is, therefore, by no means clear and more detailed investigation of this phenomenon is required before any precise conclusions can be drawn. However, the author feels that supercoiling is not necessary to obtain expression in the *in vitro* systems of all genes encoded by the template and therefore linear DNA fragments may be used with confidence. Whether linear DNAs should be used when investigating the factors involved in the control of gene expression is debatable. In such studies it may be advisable to use a circular supercoiled template to reproduce, as closely as possible, the DNA topology found *in vivo*.

5.3 Identification of Fusion Proteins

During cloning of DNA fragments into plasmid or phage λ vectors, a fusion protein may be generated across the boundary of vector and insert. In subse-

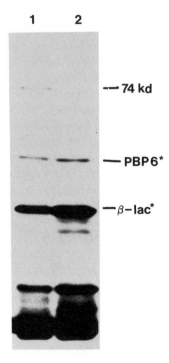

Figure 9. Identification of fusion proteins. SDS-PAGE (11% gel) and autoradiography of [35S]-methionine-labelled polypeptides synthesised *in vitro* with an S30 extract from *E. coli* MRE600. **Lane 1,** 5 μg pLG310; **lane 2,** 5 μg pLG310 digested with *Eco*RI. Plasmid pLG310 was generated by ligation of a 6.4 kb *Eco*RI fragment carrying *dacC*, the PBP6 structural gene, into pSF2124 into pSF2124 (ColE1::Tn3). Neither pSF2124 nor the *Eco*RI fragment alone encodes a protein of 74 000 mol. wt. This protein is generated by read-through across the vector:insert boundary; hence its susceptibility to *Eco*RI digestion. PBP6* and β-lac* are the precursors of penicillin-binding protein 6 and β-lactamase, respectively.

quent analyses of recombinant DNAs in one of the *in vivo* gene expression systems (see Chapter 6), the appearance of such a protein in the gel profile can complicate the interpretation of the results. This problem may be overcome using an *in vitro* system. Prior to incubation, the DNA is digested with the restriction endonucleases used for the original cloning. The polypeptide profile obtained following SDS-PAGE is then compared with that derived from intact DNA (5). Fusion proteins will always be sensitive to digestion of the DNA with these enzymes (see *Figure 9*).

5.4 Identification of Cloned Gene Products Similar in Molecular Weight to Vector-encoded Proteins

Most of the proteins encoded by plasmid cloning vectors are below 35 000 molecular weight. Therefore, while larger proteins encoded by cloned DNA may be easily identified, smaller proteins may be masked by vector proteins of similar molecular weight *Figure 10* shows the analysis of a plasmid vector and a recombinant derivative where the profiles are very similar. To identify the

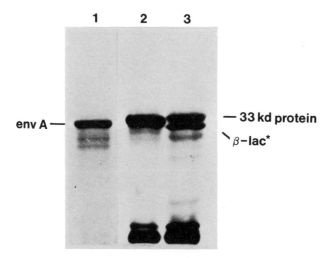

Figure 10. Identification of cloned gene products similar in size to the vector-encoded proteins. SDS-PAGE (15% gel) and autoradiography of [^{35}S]methionine-labelled polypeptides synthesised in an *in vitro* incubation with an extract from *E. coli* MRE600. Plasmid pKN410 is the vector and pLG510 is a recombinant of pKN410 carrying a 2.5-kb *Eco*RI fragment. **Lane 1**, *Eco*RI fragment from pLG510; **lane 2**, pLG510; **lane 3**, pKN410. The 33 000 mol. wt. protein is a vector-encoded protein of unknown function which spans the *Eco*RI site, β-lac* refers to the precursor of β-lactamase.

polypeptide encoded by the insert, the recombinant plasmid was digested with *Eco*RI and the cloned fragment purified from an agarose gel. When this fragment was used as a template in the Zubay system it programmed the synthesis of a 32 000 molecular weight protein (*Figure 10*, lane 1). This was subsequently identified as the *envA* gene product. The 33 000 molecular weight protein (*Figure 10*, lane 3) is a polypeptide encoded by the vector pKN410 which spans the *Eco*RI site and is therefore not present in the products synthesised using the recombinant DNA as template but is replaced by the 32 000 molecular weight protein encoded by the DNA insert. Hence the difficulty in distinguishing the polypeptides encoded by cloned DNA unless it is purified and used as template in the absence of vector DNA.

5.5 Identification of Genes Cloned with their own Promoter

Since it is possible to analyse vector and insert DNA separately in an *in vitro* transcription-translation system modified to use linear DNA templates (Section 5.1), it is easy to establish whether or not a particular gene has been cloned together with its own promoter (see *Figure 10*). If one is interested in the control of gene expression, it is particularly important that the cloned gene is using its own promoter, particularly since many cloning experiments result in at least some, if not all, transcription initiating at a promoter in the vector.

5.6 Identification of Intragenic and Intergenic Restriction Sites

Intragenic and intergenic restriction sites may be easily identified using the

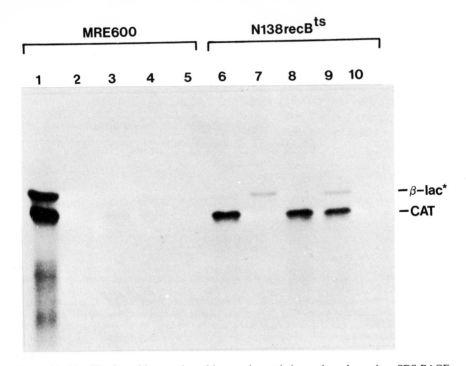

Figure 11. Identification of intragenic and intergenic restriction endonuclease sites. SDS-PAGE (15% gel) and autoradiography of [³⁵S]methionine-labelled polypeptides synthesised in an *in vitro* incubation with S30 extracts from *E. coli* MRE600 and N138 *recB*ᵗˢ. Plasmid pBR325 digested with various restriction endonucleases was the template. **Lanes 1** and **6**, pBR325; **lanes 2** and **7**, pBR325 digested with *Eco*RI; **lanes 3** and **8**, pBR325 digested with *Pst*I; **lanes 4** and **9**, pBR325 digested with *Hind*III; **lanes 5** and **10**, controls (no DNA). The abbreviations β-lac* and CAT refer to pre-β-lactamase and chloramphenicol acetyltransferase, respectively.

modified *in vitro* transcription-translation systems (Section 5.1) by treating the DNA with a range of restriction enzymes and analysing the polypeptides which survive or which are susceptible to the restriction endonuclease digestion. In this way, specific polypeptide products may be mapped to very small pieces of DNA. *Figure 11* shows the result of using pBR325, digested with various restriction enzymes, as a template for *in vitro* transcription-translation. The synthesis of CAT is sensitive to *Eco*RI digestion (lane 7) while that of β-lactamase is sensitive to *Pst*I digestion (lane 8). The tetracycline resistance protein is not visible in these incubations. Results obtained wth *recB*⁺ and *recB*ᵗˢ extracts are also compared and it is clear that with linearised DNA the *recB*ᵗˢ extract is much more efficient.

5.7 Analysis of DNA from Heterologous Sources

The author has analysed DNA prepared from *Staphylococcus aureus* (a Gram-positive bacterium) using an S30 extract prepared from *E. coli* (5). The DNA was transcribed and translated very efficiently to produce a number of polypeptide products. However, the number of polypeptides identified was significantly in excess of the number expected from DNA sequence analysis (18). Using a minicell strain of *Bacillus subtilis* (also Gram-positive), the same plasmid was analysed and far fewer proteins were synthesised than in the *in vitro* analysis (19,20). It is therefore probable that, on DNA from heterologous sources, the DNA-dependent RNA polymerase of *E. coli* is able to initiate transcription at sites not normally used. One should therefore interpret results obtained with heterologous DNAs with extreme caution. If similar transcription-translation systems could be produced from other bacterial strains this problem would be overcome. Such systems already exist for *B. subtilis* (21) and *Streptomyces lividans* (S. Rae and E. Cundliffe; personal communication).

5.8 Expression of Eukaryotic Genes

To gain expression of a eukaryotic gene in *E. coli*, the gene must either contain no introns or be cloned from cDNA, and must be provided with prokaryotic expression signals. However, due to the difference in codon usage between prokaryotes and eukaryotes, expression in *E. coli* may be inefficient. This problem may be overcome using the *in vitro* systems by altering the tRNA composition of the incubation mixture.

6. COMPARISON OF *IN VITRO* TRANSCRIPTION-TRANSLATION SYSTEMS

6.1 Advantages and Disadvantages of the Modified Zubay System

6.1.1 *Advantages*

(i)   The system is fairly easy to prepare and extremely easy to use.
(ii) The S30 is a crude extract and therefore most regulatory factors should be present.
(iii) The S30 extract can be stored in liquid nitrogen for several years without apparent loss of activity.

6.1.2 *Disadvantages*

(i)   The S30 extract may contain some residual DNA which can serve as template to give a wide range of background polypeptides. However, the residual DNA content is usually negligible.
(ii) The S30 extract contains membrane fragments, although these are inactive as acceptors during membrane protein biosynthesis (9).

6.2 Advantages and Disadvantages of the Gold and Schweiger System

6.2.1 *Advantages*

(i)   The DEAE-cellulose fractionation makes the system very low in residual DNA; background incorporation is therefore extremely low.
(ii) The Gold and Schweiger system probably contains lower concentrations of endogenous amino acids than the Zubay system, thus allowing higher levels of incorporation of radioactive amino acids.

6.2.2 *Disadvantages*

(i)   The system is complicated to prepare and use.

(ii)   The protein fraction and ribosomes are stable for only a few months.

(iii) Certain protein factors of interest may be lost during the DEAE-cellulose chromatography step in the preparation of the system.

7. ADVANTAGES OF *IN VITRO* SYSTEMS OVER *IN VIVO* SYSTEMS

In vitro transcription-translation systems have a number of advantages over the *in vivo* systems (Chapter 6):

(i)   Linear DNA fragments may be used as templates (see Section 5).

(ii)   Factors controlling gene expression may be easily investigated (22).

(iii) Membrane protein assembly and processing may be studied by the addition of membrane vesicles (9).

(iv)   Proteins which are susceptible to degradation by cellular proteases *in vivo* are more easily identified *in vitro*. Truncated polypeptides and miscompartmentalised membrane proteins not detected *in vivo* have been identified *in vitro* (J. Pratt; unpublished results).

(v)   Proteins expressed at very low levels *in vivo*, due to the action of host-encoded repressors, may be expressed at higher levels *in vitro* (4).

(vi)   The incorporation of the added labelled amino acid is extremely efficient due to the low level of endogenous amino acids in the extract.

Further comparison of the *in vitro* systems with the *in vivo* systems is given in Chapter 6, *Table 1*.

8. PROBLEMS WHICH MAY ARISE USING *IN VITRO* SYSTEMS

Although the *in vitro* transcription-translation systems described in this chapter have certain advantages over *in vivo* gene expression systems (Chapter 6), there are a number of disadvantages too. These are briefly outlined below:

(i)   The synthesis of large proteins (with molecular weights >70 000) may not be very efficient. This is mostly due to premature termination of translation of these proteins and is readily identified by a laddered appearance in the gel profile. This problem may be partly overcome by incubating for a shorter period of time (e.g., 15 min).

(ii)   Polypeptides which remain unfinished after incubation may be removed by centrifuging the samples at 5000 *g* for 2 min. This pellets the polysome-bound nascent peptides and results in a much cleaner gel profile. The detection of proteins in the low molecular weight range may also be facilitated by this procedure.

(iii) Artefacts are fairly common *in vitro*. Therefore, as far as possible, all experiments should be carried out in parallel with *in vivo* studies, or the results should be interpreted with some caution.

(iv)   [^{35}S]Methionine is usually used as the labelled amino acid in the *in vitro* incubations, but one should bear in mind that not all proteins contain methionine, especially since S30 extracts are capable of removing the

N-formyl methionine to a certain extent (23). Therefore, an alternative amino acid (e.g., [^{14}C]leucine) should be used in parallel experiments to ensure the identification of all proteins.

(v) When analysing genes cloned into phage λ using an *in vitro* system, the two λ promoters, P_L and P_R, will promote transcription of the phage λ genes as well as the cloned genes. *In vivo*, using the u.v.-irradiated host system (Chapter 6, Section 2), expression of phage λ genes can be repressed by the λ repressor. The gel profile obtained from an *in vitro* incubation may therefore be much more complicated than that obtained *in vivo*. It may be possible to prepare an S30 extract from a lysogen to overcome this problem, but this would depend on whether the repressor is stable under the conditions of preparation and storage. Alternatively, purified repressor may be added to the incubation prior to the addition of phage λ DNA to maintain repression. Neither of these possibilities has been investigated.

9. ACKNOWLEDGEMENTS

J. Pratt acknowledges the receipt of MRC Project Grant Award No. G8203714CB.

10. REFERENCES

1. Zubay,G. (1973) *Annu. Rev. Genet.,* **7**, 267.
2. DeVries,J.K. and Zubay,G. (1969) *J. Bacteriol.,* **97**, 1419.
3. Gold,L.M. and Schweiger,M. (1971) in *Methods in Enzymology,* Vol. **20**, Moldave,K. and Grossman,L. (eds.), Academic Press Inc., London and New York, p. 537.
4. Collins,J. (1979) *Gene,* **6**, 29.
5. Pratt,J.M., Boulnois,G.J., Darby,V., Orr,E., Wahle,E. and Holland,I.B. (1981) *Nucleic Acids Res.,* **9**, 4459.
6. O'Farrell,P.Z. and Gold,L.M. (1973) *J. Biol. Chem.,* **248**, 5512.
7. Konings,R.N.H. (1980) in *Methods in Enzymology,* Vol. **20**, Moldave,K. and Grossman,L. (eds.), Academic Press Inc., London and New York, p. 537.
8. Blobel,G. and Dobberstein,B. (1975) *J. Cell Biol.,* **67**, 835.
9. Pratt,J.M., Holland,I.B. and Spratt,B.G. (1981) *Nature,* **293**, 307.
10. Sutcliffe,J.G. (1978) *Proc. Natl. Acad. Sci. USA,* **75**, 3737.
11. Chang,C.N., Model,P. and Blobel,G. (1979) *Proc.Natl. Acad. Sci. USA,* **76**, 1251.
12. Jackson,M., Pratt,J.M. and Holland,I.B. (1983) *FEBS Lett.,* **163**, 221.
13. Yang,H., Ivashkiv,L., Chen,H., Zubay,G. and Cashel,M. (1980) *Proc. Natl. Acad. Sci. USA,* **77**, 7029.
14. Tomizawa,J. and Ogawa,H. (1972) *Nature New Biol.,* **239**, 14.
15. Yang,H., Heller,K., Gellert,M. and Zubay,G. (1979) *Proc. Natl. Acad. Sci. USA,* **76**, 3304.
16. Seeburg,P.H., Nüsslein,C. and Schaller,H. (1977) *Eur. J. Biochem.,* **74**, 107.
17. Gellert,M. (1981) *Annu. Rev. Biochem.,* **50**, 879.
18. Horinouchi,S. and Weisblum,B. (1982) *J. Bacteriol.,* **150**, 815.
19. Shivakumar,J.H. and Dubnau,D. (1979) *Plasmid,* **2**, 279.
20. Shaw,W.V. (1983) *CRC Crit. Rev. Biochem.,* **14**, 1.
21. Leventhal,J.M. and Chambliss,G.H. (1979) *Biochim. Biophys. Acta,* **564**, 162.
22. De Crombrugghe,B., Pastan,I., Shaw,W.V. and Rosner,J.L. (1973) *Nature New Biol.,* **241**, 237.
23. Jerez,C. and Weissbach,H. (1980) *J. Biol. Chem.,* **255**, 8706.

CHAPTER 8

Purification of Eukaryotic Messenger RNA

MICHAEL J. CLEMENS

1. INTRODUCTION

There has been much expansion of activity in the fields of mRNA purification and *in vitro* translation over the last decade. This work has laid the basis for the development of recombinant DNA technology which is now providing valuable information in many areas of biology and medicine. In addition, the characterisation of isolated mRNA species and their translation products has contributed to our understanding of the regulation of gene expression at different levels within the cell. This work has been made possible by the development of procedures for obtaining undegraded, biologically active mRNAs from a variety of sources. This chapter will describe the techniques involved in the isolation of mRNA from eukaryotic cells. The isolation of mRNA from bacteria is not covered here but has been described in a number of earlier reviews (1,2). However, for many purposes the need to prepare such mRNA has been eliminated by the development of coupled transcription-translation systems (see Chapter 7).

2. EXTRACTION AND PURIFICATION OF MESSENGER RNA

Because of the great diversity of cells and tissues from which eukaryotic mRNAs have been extracted, there are many different procedures described in the literature. These have certain features and requirements in common, the most important of which are:
(i)   The need to separate RNA from protein, and in some cases DNA, in as quantitative a manner as possible.
(ii) The need to prevent ribonuclease activity, both during the isolation of the RNA and subsequently, in order to obtain intact, and thus biologically active, mRNA.
(iii) The requirement, in many procedures, to separate certain mRNA species from other mRNAs and from other classes of RNA.

The nature of the investigation to be undertaken and the properties of the biological material to be used will dictate which particular combination of techniques is best.

2.1 Apparatus and Solutions

Workers in the field of mRNA purification must be extremely careful to avoid ribonuclease contamination of their samples. Ribonucleases are very robust

Table 1. Precautions against Ribonuclease during RNA Isolation[a].

Bake or autoclave glassware; autoclave plasticware
Use autoclaved water for preparation of solutions
Use pure reagents
Treat all solutions with 0.2% (final concentration) diethyl pyrocarbonate to inactivate ribo-
nucleases and preferably filter sterilise them to eliminate microorganisms.
Use ribonuclease inhibitors (see Section 2.7) in solutions where possible
Wear disposable plastic or latex gloves
Protect solutions and RNA preparations from dust, fingerprints, etc.

---

[a]For details see the text.

enzymes and only minute amounts are sufficient to destroy mRNA activity. The problem of endogenous ribonuclease activity in the biological material itself can be overcome as described in Section 2.7 below. The introduction of ribonuclease contamination from other sources is avoided by good laboratory practice as summarised in *Table 1* and described in more detail below.

2.1.1 *Glassware and Plasticware*

All glass and sterile, disposable plasticware requirements should be anticipated (and preferably somewhat over-estimated) and the necessary non-sterile items treated by heating in an oven (160°C, 4 h) or autoclaving (15 p.s.i, 15 min). Be careful to check the maximum temperature which plastic apparatus can withstand. Where possible, use disposable sterile tubes for handling RNA solutions.

2.1.2 *Solutions*

Solutions should be prepared from the purest possible reagents which are available, using autoclaved, double glass-distilled water. Many specialist suppliers now sell ultra-pure, ribonuclease-free reagents such as sucrose *(Table 2)*. Ideally, solutions should be sterilised immediately after preparation either by passage through a membrane filter (0.22 μ pore-size) or by autoclaving. Any residual ribonuclease contamination can be eliminated by treating solutions with 0.2% (final concentration) diethyl pyrocarbonate (3) for 12 h, but this reagent should be removed by heating (100°C, 15 min) before use of the solution because it can react with nucleic acids. Diethyl pyrocarbonate cannot be used to treat Tris buffers because it reacts with primary amines.

Phenol is frequently used in mRNA extraction procedures and should also be as pure as possible. It can readily be re-distilled in a fume cupboard, taking care to maintain a suitable temperature (160°C) in the heating mantle and using an air-cooled condenser to collect the liquid distillate. Strict safety precautions (e.g., wearing safety goggles) should be observed during this procedure. The liquid distillate should be run into sterile water and stored at -20°C as a water-saturated solution. Use reasonably pure phenol to start with and stop the distillation before the residue is too small (to minimise the risk of explosion). Phenol should never be allowed to come into contact with the skin since

Table 2. Sources of Reagents for Messenger RNA Isolation.

Ultra-pure reagents	
Bovine serum albumin (ribonuclease-free)	Bethesda Research Laboratories Inc. Enzo Biochem Inc.
DTT	Bethesda Research Laboratories Inc. Cambridge Biotechnology Laboratories
EDTA	Cambridge Biotechnology Laboratories
Formamide	Bethesda Research Laboratories Inc.
Guanidinium chloride	Bethesda Research Laboratories Inc.
Guanidinium thiocyanate	Fluka AG
Nonidet P-40	Bethesda Research Laboratories Inc. Sigma Chemical Co.
Phenol	Bethesda Research Laboratories Inc.
SDS	Bethesda Research Laboratories Inc. Cambridge Biotechnology Laboratories
Sucrose (ribonuclease-free)	Bethesda Research Laboratories Inc. Cambridge Biotechnology Laboratories
Triton X-100	Cambridge Biotechnology Laboratories Inc.
Urea	Bethesda Laboratories Inc. Cambridge Biotechnology Laboratories
Affinity matrices	
Oligo(dT)-cellulose	Bethesda Research Laboratories Inc. Collaborative Research Sigma Chemical Co.
Poly(U)-Sepharose	Bethesda Research Laboratories Inc. Pharmacia Sigma Chemical Co.
Ribonuclease inhibitors	
Heparin	Sigma Chemical Co.
Human placental ribonuclease inhibitor	Bethesda Research Laboratories Inc. Enzo Biochem Inc. Sigma Chemical Co. P & S Biochemicals
Vanadyl ribonucleoside complex	Bethesda Research Laboratories Inc.

phenol burns can be serious and may require emergency treatment. Disposable plastic gloves should be worn at all times when working with RNA, with the dual purpose of protecting the investigator from phenol and of protecting the solutions from skin ribonucleases.

2.2 Choice of RNA Extraction Method

Numerous procedures have been described for the extraction of RNA from animal and plant cells and tissues. The best approach to use will depend on the precise requirements of the experiment and the properties of the biological starting material. An initial decision is whether to prepare RNA from whole cells or from a subcellular fraction (e.g., nucleus, total cytoplasm, polysomes or post-ribosomal supernatant). The next section deals with methods for RNA extraction from whole cell (or tissue) homogenates and cytoplasmic fractions. The specialised techniques for obtaining RNA from total polysomes or specific subclasses of polysomes are covered in Section 2.4. The choice of protocol will be governed by the nature of the questions to be answered. Several of these aspects have been discussed in the review by Taylor (4). In general, similar methods are applicable to plant tissues as to animal tissues, provided the necessary measures to prevent ribonuclease activity are taken (Section 2.7).

2.3 Preparation of RNA from Whole Cells or Cytoplasmic Fractions

2.3.1 *Phenol Methods*

The majority of RNA extraction methods starting from crude cell homogenates or extracts use phenol as a means of denaturing protein and separating this material from the nucleic acids. In general, the tissue is homogenised in a buffered salt solution, sometimes containing 0.35 M sucrose to prevent disruption of lysosomes and the consequent release of ribonucleases. A non-ionic detergent such as Triton X-100 or Nonidet P-40 is often present to aid the disruption of subcellular membranes. Some difficult tissues may need to be pulverised in liquid nitrogen before addition of the homogenisation medium. Large volumes of the medium (10−20 x tissue weight) can be used to dilute out ribonucleases and the use of high pH (pH 8.5−9.0) and ionic strength (0.2−0.5 M salt) further inhibits these enzymes. The latter conditions are particularly important when dealing with plant tissues (5). If only cytoplasmic RNA is required, the homogenate is first centrifuged to remove nuclei and an ionic detergent such as SDS or Sarkosyl (sodium N-laurylsarcosine) is then added (usually together with EDTA) to dissociate nucleoprotein complexes and further inhibit ribonuclease activity. Extraction is carried out by vigorous shaking with phenol or phenol:chloroform (1:1), pre-saturated with the homogenisation buffer. The organic and aqueous phases are separated by centrifugation, during which denatured protein accumulates at the interface. The RNA partitions into the upper aqueous layer, but some will also be trapped in the interface material. The latter should therefore be re-extracted with more buffer and the aqueous phases combined. Alternatively, brief treatment of the cell extract with Proteinase K (200 μg/ml, 37°C, 30 min) before phenol extraction will reduce the protein interface and improve the RNA yield. The presence of chloroform also aids the partitioning of RNA into the aqueous layer. Other components are often included in the organic extraction mixture, such as 1% (v/v) isoamyl alcohol (to minimise foaming) and 0.1% (w/v) 8-hydroxyquinoline (as an anti-oxidant for the phenol and as a chelating agent to destabilise RNA-protein interactions).

Total nucleic acid is precipitated from the aqueous layer by the addition of 2.5 volumes of cold (−20°C) ethanol in the presence of 0.2 M salt (usually sodium or potassium acetate) at pH 5.0, a pH chosen to minimise chemical hydrolysis of RNA. The time required for precipitation depends on how much RNA is present, and can vary from 2 h to overnight at −20°C. The RNA is recovered by centrifugation and is then washed in cold 70% ethanol, 0.1 M sodium acetate to remove traces of phenol. Resuspension and centrifugation in 3 M sodium acetate will remove DNA and small RNAs (tRNA and 5S rRNA) as well as polysaccharide or proteoglycan material. Finally, the RNA preparation is washed again in cold 70% ethanol, 0.1 M sodium acetate and then is dried by gently blowing a stream of nitrogen over the pellet. It is then readily dissolved in sterile water or buffer for further use. The yield and purity of an RNA preparation may be assessed by measuring the absorbance of a diluted sample at 260 nm and 280 nm; 1 mg of RNA is equivalent to 23 A_{260} units and a pure product will have an A_{260}:A_{280} ratio of 2.0.

Two examples of RNA extraction protocols which use the above procedures starting with whole tissue or cultured cells are summarised in *Tables 3* and *4*, respectively.

Table 3. A Phenol Extraction Procedure for Preparation of Cytoplasmic RNA from Liver[a].

1. Rinse the tissue and homogenise it in 15 volumes of cold buffer:
    0.35 M sucrose
    50 mM KCl
    10 mM magnesium acetate
    1.3% Triton X-100
    0.2 M Tris-acetate (pH 8.5)
2. Centrifuge at 2000 *g* for 5 min to pellet nuclei.
3. Collect the supernatant, add SDS to 1% and EDTA to 2 mM and bring to room temperature.
4. Shake for 10 min with 2 volumes of phenol:chloroform (1:1, v/v) pre-equilibrated with the above components.
5. Centrifuge at 10 000 *g* for 10 min at 20°C.
6. Collect the upper (aqueous) phase and store on ice. Do not contaminate with interface material.
7. Shake the phenol layer and interface material with an equal volume of 0.1 M sodium acetate, 2 mM EDTA, 0.1 M Tris-acetate (pH 9.0).
8. Centrifuge as in step 5.
9. Combine the aqueous layer with the previous aqueous phase (from step 6) and re-extract these by shaking with an equal volume of phenol:chloroform.
10. Centrifuge as in step 5.
11. Collect the aqueous phase. Add potassium acetate (pH 5.5) to 0.2 M and precipitate the RNA by adding 2.5 volumes of cold (−20°C) ethanol. Mix and leave at −20°C for 16 h.
12. Centrifuge at 2000 *g* for 10 min at 0°C.
13. Discard the supernatant and wash the pellet by resuspension in 3 M sodium acetate (pH 6.0).
14. Repeat steps 12 and 13.
15. Resuspend the pellet in cold 70% ethanol, 0.1 M sodium acetate and centrifuge as in step 12.
16. Decant the supernatant and dry the pellet in a gentle stream of clean nitrogen.
17. Dissolve the RNA in water or buffer, as required, and store at −70°C.

[a]Adapted from ref. 6.

Table 4. A Phenol Extraction Procedure for Preparation of Cytoplasmic RNA from Cultured Cells[a].

---

1. Centrifuge the cell suspension (1000 g, 5 min) and wash the cells by resuspending in 10 volumes of cold phosphate-buffered saline (0.14 M NaCl, 2.7 mM KCl, 6.5 mM Na_2HPO_4, 1.5 mM KH_2PO_4, pH 7.2).

2. Repeat this centrifugation and washing three times more.

3. Resuspend the cells in cold lysis buffer (5 ml per 10^8 cells):

   0.14 M NaCl
   1.5 mM $MgCl_2$
   0.5% Nonidet P-40
   10 mM Tris-HCl (pH 8.6)
   plus 10 mM vanadyl ribonucleoside complex
   or 1000 units/ml human placental ribonuclease inhibitor

   Vortex the suspension briefly.

4. Layer the lysate onto an equal volume of 24% (w/v) sucrose, 1% Nonidet P-40 in cold lysis buffer and leave for 5 min at 0°C.

5. Centrifuge at 8500 g for 20 min in a swing-out rotor (e.g., Sorvall HB-4 rotor).

6. After centrifugation, collect the upper layer and add an equal volume of:

   0.3 M NaCl
   25 mM EDTA
   2% SDS
   0.4 mg/ml Proteinase K (BRL or Sigma Chemical Co.)
   0.2 M Tris-HCl (pH 7.5)

   Incubate at 37°C for 30 min.

7. Shake the sample briefly with an equal volume of phenol:chloroform:isoamyl alcohol (50:50:1, by vol.) containing 0.1% 8-hydroxyquinoline, pre-equilibrated with 0.15 M NaCl, 1 mM EDTA, 50 mM Tris-HCl (pH 7.5).

8. Centrifuge the mixture at 10 000 g for 10 min at 18°C in Corex (Dupont) glass tubes.

9. Collect the upper (aqueous) phase and add 2.5 volumes of cold (-20°C) ethanol. Allow the RNA to precipitate at -20°C for $2-18$ h.

10. Centrifuge the sample at 10 000 g for 10 min at 4°C in Corex glass tubes.

11. Discard the supernatant and wash the pellet by resuspension in cold 75% ethanol, 0.1 M sodium acetate (pH 5.3).

12. Centrifuge as in step 10. Discard the supernatant and dry the pellet in a gentle stream of nitrogen.

13. Dissolve in sterile water or buffer, as required, and store at -70°C.

---

[a]Adapted from ref. 7.

2.3.2 *Guanidinium Salt Methods*

In recent years a number of related methods have been developed for the preparation of RNA which avoid the use of phenol ($8-10$). These are based on the ability of high concentrations of guanidinium salts and other denaturing agents to dissociate ribonucleoprotein (RNP) complexes and are particularly useful with cells which have high levels of endogenous ribonuclease activity. For example, Chirgwin *et al.* (8) have described the optimal conditions for preparation of intact RNA from rat pancreas. In this procedure the tissue is rapidly disrupted in 0.1% 2-mercaptoethanol, 4 M guanidinium thiocyanate (to reduce and denature ribonuclease), centrifuged and the RNA is then precipitated from the supernatant with ethanol acidified with acetic acid.

Table 5. Extraction of RNA by the Guanidinium Thiocyanate Procedure[a].

1. Homogenise the tissue rapidly and thoroughly for 30 – 60 sec in 20 volumes of the following buffer at 20°C, using a high-speed motor-driven pestle:

    4 M guanidinium thiocyanate
    0.5% (w/v) Sarkosyl
    0.1 M 2-mercaptoethanol
    0.1% (v/v) Antifoam A (Sigma)
    25 mM sodium citrate (pH 7.0)

2. Centrifuge at 8500 g for 10 min at 10°C.

3. Collect the supernatant, add 0.025 volume of 1 M acetic acid and 0.75 volume of ethanol, shake thoroughly and leave at −20°C overnight.

4. Centrifuge at 5000 g for 10 min at −10°C.

5. Drain off the supernatant and resuspend the pellet vigorously by homogenising in 0.5 volume (relative to the volume of homogenisation buffer) of 7.5 M guanidinium chloride, 5 mM DTT, 25 mM sodium citrate (pH 7.0). Brief warming to 68°C may help.

6. Re-precipitate the RNA by adding 0.025 volume (relative to the volume of guanidinium chloride) of 1 M acetic acid and 0.5 volume of ethanol. Leave the mixture at −20°C for at least 3 h.

7. Centrifuge as in step 4.

8. Repeat steps 5 – 7 using half the volumes previously used.

9. Remove the supernatant, resuspend the pellet thoroughly in ethanol at room temperature and centrifuge as in step 4.

10. Drain the pellet and dry it under a gentle stream of clean nitrogen.

11. Dissolve the RNA in cold water (1 ml per gram of original tissue) and clarify the solution by centrifugation at 22 500 g for 10 min at 10°C.

12. Save the supernatant and resuspend the pellet in 0.5 ml of water per gram of tissue. Next, centrifuge the pellet as in step 11. Recover the supernatant.

13. Combine the supernatants from step 12, add 0.1 volume of 2 M potassium acetate (pH 5.0) and 2 volumes of ethanol, mix and then leave overnight at −20°C.

14. Centrifuge the sample at 13 000 g for 20 min at −10°C.

15. Discard the supernatant, resuspend the pellet in 95% ethanol and re-centrifuge as in step 14.

16. Dry the pellet with a stream of nitrogen and dissolve it in water or buffer (1 ml per gram of tissue) as required. Store at −70°C.

[a]Adapted from ref. 8.

The pellet is dissolved in 7.5 M guanidinium chloride and the RNA is re-precipitated with ethanol. This is repeated and the RNA is then further washed with ethanol. *Table 5* summarises the procedure, which has also been used successfully with other tissues and with cultured cells. The method may be modified by layering the initial homogenate onto a solution of caesium chloride of sufficient density to prevent DNA from being pelleted during a subsequent centrifugation (8). The RNA, which is pelleted under these conditions, is then resuspended in guanidinium chloride and washed by ethanol precipitation. This adaptation is particularly useful for preparing small amounts of RNA and is reported to give a preparation from liver which is five times more active in translation than mRNA obtained by phenol extraction (9).

Table 6. Extraction of RNA by the Lithium Chloride/Urea Procedure[a].

1.	Homogenise the frozen tissue in 10 ml of 3 M LiCl, 6 M urea per gram of tissue for 2 min at 0°C in a Waring blender (full speed). Keep the homogenate overnight at 0°C.
2.	Centrifuge the homogenate at 16 000 g for 20 min at 0°C. Discard the supernatant.
3.	Dissolve the pellet in 0.5% SDS, 10 mM Tris-HCl (pH 7.6) (10 ml per gram of original tissue) and extract this solution by shaking it for 10 min with an equal volume of chloroform:iso-amyl alcohol (24:1, v/v).
4.	Centrifuge the mixture at 10 000 g for 10 min at 18°C in Corex glass tubes using a swing-out rotor.
5.	Collect the upper (aqueous) phase and then proceed as described in steps 13 – 16 of *Table 5*.

[a]Adapted from ref. 10.

2.3.3 *LiCl/urea method*

Another method which can be used to obtain intact, translationally-active mRNAs from a variety of sources involves homogenisation of frozen tissue in 3 M LiCl, 6 M urea (10). These conditions inhibit ribonucleases and result in precipitation of RNA species which are largely free of DNA fragments, poly-saccharides and proteins. *Table 6* summarises the procedure, which was originally developed for the preparation of immunoglobulin mRNA from mouse myeloma tumours (10).

2.4 **Preparation of Polysomal RNA**

2.4.1 *High Speed Centrifugation of Polysomes*

If it is necessary to examine the nature and composition of the translated mRNA population in cells (as opposed to the total cellular mRNA) then isola-tion of polysomes must precede the RNA extraction itself. The preparation of polysomes is often also a prerequisite for the purification of individual species of mRNA. There are numerous descriptions in the literature of methods of polysome preparation by differential centrifugation. Classically, a post-mito-chondrial supernatant of a cell lysate or homogenate is layered onto a 1 M sucrose cushion in a neutral or alkaline pH buffer (pH 7.6 – 8.5) and centrifug-ed for 300 000 – 400 000 g.hours. The polysome pellets are then resuspended in buffer for further fractionation or analysis. *Table 7* describes a general pro-cedure for polysome preparation. Although this often works well, there can be problems in some cases. The relatively long time involved in polysome preparation may allow some degradation of the mRNAs by endogenous ribo-nucleases. There may be some shearing, particularly of long mRNAs, by mechanical damage during pelleting and resuspension of the polysomes. Some non-polysomal RNP particles may also contaminate the pellet. The latter two problems can be overcome by preparing the polysomes by sucrose density gra-dient centrifugation (*Table 7*, second part) rather than by differential cen-trifugation. Finally, the recovery of polysomes from some tissues is often less than quantitative, with considerable (perhaps differential) losses during the in-itial low speed centrifugations of the homogenate.

Table 7. General Procedures for Preparation and Size Fractionation of Polysomes from Eukaryotic Cells.

Preparation of Polysomes

1. Suspend the tissue or cells in 3 volumes of cold polysome buffer[a]:

    0.25 M sucrose
    0.1 M KCl
    5 mM magnesium acetate
    6 mM 2-mercaptoethanol
    20 mM Tris-HCl (pH 7.6)

    Homogenise using 3 − 20 strokes of a Teflon/glass or an all-glass (Dounce-type) homogeniser[b], with the temperature kept at 4°C.

2. Centrifuge at 10 000 *g* for 10 min to remove debris, plasma membranes, nuclei and mitochondria.

3. Add Triton X-100[c] to the post-mitochondrial supernatant to 1% (v/v) final concentration. Layer this mixture over an equal volume of polysome buffer containing 1.0 M sucrose.

4. Centrifuge at 260 000 *g* for 2 h at 4°C[d].

5. Aspirate the supernatant and discard this. Gently rinse the walls of the centrifuge tube with cold polysome buffer lacking sucrose. Drain the polysome pellet for a few minutes over tissue paper. Then gently resuspend the polysomes in cold sucrose-free polysome buffer (0.5 ml per gram of tissue) using an all-glass homogeniser.

Size Fractionation of Polysomes

1. For size-fractionation of polysomes, prepare sucrose gradients (e.g., 10 − 40%, 15 − 45%, 20 − 50%[e]) in cold 0.1 M KCl[f], 3 mM magnesium acetate, 20 mM Tris-HCl (pH 7.6).

2. Layer either detergent-treated, post-mitochondrial supernatant (step 3 above) or resuspended polysomes (step 5 above) on each gradient in as small a volume as possible. A typical load is 3 − 10 A_{260} units of polysomes per 10 ml of gradient. Centrifuge for ∼150 000 *g* for 1 h at 4°C[e].

3. Fractionate the gradients by pumping them through a flow-cell of a recording spectrophotometer set at 260 nm, using a flow rate of 1 − 5 ml/min (depending on the size of the gradients). After passage through the flow cell, collect 20 − 30 fractions per gradient for subsequent analysis or purification.

---

[a]The method described is appropriate for a tissue such as liver. The composition of the polysome buffer may be varied according to the specific properties of the tissue used. For example, for muscle a higher salt concentration is necessary to prevent precipitation of contractile proteins and the other conditions used may vary considerably from those described here (11). For many types of cultured mammalian cells, an alternative procedure to step 1, involving hypotonic swelling and detergent lysis of the cells (Chapter 9, *Table 11*), may be used to obtain extracts for preparation of polysomes.
[b]The type of homogeniser, its clearance and the number of strokes to be used will depend on the tissue and may need to be determined experimentally.
[c]Triton X-100 dissolves membranes of the endoplasmic reticulum, releasing bound polysomes. Addition of Triton X-100 is unnecessary if detergent has been used to lyse the cells in step 1.
[d]Some protocols involve centrifugation of polysomes through 2 M sucrose. In this case the time or speed of centrifugation should be increased to maximise the yield of polysomes. This can be determined by assaying the RNA content of the final polysome preparation.
[e]The sucrose concentrations used to construct the gradient, and the time and speed of centrifugation employed will depend on the nature of the sample and the size distribution of the polysomes of interest. In general, crude post-mitochondrial supernatants should be fractionated on 20 − 50% sucrose gradients to minimise contamination of the polysomes by proteins diffusing from the top of the gradient. This problem does not arise if purified polysomes are used.
[f]For muscle polysomes, use 0.25 M NaCl or 0.25 M KCl (11).

Table 8. Rapid Preparation of Polysomes by Magnesium Precipitation[a].

1.	Mince the tissue and then homogenise it thoroughly in 9 volumes of ice-cold buffer containing:

$$25 \text{ mM NaCl}$$
$$5 \text{ mM MgCl}_2$$
$$1 \text{ mg/ml heparin}$$
$$2\% \text{ Triton X-100}$$
$$25 \text{ mM Tris-HCl (pH 7.5)}$$

2.	Centrifuge the homogenate at 27 000 *g* for 5 min and collect the supernatant.
3.	Add an equal volume of the homogenisation buffer (but containing 0.2 M MgCl$_2$) and incubate at 4°C for 1 h.
4.	Layer 8 ml of this sample over 4 ml of 1 M sucrose in 25 mM NaCl, 0.1 M MgCl$_2$, 25 mM Tris-HCl (pH 7.5) and centrifuge at 27 000 *g* for 10 min.
5.	Aspirate the supernatant and part of the sucrose layer and discard.
6.	Rinse the walls of the tube with water, aspirate this and then decant the remaining sucrose.
7.	Drain the pellet (polysomes), wipe the walls of the tube with tissue paper and then resuspend the polysomes in an appropriate buffer.

[a]Adapted from ref. 12.

2.4.2 *Magnesium Precipitation of Polysomes*

As an alternative to the high speed centrifugation of polysomes, it is possible to aggregate these structures reversibly by the addition of magnesium salts to 0.1 M final concentration. This allows polysomes to be pelleted from crude homogenates by low-speed centrifugation (12). A suitable procedure is described in *Table 8*. The method has the advantage of being relatively fast and gives intact polysomes, provided the usual precautions are taken to avoid ribonuclease activity. Non-polysomal mRNP complexes can also be precipitated with magnesium ions, but free RNA cannot.

2.4.3 *Fractionation of Polysomes*

It is often desirable or essential to separate different classes of polysomes before extraction of the mRNA species they contain. This may be necessary in order to determine whether particular proteins are synthesised by specific populations of polysomes (e.g., membrane-bound versus free polysomes) or as an initial step in purifying individual species of mRNA by exploiting different properties of the polysomes in which they occur. One commonly-used procedure is the fractionation of polysomes by size on sucrose gradients. This is most useful for obtaining mRNA species which are abundant in the total population and which are considerably larger or smaller than the average mRNAs. Alternatively, if a specific antibody is available against the protein coded by a particular mRNA, an extremely powerful technique is to utilise the antigenic specificity of the nascent polypeptide chains in the polysomes containing that mRNA. These techniques are described below.

(i) *Separation of bound and free polysomes.* A number of good procedures have been published for the near-quantitative fractionation of polysomes into

Table 9. Separation of Bound and Free Polysomes from Mouse Myeloma Cells[a].

1. Suspend washed cells at a concentration of 5×10^8 cells/ml in cold hypotonic buffer (10 mM KCl, 1.5 mM $MgCl_2$, 10 mM Tris-HCl, pH 7.4). Allow the cells to swell for 5 min.

2. Homogenise with 10 strokes of tight-fitting Dounce glass homogeniser.

3. Dilute the cell homogenate 5-fold with:
   2.5 M sucrose
   0.15 M KCl
   5 mM $MgCl_2$
   50 mM Tris-HCl (pH 7.4)

4. Layer this sample over 2 volumes of 2.5 M sucrose in this buffer in a centrifuge tube. Then overlay successively with 3 volumes of 2.05 M sucrose and 1 volume of 1.2 M sucrose in the same buffer.

5. Centrifuge at 82 000 g for 5 h at 4°C using a Beckman SW28.1 rotor or its equivalent.

6. Collect fractions from the bottom by puncturing the tube. The nuclei sediment into the 2.5 M sucrose layer. The free polysomes remain in the region of the loaded material, whereas the bound polysomes float up to the 1.20 M/2.05 M sucrose interface.

7. Treat the bound polysomes with 0.5% (w/v) sodium deoxycholate and 0.5% (w/v) Brij 58 to dissolve the membranes.

8. Centrifuge the free and detergent-treated bound polysomes separately through a 15 − 30% sucrose gradient prepared in 80 mM KCl, 5 mM $MgCl_2$, 50 mM Tris-HCl (pH 7.4), onto a 4 ml cushion of 69% sucrose in the same buffer. Use a Beckman SW28 rotor, or its equivalent, run at 27 000 r.p.m. (96 000 g) for 8.5 h.

9. Collect the polysomes from the 30%/69% sucrose interface and dilute them with 1 volume of distilled water prior to storage (-70°C) or use.

[a]From refs. 16 and 17.

membrane-bound and free populations and for the extraction of intact mRNAs from these. The most thoroughly studied systems have been tissues and cells which are active in protein secretion, particularly rat liver (13 − 15) and cultured mouse myeloma cells (16 − 18). One problem is the potential contamination of the bound polysomes with lysosomes, which are rich in ribonuclease activity. Therefore, the methods have been developed to separate rough microsomal membranes from lysosomes without disrupting the latter (13 − 18). An example of a suitable protocol, developed for myeloma cells, is given in *Table 9*.

(ii) *Size fractionation of polysomes.* Provided that suitable precautions are taken to prevent the breakdown of polysomes, which can result from either ribonuclease action (Section 2.7) or ribosome 'run-off' during cell lysis (Chapter 9, Section 7.1.2), different size classes of mRNA can be fractionated by centrifugation of the corresponding polysomes through continuous sucrose density gradients. In general, the largest mRNAs form the largest polysomes and *vice-versa* although exceptions to this rule may occur if a particular mRNA species can initiate protein synthesis more or less efficiently than average, resulting in more or less dense packing of ribosomes, respectively. Because most mRNAs fall within a similar size range, this method of mRNA fractionation is only useful for partial purification of especially small mRNAs, such as those coding for histones (19) or ribosomal proteins (20), or especially large mRNAs such as myosin heavy chain mRNA (11).

The usual conditions for size fractionation of polysomes are centrifugation at 4°C through linear gradients of $10-40\%$, $15-45\%$ or $20-50\%$ sucrose in a buffer such as 0.1 M KCl, 3 mM magnesium acetate, 20 mM Tris-HCl (pH 7.6) (*Table 7*). In some cases, for example muscle polysomes (11), a higher salt concentration is necessary to prevent protein aggregation and precipitation. The time period and speed of centrifugation and the type of sucrose gradient to be used depend on the size of the polysomes of interest and must be determined empirically.

(iii) *Immunoprecipitation of polysomes.* A number of alternative strategies may be used, employing direct or indirect immunoprecipitation (4). In the former, a specific antibody is bound to the polysomes and carrier antigen and antibody are then added in amounts appropriate to form an immunoprecipitate which can be centrifuged through a sucrose cushion. When indirect immunoprecipitation is used, an insoluble complex is formed by the addition of a second antibody (anti-immunoglobulin) to the polysome-first antibody complex $(21-23)$. A typical protocol is given in *Table 10*. As an alternative to precipitation, the polysome-antibody complexes can be adsorbed to an affinity column containing a matrix of insoluble cross-linked antigen (25).

These methods have been applied successfully in many systems, notably for the purification of egg white protein mRNAs from chick oviduct and for the preparation of serum albumin mRNA from liver (4,22). The RNAs were obtained intact and in high yield after extraction of the immunoprecipitates. Payvar and Schimke (24) have defined some of the most important features

Table 10. Indirect Immunoprecipitation of Polysomes by the Double Antibody Method[a].

1. Siliconise all glassware to be used in the procedure.

2. Suspend the polysomes at $10-25$ A_{260}/ml in cold antibody buffer:
   - 0.15 M NaCl
   - 5 mM $MgCl_2$
   - 2 mg/ml heparin
   - 25 mM Tris-HCl (pH 7.1)

3. Centrifuge three times at 19 000 g for 10 min, each time retaining the supernatant. This removes insoluble polysome aggregates.

4. Add purified antibody (free of ribonuclease)[b] to ~150 μg/ml; the optimum concentration must be determined experimentally. Incubate for 40 min at 2°C.

5. Add purified second antibody[b] (directed against the first antibody) in 50- to 70-fold excess and continue the incubation for 2 h at 2°C.

6. Layer the incubation mixture over 3 ml of 0.5 M sucrose which is itself layered over 6 ml of 1.0 M sucrose, both in antibody buffer containing 0.4% Triton X-100. Centrifuge at 16 000 g for 20 min.

7. Aspirate the supernatant and sucrose solutions and discard these. Wash the walls of the tubes gently 3 times with antibody buffer, taking care not to disturb the pellets. This washing reduces non-specific contamination of the immunoprecipitates.

8. Suspend each pellet in 1 ml of antibody buffer and re-sediment through sucrose as in step 6.

[a]From refs. 23 and 25.
[b]For purification of antibodies, see refs. 20 and 23.

required for successful immunoprecipitation of mRNAs. These include high purity of the original antigen used to raise the antibodies, preparation of specific antibodies free from contamination with ribonuclease and reduction of non-specific trapping of polysomes in immunoprecipitates. When careful attention is paid to these points, it is possible to purify mRNAs which constitute as little as 1% of the total mRNA population.

2.4.4 *Extraction of RNA from Polysomes*

RNA is readily extracted from preparations of total polysomes, or from polysomal fractions separated on the basis of their association with membranes, their size, or their ability to be immunoprecipitated with specific antibodies. The principles involved are similar to those described for the isolation

Table 11. Extraction of RNA from Purified Polysomes.

SDS-Sucrose Gradient Procedure[a]

1. To suspensions of polysomes or polysome-antibody complexes, add EDTA to 50 mM, NaCl to 0.1 M and SDS to 1%. If necessary, adjust the pH to pH 7.0 by adding 0.2 volumes of 1.0 M Tris-HCl (pH 7.0).

2. Add 2 volumes of ethanol and allow the RNA to precipitate at $-20°C$ for at least 6 h.

3. Centrifuge at 14 000 g for 20 min at 0°C.

4. Dissolve the precipitate in 0.5% SDS, 5 mM EDTA, 20 mM sodium acetate (pH 5.0). Layer the solution onto a 5 – 20% sucrose gradient prepared in the same buffer and centrifuge at 195 000 g for 6 h at 20°C (using the Beckman SW41 rotor or its equivalent).

5. Analyse the gradient by pumping it through a flow cell of a recording spectrophotometer (set at 260 nm) and pool the material in the lower part of the gradient (>10S). Precipitate this with ethanol as in steps 1 and 2[b].

Proteinase K-Phenol Procedure[c]

1. Add the polysomal suspension to an equal volume of hot (100°C) SDS buffer:

    1% SDS
    0.2 M NaCl
    40 mM EDTA
    20 mM Tris-HCl (pH 7.4)

    Incubate at 100°C for 2 min.

2. Rapidly cool the mixture to 30°C and then add Proteinase K (0.5 mg/ml). Incubate for 10 min at this temperature.

3. Add concentrated Tris-HCl (pH 9.0) to 0.1 M and SDS to 1% and extract three times with an equal volume of phenol:chloroform (1:1, v/v) as described in *Table 3* (steps 4 and 5), discarding the phenol layer each time.

4. Precipitate the RNA from the aqueous phase by adding 2.5 volume of ethanol and 0.1 volumes of 2 M sodium acetate (pH 5.2) and storing at $-20°C$ overnight. Recover the precipitated RNA by centrifugation.

[a]From refs. 23 and 26.
[b]The sucrose gradient step separates mRNA and rRNA from tRNA, denatured proteins and heparin which remain near the top of the gradient. Contamination with these molecules can be further reduced by subsequent washing of the ethanol-precipitated RNA with 2.0 M LiCl, 5 mM EDTA, then with 3 M sodium acetate (pH 5.5), 5mM EDTA, and finally 70% ethanol containing 0.15 M NaCl.
[c]From ref. 16.

of total cellular or cytoplasmic RNA (Section 2.3), that is, deproteinisation of the polysomes, separation of the protein and RNA moieties, and then precipitation and washing of the RNA. Two examples of suitable procedures are outlined in *Table 11*.

2.5 Preparation of Viral RNA

Broadly similar procedures to those used for cellular RNA (Section 2.3) can be applied to the extraction of eukaryotic virus RNA species.

If purified virus preparations are available, it is relatively easy to obtain nucleic acid from the virions using a phenol method (*Table 12*) or a salt extraction procedure (26). Virus particles are disrupted by treatment with detergent (SDS or sodium deoxycholate) in the presence of EDTA before the phenol is added. Following the usual mixing and separation of phases by centrifugation, the phenol layer is re-extracted with fresh buffer and the aqueous layers combined. A further extraction of the latter with more phenol is often also carried out. Some procedures employ an elevated temperature (frequently 42°C) during these extractions. The viral RNA is precipitated with ethanol, washed and handled subsequently in exactly the same way as other kinds of RNA.

The preparation of viral RNAs from infected or virus-transformed cells is essentially the same as for RNA isolation from the corresponding uninfected cells (Section 2.3). Indeed, many species of viral RNAs can only be obtained in

Table 12. Extraction of RNA from Purified Encephalomyocarditis Virus[a].

1.	Suspend the purified virus at 1 mg/ml in 10 mM 2-mercaptoethanol, 50 mM Tris-HCl (pH 7.6) and add 112 μl of 5% sodium deoxycholate and 125 μl of 0.1 M EDTA per ml of virus suspension.
2.	Incubate at 42°C for 2 min and then add an equal volume of warm (42°C) phenol pre-equilibrated with 50 mM Tris-HCl (pH 7.6). Mix for 6 min, maintaining the temperature at 42°C.
3.	Cool for 10 min on ice and then centrifuge at 10 000 g for 10 min at 4°C. Collect the aqueous layer, avoiding any interface material.
4.	Re-extract the phenol layer with a half volume of 50 mM Tris-HCl (pH 7.6) and re-centrifuge as in steps 2 and 3. Combine the aqueous layer with that from step 3.
5.	Precipitate the RNA from the combined aqueous layers by adding 2.5 volumes of ethanol and KCl to 0.1 M followed by storage at -20°C overnight.
6.	Recover the RNA by centrifugation at 2000 g for 5 min at 4°C.
7.	Drain the RNA pellet thoroughly at -20°C and then resuspend it in 1 ml of 70% ethanol. Centrifuge as in step 6.
8.	Repeat step 7.
9.	Drain the RNA pellet and dissolve it in a small volume of sterile water. Dilute a small aliquot and measure its absorbance at 260 nm to determine the RNA concentration as described in the text (Section 2.3.1).
10.	Store the RNA solution in small aliquots at -70°C.

[a]This procedure is given as just one example of numerous protocols which have been developed for use with various types of virus. The original literature should be consulted for techniques applicable to other viruses.

this way since the RNA species involved in the processes of viral replication or cellular transformation are frequently not the same as those found in the virions. During the later stages of lytic infections, host cell protein synthesis is commonly shut off so that the translation of viral polypeptides predominates. Preparation of polysomal RNA (see Section 2.4) at the appropriate time post-infection may therefore give an mRNA fraction significantly enriched in certain viral sequences. However, since there is usually some cellular mRNA remaining in the non-polysomal (or even the polysomal) fraction, subsequent purification of the viral RNAs may be necessary. Suitable techniques are described in the next section.

2.6 Fractionation of RNA

Having isolated either total cellular RNA, cytoplasmic RNA, polysomal RNA, non-polysomal RNA, or immunoprecipitated polysomal RNA by one or more of the above procedures, the next stage is the removal of unwanted RNA species from the mRNA of interest. Several methods are available which can be broadly classified into two groups; fractionation of RNA by size and fractionation by nucleotide sequence.

2.6.1 *Fractionation by Size*

Sucrose density gradient centrifugation.   Much use has been made of sucrose density gradient centrifugation to separate RNA species with different molecular weights and sedimentation properties. Messenger RNAs have sedimentation coefficients ranging from 6S to 35S or more, with the vast majority in the 16–20S range. Sucrose gradient centrifugation is therefore particularly useful for purifying very large or very small mRNAs, and for separating many mRNAs from tRNA and the larger (28S) rRNA. A problem to be avoided is aggregation of the RNA during sedimentation through sucrose. This can be minimised by heating the RNA at 65°C for 10 min prior to fractionation or by carrying out the sucrose gradient fractionation under denaturing conditions. In the latter case, the RNA is dissolved in buffer containing a denaturant such as 0.5% SDS or 99% dimethylsulphoxide (DMSO) or 70% formamide and this is also included in the gradient itself. A protocol using 70% formamide is given in *Table 13*. After centrifugation, the RNA is recovered from appropriate fractions by ethanol precipitation.

Gel electrophoresis.   Preparative polyacrylamide or agarose gels, employing denaturing conditions such as methylmercuric hydroxide, are also widely used to fractionate RNA species and can give large quantities of pure, translationally-active mRNAs. These techniques have been described in detail in a recent volume in this Series (27) and will therefore not be considered further here.

2.6.2 *Fractionation by Nucleotide Sequence*

With the advent of knowledge concerning the nucleotide sequence of eukaryotic cell mRNAs, methods have been developed for fractionation of

Table 13. Sucrose Gradient Fractionation of RNA under Denaturing Conditions.

1. Prepare 5 – 20% sucrose gradients in 3 mM EDTA, 70% formamide, 3 mM Tris-HCl (pH 7.9)[a].

2. Dissolve the RNA in the above buffer, layer it onto the gradient and centrifuge at 186 000 g for up to 20 h at 25°C[b]. The Beckman SW41 rotor or its equivalent is suitable for this.

3. After centrifugation, pump the gradients through a flow-cell of a recording spectrophotometer set to monitor at 260 nm. After passage through the flow cell, collect fractions. Precipitate the RNA from the desired fractions by addition of 2.5 volumes of ethanol in the presence of 0.2 M NaCl. Leave at – 20°C overnight and then recover the RNA by centrifugation.

[a]Alternative protocols employ buffers containing 5 mM EDTA (pH 7.4), 0.5% SDS or 10 mM LiCl, 1 mM EDTA, 99% DMSO instead of the 70% formamide buffer.
[b]The optimal time of centrifugation will depend on the size of the RNA species of interest.

Table 14. Purification of Messenger RNA by Chromatography on Oligo(dT)-Cellulose[a].

1. Suspend the oligo (dT)-cellulose in sterile (high salt) loading buffer:
   0.5 M NaCl (or KCl)
   1 mM EDTA
   0.1% SDS
   10 mM Tris-HCl (pH 7.5).

2. Pour a 1 – 2 ml column of this oligo(dT)-cellulose in a sterile syringe or Pasteur pipette and wash it successively with:
   (i) H_2O
   (ii) 0.1 M NaOH, 5 mM EDTA
   (iii) H_2O
   Continue washing the oligo(dT)-cellulose with water until the pH of the effluent is close to pH 7.0.

3. Re-equilibrate the column in sterile loading buffer.

4. Heat the RNA solution in water at 65°C for 5 min. Add an equal volume of 2-fold concentrated loading buffer, cool and load this sample onto the column.

5. Collect the unbound RNA which elutes. Heat this to 65°C, cool and re-apply it to the column as in step 4.

6. Wash the column with loading buffer (at least 5 column volumes) until the A_{260} of the effluent is close to zero.

7. Elute poly(A)[+] RNA with sterile low salt buffer:
   1 mM EDTA
   0.05% SDS
   10 mM Tris-HCl (pH 7.5)

8. Re-adjust the poly(A)[+] RNA solution to 0.5 M NaCl or KCl (final concentration) and then repeat steps 3 – 7.

9. Add sodium acetate (pH 5.2) to the poly(A)[+] RNA sample to 0.3 M final concentration and precipitate the RNA with 2.5 volumes of ethanol at – 20°C overnight.

10. Collect the RNA by centrifugation (10 000 g, 10 min), rinse the pellet in 70% ethanol, dry it and dissolve it in sterile water. Store at – 70°C.

11. Regenerate the oligo(dT)-cellulose by washing it successively with 0.1 M NaOH, 5 mM EDTA then water and finally loading buffer as in steps 2 and 3.

[a]The method may be adapted for use in a batch procedure, employing 0.6 g dry weight of oligo(dT)-cellulose per mg of RNA. Several centrifugations are required to wash the bound RNA with loading buffer before eluting the RNA with low salt buffer. This procedure is useful for handling multiple samples.

226

mRNAs on the basis of their characteristic primary structures.

Affinity chromatography on oligo(dT)-cellulose. This is the most widely used method for the purification of mRNA molecules which have a 3' poly(A) tract, that is, polyadenylated [poly(A)$^+$] mRNAs. The details of the procedure vary somewhat between laboratories but, typically, up to 10 mg of total RNA is loaded slowly on to a small (1−2 ml) column at room temperature in a buffer containing 0.5 M NaCl or KCl, 1 mM EDTA, 0.1% SDS, 10 mM Tris-HCl (pH 7.5) and the column is then washed extensively with this solution (*Table 14*). RNA species lacking poly(A) (principally rRNA and tRNA) fail to bind to oligo(dT)-cellulose and hence pass through the column. The poly(A)$^+$ mRNA does bind and can then be eluted by lowering the ionic strength. Usually a second passage through the column is necessary to remove most contaminating rRNA from poly(A)$^+$ mRNAs. The latter are then recovered by precipitation with ethanol in the presence of 0.3 M sodium acetate (pH 5.2) *(Table 14)*. Krystosek *et al.* (28) have shown that polysomes can be dissolved in buffer containing SDS and the RNA successfully fractionated on oligo(dT)-cellulose directly, without prior deproteinisation.

Affinity chromatography on poly(U)-Sepharose. Messenger RNA molecules which have only short 3' poly(A) sequences (<20 nucleotides) do not bind well to oligo(dT)-cellulose but often can be retained on poly(U)-Sepharose (which has a longer ligand). In this case, elution of poly(A)$^+$ mRNA requires stronger denaturing conditions (70−90% formamide) *(Table 15)*. Messenger RNA can also be recovered from these affinity columns by thermal elution rather than by lowering the ionic strength (30,31). This has been extended to the fractionation of poly(A)$^+$ mRNA on the basis of poly(A) length by elution with stepwise temperature increases (29). A similar technique has been used for

Table 15. Purification of Messenger RNA by Chromatography on Poly(U)-Sepharose[a].

1.	Suspend the poly(U)-Sepharose in sterile elution buffer:
	10 mM EDTA
	0.2% Sarkosyl
	90% formamide
	10 mM potassium phosphate (pH 7.5)
2.	Pour this suspension into a sterile syringe to give a column with dimensions 0.5 cm x 2.0 cm.
3.	Wash the column with 3 ml of elution buffer.
4.	Wash the column with 3 ml of sterile loading buffer;
	0.5 M NaCl
	10 mM EDTA
	25% formamide
	50 mM Tris-HCl (pH 7.5)
5.	Dissolve the RNA sample in loading buffer and apply this to the column.
6.	Elute the unbound RNA with three successive 1 ml portions of loading buffer.
7.	Elute the bound poly(A)$^+$ RNA with six successive 0.25 ml portions of elution buffer (step 1).
8.	Precipitate the RNA with ethanol and recover it as described in *Table 14*, steps 9 and 10.
9.	Regenerate the poly(U)-Sepharose by re-equilibrating it in loading buffer (step 3).

[a]From ref. 30.

chromatography of mRNP particles on oligo(dT)-cellulose (32).

Purification of poly(A)$^+$ mRNAs by other methods. The presence of poly(A) tracts in mRNAs confers a number of other properties on these molecules which can be useful during their purification. Poly(A)$^+$ mRNAs partition in the phenol phase during phenol extraction at neutral pH, but can be recovered upon re-extraction at pH 9.0 (33). They also bind to unmodified cellulose (34) and to Millipore-type cellulose ester filters (35) in 0.5 M KCl and can be subsequently eluted with 0.05% SDS at low ionic strength. However, none of these methods is as specific as affinity chromatography on oligo(dT)-cellulose or poly(U)-Sepharose.

Purification of poly(A)$^-$ mRNA. The purification of cellular mRNAs which apparently lack 3' poly(A) tails presents more of a problem. However, Greenberg (36) has described a method based on the differential buoyant densities of EDTA-dissociated ribosomes and mRNP particles in caesium sulphate gradients containing 15% DMSO. The mRNP fraction can subsequently be deproteinised and then fractionated into its poly(A)$^+$ and poly(A)$^-$ mRNA populations as described above.

Purification of specific mRNAs. The isolation of particular mRNA species from total RNA preparations is possible if a DNA probe complementary to the mRNA sequences is available for hybridisation-selection. The DNA is immobilised on nitrocellulose or chemically-activated paper and then used to bind individual mRNAs which, when eluted, are translated *in vitro* to give specific polypeptide products (37,38). This method has great potential and is likely to achieve wide usage as cloned DNA probes become more generally available. A protocol that may be applied to the isolation of mRNA is described in Chapter 2, *Table 4*.

2.7 Ribonuclease Inhibitors

Several methods have already been described in this chapter for minimising ribonuclease activity (Sections 2.1 – 2.4 and *Table 1*). In addition to these general procedures, there are a number of specific ways of inhibiting ribonuclease activity during mRNA purification which are briefly summarised here.

Heparin. This sulphated polysaccharide is widely used as a ribonuclease inhibitor. It adsorbs the nucleases and competitively inhibits them. Addition of this compound to buffers at a concentration of approximately 1 mg/ml during subcellular fractionation and polysome preparation can substantially improve the yield and translational activity of purified mRNA. However, the heparin must be removed from the RNA (with which it co-purifies) before translational assays are carried out since it is a potential inhibitor of polypeptide chain initiation. This can be achieved by washing the ethanol-precipitated RNA with 3 M sodium acetate or 2 M lithium chloride. Other poly-anionic compounds, such as polyvinylsulphate, may act similarly to heparin (39), presumably by binding ribonucleases in competition with the

RNA. In early experiments, the diatomaceous earth, bentonite, was used to adsorb ribonucleases but it is not completely effective.

Ribonuclease inhibitor proteins. Proteins which bind ribonuclease and inhibit it (reversibly) have been characterised and purified from rat liver and human placenta. The latter inhibitor is available commercially (*Table 2*) and is useful not only for preventing polysome degradation during subcellular fractionation, and RNA degradation during extraction (*Table 4*), but also for improving the yield of translation products when added with exogenous mRNA to the wheat germ cell-free system (Chapter 9, Section 3.2) (40).

Proteinase K. The enzyme, Proteinase K, is useful since it it active in the presence of SDS and rapidly inactivates nucleases from many sources. It is also advantageous in reducing the amount of the protein interface obtained after phenol extraction (*Table 4*).

Agents which minimise ribonuclease activity. Use of vanadyl ribonucleoside complexes, which bind to ribonucleases and inhibit them, may be helpful in some cases (e.g., *Table 4*). In other situations, the simple precaution of raising the Mg^{2+} concentration to 50 mM or more may protect polysomes from degradation during their isolation (41).

3. ACKNOWLEDGEMENTS

The author is supported by a Career Development Award from the Cancer Research Campaign.

4. REFERENCES

1. Lodish,H.F. (1976) *Annu. Rev. Biochem.,* **45**, 39.
2. Moldave,K. and Grossman,L., eds. (1979) *Methods in Enzymology,* Vol. **60**, published by Academic Press, Inc., NY and London.
3. Ehrenberg,L., Fedorcsak,I. and Solymosy,F. (1976) in *Progress in Nucleic Acid Research and Molecular Biology,* Vol. **16**, Cohn,W.E. (ed.), Academic Press, Inc., NY and London, p. 189.
4. Taylor,J.M. (1979) *Annu. Rev. Biochem.,* **48**, 681.
5. Verma,D.P.S., Maclachlan,G.A., Byrne,H. and Ewings,D. (1975) *J. Biol. Chem.,* **250**, 1019.
6. Shore,G.C. and Tata,J.R. (1977) *J. Cell Biol.,* **72**, 726.
7. Favaloro,J., Treisman,R. and Kamen,R. (1980) in *Methods in Enzymology,* Vol. **65**, Grossman,L. and Moldave,K. (eds.), Academic Press, NY and London, p. 718.
8. Chirgwin,J.M., Przybyla,A.E., MacDonald,R.J. and Rutter,W.J. (1979) *Biochemistry (Wash.),* **18**, 5294.
9. Raymond,Y. and Shore,G.C. (1979) *J. Biol. Chem.,* **254**, 9335.
10. Auffray,C. and Rougeon,F. (1979) *Eur. J. Biochem.,* **107**, 303.
11. Buckingham,M.E., Cohen,A. and Gros,F. (1976) *J. Mol. Biol.,* **103**, 611.
12. Palmiter,R.D. (1974) *Biochemistry (Wash.),* **13**, 3606.
13. Ramsey,J.C. and Steele,W.J. (1976) *Biochemistry (Wash.),* **15**, 1704.
14. Dissous,C., Lempereur,C., Verwaerde,C. and Krembel,J. (1978) *Eur. J. Biochem.,* **83**, 17.
15. Adesnik,M. and Maschio,F. (1981) *Eur. J. Biochem.,* **114**, 271.
16. Mechler,B. and Rabbitts,T.H. (1981) *J. Cell. Biol.,* **88**, 29.
17. Mechler,B. (1981) *J. Cell. Biol.,* **88**, 37.
18. Mechler,B. (1981) *J. Cell. Biol.,* **88**, 42.
19. Schochetman,G. and Perry,R.P. (1972) *J. Mol. Biol.,* **63**, 591.
20. Nabeshima,Y.-I., Imai,K. and Ogata,K. (1979) *Biochim. Biophys. Acta,* **564**, 105.
21. Schechter,I. (1974) *Biochemistry (Wash.),* **13**, 1875.
22. Shapiro,D.J., Taylor,J.M., McKnight,G.S., Palacios,R., Gonzalez,C., Kiely,M.L. and Schimke,R.T. (1974) *J. Biol. Chem.,* **249**, 3665.

23. Shapiro,D.J. and Schimke,R.T. (1975) *J. Biol. Chem.,* **250**, 1759.
24. Payvar,F. and Schimke,R.T. (1979) *Eur. J. Biochem.,* **101**, 271.
25. Palacios,R., Sullivan,D., Summers,N.M., Kiely,M.L. and Schimke,R.T. (1973) *J. Biol. Chem.,* **248**, 540.
26. Villa-Komaroff,L., McDowell,M., Baltimore,D. and Lodish,H.F. (1974) in *Methods in Enzymology,* Vol. **30**, Moldave,K. and Grossman,L. (eds.), Academic Press, Inc., New York and London, p. 709.
27. Rickwood,D. and Hames,B.D., eds. (1982) *Gel Electrophoresis of Nucleic Acids,* published by IRL Press Ltd., Oxford and Washington, D.C.
28. Krystosek,A., Cawthon,M.L. and Kabat,D. (1975) *J. Biol. Chem.,* **250**, 6077.
29. Lindberg,U. and Persson,T. (1972) *Eur. J. Biochem.,* **31**, 246.
30. Firtel,R.A. and Lodish,H.F. (1973) *J. Mol. Biol.,* **79**, 295.
31. Rhoads,R.E. (1975) *J. Biol. Chem.,* **250**, 8088.
32. Jain,S.K., Pluskal,M.G. and Sarkar,S. (1979) *FEBS Lett.,* **97**, 84.
33. Brawerman,G., Mendecki,J. and Lee,S.Y. (1972) *Biochemistry (Wash.),* **11**, 637.
34. Schutz,G., Beato,M. and Feigelson,P. (1972) *Biochem. Biophys. Res. Commun.,* **49**, 680.
35. Gorski,J., Morrison,M.R., Merkel,C.G. and Lingrel,J.B. (1974) *J. Mol. Biol.,* **86**, 363.
36. Greenberg,J.R. (1976) *Biochemistry (Wash.),* **15**, 3516.
37. Ricciardi,R.P., Miller,J.S. and Roberts,B.E. (1979) *Proc. Natl. Acad. Sci. USA,* **76**, 4927.
38. Bedard,L. (1983) *J. Virol.,* **46**, 656.
39. Fellig,J. and Wiley,C.E. (1959) *Arch. Biochem. Biophys.,* **85**, 313.
40. Scheele,G. and Blackburn,P. (1979) *Proc. Natl. Acad. Sci. USA,* **76**, 4898.
41. Morton,B., Nwizu,C., Henshaw,E.C., Hirsch,C.A. and Hiatt,H.H. (1975) *Biochim. Biophys. Acta,* **395**, 28.

CHAPTER 9

Translation of Eukaryotic Messenger RNA in Cell-free Extracts

MICHAEL J. CLEMENS

1. INTRODUCTION

Several cell-free protein-synthesising systems have been used in recent years for the translation of eukaryotic mRNAs. Of these, the rabbit reticulocyte lysate and the wheat germ extract have received the most attention. Both have a number of advantages over other systems and are easy to prepare, provided that good laboratory practice is adopted to avoid contamination with ribonucleases and other extraneous factors. In addition to these systems, extracts from a variety of cultured cells and transplantable ascites tumour cells have been used which initiate translation and produce full-size polypeptide products. This chapter describes the preparation and properties of these eukaryotic mRNA translation systems. It also describes the use of micrococcal nuclease treatment to eliminate endogenous mRNA-coded protein synthesis, considers the variety of ways in which the products of exogenous mRNA translation can be characterised and discusses how cell-free systems may be used to provide information on the fate of primary translation products and on the mechanisms which regulate protein synthesis. Prokaryotic translation systems are not covered here since coupled transcription-translation systems (Chapter 7) are now commonly used instead for studies on many aspects of bacterial and phage gene expression.

2. THE RETICULOCYTE LYSATE SYSTEM

Although reticulocyte lysate may be obtained from a number of commercial sources (*Table 1*) which are undoubtedly convenient and provide preparations already optimised for translation of mRNA (Section 2.2), this is an extremely expensive way of obtaining lysate. At a cost of at least £24 per ml at the time of writing, the lysate obtained from one rabbit has a commercial value of approximately £500. Therefore, it makes considerable economic sense to prepare the lysate oneself.

2.1 Preparation and Storage of Lysate

A reticulocyte lysate is made by lysing red blood cells obtained from rabbits which are recovering from experimentally-induced anaemia. The recommended procedure is summarised in *Table 2*. A number of different protocols have

Translation in Eukaryotic Cell-free Extracts

Table 1. Sources of Reagents for Cell-free Translation Studies.

Translation systems

Reticulocyte lysate	Collaborative Research
Nuclease-treated reticulocyte lysate	Amersham International
	Bethesda Research Laboratories Inc.
	Collaborative Research
	New England Nuclear
	P & S Biochemicals
Wheat germ cell-free system	Bethesda Research Laboratories Inc.
Nuclease-treated wheat germ system	Bethesda Research Laboratories Inc.

Reagents

E. coli aminoacyl-tRNA synthetases	Sigma Chemical Co.
Edeine	Calbiochem
Globin mRNA	Bethesda Research Laboratories Inc.
Mammalian tRNA	Boehringer
Micrococcal nuclease	Boehringer
	P-L Biochemicals
Microsomal membranes	New England Nuclear
TMV-RNA	Amersham International

Table 2. Preparation of Reticulocyte Lysate.

1. Inject each of several rabbits subcutaneously with 2.5 ml of acetylphenylhydrazine (10 mg/ml in water) daily for 4 days.

2. Wait another 5 days before bleeding the rabbit as follows.

3. Inject $1.0-1.5$ ml of Nembutal plus 1 ml of 1% heparin in 0.9% NaCl into an ear vein. Check for complete anaesthesia.

4. Collect the blood by cardiac puncture and stand it in ice.

5. Centrifuge the blood in a swing-out rotor at 1200 g for 10 min at 4°C.

6. Aspirate the supernatant and resuspend the cells with a glass rod in 200 ml of:

    0.14 M NaCl
    5 mM KCl
    5 mM magnesium acetate
    5 mM glucose
    5 mM Hepes (pH 7.2 with KOH).

7. Centrifuge as in step 5.

8. Repeat steps 6 and 7 twice more. Finally centrifuge the preparation at 1600 g for 15 min.

9. Aspirate the supernatant, measure the packed cell volume and lyse the cells by addition of $1.0-1.5$ volumes of cold double-distilled water. Mix thoroughly.

10. Centrifuge at 21 000 g for 20 min at 4°C in Corex (Dupont) or transparent plastic tubes[a].

11. Collect the supernatant taking care not to contaminate it with fragments of the pellet (stroma). Dispense in 1 ml aliquots, freeze rapidly and store under liquid nitrogen.

[a]To aid in the recovery of uncontaminated supernatant in step 11.

been used for the treatment of the animals. One which works well consists of four daily injections of 25 mg of fresh acetylphenylhydrazine (10 mg/ml in water) subcutaneously into $2.0 - 2.5$ kg male rabbits to cause anaemia, followed by five days of recovery before blood is taken (*Table 2*). This schedule allows the blood haematocrit to rise, thus giving more red blood cells and hence more lysate per rabbit, and it also gives time for the acetylphenylhydrazine and its breakdown products to disappear from the circulation. It is convenient to treat several rabbits at once to save time. The animals should remain healthy (although anaemic) throughout.

On day 9 after the initial injection, the rabbits are anaesthetised by the injection of Nembutal mixed with heparin into an ear vein (*Table 2*). The vein is readily visible at the ear margin when a little fur is shaved off. The injection requires some experience and is easier if an assistant holds the rabbit firmly to keep it calm. Anaesthesia occurs rapidly but the heart should still beat firmly. With the anaesthetised animal on its back, bleeding is then performed by cardiac puncture using a large needle attached to flexible plastic tubing. The heart should pump the blood out quite rapidly. If this does not work, or the heart stops, the blood can still be collected by opening the thorax, cutting the aorta and then sucking up the blood using a syringe. It will not clot because of the heparin present in the injection and should be collected and kept on ice at this stage. There is often considerable variation in the relative activities of reticulocyte lysates prepared from different rabbits. Therefore, it is advisable to keep the blood from each animal separate in case the least active preparations contain inhibitors which would impair the activity of the better lysates. The yield of blood should be up to 100 ml per rabbit. The reticulocyte count can be determined by mixing a few drops of blood with 1% brilliant cresyl blue in saline and microscopically examining a smear under an oil immersion lens. It is a common experience, however, that the reticulocyte count does not always correlate with the translational activity of the lysate obtained.

The cells are harvested by centrifugation, washed and then lysed with cold water (*Table 2*). After centrifugation to remove the stroma, the lysate can either be frozen immediately in 1 ml aliquots in liquid nitrogen and stored in this or first treated with micrococcal nuclease (Section 2.4) before storage. Reticulocyte lysates retain their activity during storage in liquid nitrogen for very long periods (at least two years). At $-70°C$ they keep for several months, but storage at $-20°C$ is not recommended. Lysates should be sealed in plastic bags before transportation in dry ice, to prevent loss of activity due to exposure to high concentrations of carbon dioxide. Once thawed for use, it is best not to re-freeze lysates, although this may be done once without too much loss of activity.

2.2 Optimisation of Activity

Each reticulocyte lysate that is prepared should be characterised by monitoring protein synthetic activity in response to different concentrations of haemin, a compound which prevents activation of a translational repressor in this system (1). Such an experiment will indicate not only how much amino acid incor-

Table 3. Preparation of a Haemin Stock Solution.

1.	Dissolve 6.5 mg of haemin in 0.25 ml of 1 M KOH.
2.	Add in order, mixing after each addition:
	0.5 ml of 0.2 M Tris-HCl (pH 7.8).
	8.9 ml of ethylene glycol.
	0.25 ml of 1 M HCl.
3.	Store at $-20°C$. The final concentration of haemin is 1 mM.

poration occurs during translation of endogenous mRNA under standard conditions, but also the extent to which the lysate is stimulated by haemin and the optimal haemin concentration which produces this effect. In general, the most active lysates also show the greatest response to haemin.

Haemin stock solutions are difficult to make and store unless the correct procedure is followed. The solutions are stable only if prepared in 90% ethylene glycol buffered with Tris-HCl (pH 7.8). *Table 3* summarises the correct procedure. In order to determine the haemin optimum, a lysate is incubated with various amounts of this haemin solution to give haemin concentrations up to 40 μM under conditions suitable for protein synthesis to take place (Section 2.3) using [^{14}C]leucine as the labelled amino acid. After 60 min at 30°C, the amount of radioactive leucine incorporated is measured. In most cases, good lysates show an approximately 10-fold stimulation of translation by haemin, with an optimum of $10-20$ μM. *Figure 1* illustrates a typical result. The calculated rate of incorporation of added leucine under optimal conditions is of the order of 50 pmol/min/20 μl incubation mixture for good, non-nuclease-treated lysates, although the true rate is higher because the lysate itself contains a significant pool of leucine.

2.3 Incubation Conditions for Protein Synthesis

Reticulocyte lysate contains substantial amounts of endogenous mRNAs (mainly those for α- and β-globins) which can be translated under suitable conditions. In this mode, the reticulocyte lysate is now mainly used to study the mechanism of protein synthesis itself (Section 7.1). For the translation of exogenous mRNAs, purified from other sources (Chapter 8), the lysate is usually first treated with micrococcal nuclease (Section 2.4) to reduce the translation due to the endogenous globin mRNAs. The incubation conditions described below are optimal both for *endogenous* mRNA translation and, with minor modifications (Section 2.4.2), for the translation of *exogenous* mRNAs by nuclease-treated lysate.

When the reticulocyte lysate is used in a cell-free protein synthesis assay, it is important that the lysate itself should remain as concentrated as possible to ensure maximum translational activity. Once the optimal haemin concentration (Section 2.2) is known, the frozen lysate is best thawed in the presence of twice this haemin concentration and the lysate should comprise 50% of the final volume of the assay mixture. A convenient way of setting up the incubations is described in *Table 4*. The final concentration of each component added to the

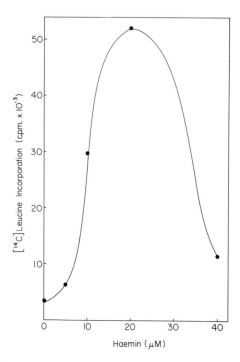

Figure 1. Determination of the haemin optimum for protein synthesis in the reticulocyte lysate. A freshly prepared lysate was incubated with the indicated concentrations of haemin for 60 min at 30°C under the conditions described in *Table 4*. Aliquots (20 μl) were taken for determination of the radioactivity incorporated into protein as described in *Table 6*.

lysate is given in *Table 5* but, in addition, the lysate itself contributes K^+ (up to 20 mM) and Mg^{2+} (1 mM). Some investigators omit ATP and GTP from the salts-amino acid-energy mixture (*Table 4*) since these are already present in the lysate. If this is done, the amount of magnesium acetate added is reduced so that its final concentration is 0.5 mM.

When *exogenous* mRNAs are to be translated, these are added to the reticulocyte lysate (usually nuclease-treated: see Section 2.4) in place of an equivalent volume of water (*Table 4*), to give a final concentration usually in the range of 5 – 100 μg/ml. In these cases, the optimal salt conditions may vary from those given in *Table 5* and therefore should be determined in preliminary experiments.

Incubations are conveniently carried out in disposable microcentrifuge tubes, using Finn pipettes and Drummond pipettes (or their equivalents) to make additions and removals. It is important to ensure thorough mixing in the tubes (without frothing) since the lysate is very dense. All components should be kept on ice before commencing the incubations, which are best performed at 30°C for prolonged protein synthesis. Under optimal conditions, amino acid incorporation should be linear for at least 60 min. If it ceases before this, it is worth checking the fraction of the labelled amino acid which has been incorporated because very active lysates can rapidly exhaust the amino acid pool

Table 4. Incubation Conditions for the Reticulocyte Lysate.

Stock solutions

Salts-amino acids-energy mixture (5x stock)[a]:

   0.375 M KCl[b]
   10 mM magnesium acetate
   15 mM glucose
   0.25 – 1.0 mM amino acids
   20 μCi/ml [^{14}C]leucine (20 – 30 mCi/mmol) *or* up to 2.5 mCi/ml [^{35}S]methionine (~ 1200 Ci/mmol).
   5 mM ATP
   1 mM GTP
   50 mM Tris-HCl (pH 7.6)

Creatine phosphate – creatine phosphokinase (10x stock)[c]:
   70 mM creatine phosphate
   10 mg/ml creatine phosphokinase

Reaction mixture

For each incubation (final volume 100 μl), mix the following components carefully on ice in a sterile microcentrifuge tube:

   50 μl of reticulocyte lysate (thawed immediately before use and with haemin at twice optimal concentration)
   20 μl of salts — amino acid — energy mixture (5x stock)
   10 μl of creatine phosphate – creatine phosphokinase (10x stock)
   20 μl of water or other components (e.g., exogenous mRNA, if applicable, to ~ 5 – 100 μg/ml final concentration).

Incubate at 30°C.

---

[a]Stable for prolonged periods at −20°C (if [^{14}C]leucine is used) or −70°C (if [^{35}S]methionine is used).
[b]Substitution of potassium acetate for potassium chloride will allow a higher K^+ ion concentration to be used for translation of certain exogenous mRNAs since high chloride ion concentrations inhibit initiation.
[c]Make fresh just before use.

Table 5. Concentrations of Added Components in Incubations with Reticulocyte Lysate.

---

75 mM KCl[a]

2 mM magnesium acetate

3 mM glucose

50 – 200 μM amino acids

4 μCi/ml [^{14}C]leucine (*or* up to 500 μCi/ml [^{35}S]methionine)

1 mM ATP

0.2 mM GTP

10 mM Tris-HCl (pH 7.6)

7 mM creatine phosphate

1 mg/ml creatine phosphokinase

10 – 20 μM haemin

Up to 100 μg/ml mRNA[b]

---

[a]See the second footnote to *Table 4*.
[b]When the translation of exogenous mRNA is being studied.

unless sufficient amino acids are added. Thus the use of high specific radio-activity precursors at low concentrations is not always an advantage in this system, although this is more of a problem with radiolabelled leucine than with [^{35}S]methionine.

Typically, the lysate may contribute up to 5 μM methionine, 10 μM leucine and more of some other amino acids. However, reticulocyte lysates are not normally fractionated by gel filtration to lower the concentration of these endogenous amino acids since this treatment also removes other low molecular weight components (polyamines, sugar phosphates and substrates which generate reducing power) and so severely impairs protein synthesis. This can only be reversed by the re-addition of these compounds to the lysate (2). This is easily achieved but may require further optimisation experiments.

The extent of amino acid incorporation at the end of an incubation is easily assessed by pipetting small aliquots (2 – 10 μl) onto numbered Whatman No. 1 filter paper discs followed by precipitation with TCA (*Table 6*). Charged tRNA is hydrolysed by heating the discs in hot TCA and the samples are then bleached with hydrogen peroxide before determination of radioactivity using a liquid scintillation counter. As an alternative to this disc method, some investigators transfer a sample of each incubation mixture into 0.5 ml of water to which is added NaOH and hydrogen peroxide (*Table 6*). In this case, the NaOH hydrolyses charged tRNA and bleaching with hydrogen peroxide occurs simultaneously. The sample proteins are recovered by precipitation with TCA followed by filtration onto glass fibre filters and then determination of radioactivity as before.

As a guide to the endogenous protein synthetic activity of a typical lysate, TCA precipitation of 5 μl from an assay containing 4 μCi/ml of [^{14}C]leucine (20 mCi/mmol) should give up to 20 000 c.p.m. after a 60 min incubation. When calculating the amount of protein synthesised, remember the contribution of the endogenous amino acids to the final specific radioactivity of the labelled precursor.

2.4 Micrococcal Nuclease Treatment

A very important development in the use of cell-free protein synthesising systems for the translation of exogenous mRNAs came in 1976 when Pelham and Jackson described a method for eliminating endogenous mRNA by treatment of reticulocyte lysates with micrococcal nuclease (3). The elegant simplicity of this procedure lies in the fact that this enzyme (from *Staphylococcus aureus*) is entirely calcium-dependent for its activity and may therefore be inactivated by subsequent addition of the Ca^{2+}-chelating agent, EGTA. Furthermore, conditions exist in which the nuclease destroys mRNAs without significantly damaging the biological activity of tRNA or ribosomes. Thus a cell-free system can be obtained in which the background level of amino acid incorporation is virtually eliminated but which will translate added mRNAs with almost undiminished activity.

Table 6. Determination of Amino Acid Incorporation in the Reticulocyte Lysate System[a].

Filter-Paper Disc Method

1. Pipette $2-10$ μl of each incubation mixture onto separate, numbered Whatman No.1 filter paper discs (2.5 cm diameter) laid on a sheet of aluminium foil.

2. Place the discs into a beaker of ice-cold 5% TCA and wash them with gentle swirling for 15 min.

3. Transfer the discs to 5% TCA at 90°C and incubate at this temperature for 15 min to hydrolyse charged tRNA.

4. Transfer the discs to fresh 5% TCA at room temperature and wash them by swirling gently for 15 min.

5. Wash the discs successively in excess absolute ethanol and then acetone with gentle swirling, each for about 1 min.

6. Allow the discs to dry at room temperature on aluminium foil.

7. Add 40 μl of 10% (w/v) hydrogen peroxide to each disc and leave at room temperature for ~ 1 h. This bleaches the samples and thus prevents colour quenching during scintillation counting.

8. After the 1 h incubation in step 7, the hydrogen peroxide will have dried. Place each disc in a vial or vial insert and determine the radioactivity using a toluene-based scintillation fluid.

Filtration Method

1. Transfer $2-10$ μl of each incubation mixture into 0.5 ml of water and then add 0.5 ml of 1.0 M NaOH (to hydrolyse aminoacyl-tRNA) and 50 μl of 30% (w/v) hydrogen peroxide (to bleach the samples).

2. As soon as the colour of each sample is bleached, add 0.25 ml of 50% TCA to give a final concentration of 10% TCA. Mix and then keep the samples on ice for 30 min to precipitate the protein.

3. Filter each sample through a glass fibre (GF/C) disc (2.5 cm diameter) using a filtration manifold attached to a vacuum line.

4. Wash each disc whilst on the filter manifold three times each with 10 ml of 5% TCA.

5. Wash each disc with 10 ml of absolute ethanol.

6. Dry the discs under an infra-red lamp and then determine the radioactivity using a toluene-based scintillation fluid.

[a]These methods can also be used for determining the amino acid incorporation in translation systems from other sources but in these cases bleaching with hydrogen peroxide is not necessary.

2.4.1 *Preparation of Nuclease-treated Reticulocyte Lysate*

Table 7 summarises the simple procedure for obtaining nuclease-treated, mRNA-dependent reticulocyte lysate. In essence, the method involves just two steps. In the first, the lysate is incubated with micrococcal nuclease in the presence of 1 mM $CaCl_2$ for $15-20$ min at 20°C. Then EGTA is added to a concentration of 2 mM and the mixture is placed on ice. It can either be used immediately for a translation assay or stored in aliquots in liquid nitrogen. Haemin (40 μM) should be included in the incubation with nuclease to prevent activation of the haem-controlled translational repressor (Section 2.2). This gives a final haemin concentration of 20 μM when the nuclease-treated

Table 7. Preparation of Micrococcal Nuclease-treated Reticulocyte Lysate.

1.	Add sufficient 0.2 M $CaCl_2$ and 1 mM haemin stock solution (*Table 3*) to the lysate to give final concentrations of 1 mM and 40 μM[a] respectively.
2.	Add micrococcal nuclease to 25 – 100 units/ml and incubate for 15 – 20 min at 20°C.
3.	Add sufficient 0.1 M EGTA (pH 7.0) to give a final concentration of 2 mM and return the lysate to ice.
4.	The nuclease-treated lysate can either be used immediately or stored as aliquots in liquid nitrogen.

[a]This yields a final haemin concentration of 20 μM during *in vitro* translation since the lysate is diluted 2-fold for the incubation (*Table 4*). If the haemin optimum of the untreated lysate (Section 2.2) is less than 20 μM, use a correspondingly lower concentration of haemin during nuclease treatment.

reticulocyte lysate is used for *in vitro* translation since the lysate is diluted so as to comprise 50% of the final volume of the incubation mixture (*Table 4*). If the haemin optimum of the untreated lysate (Section 2.2) is less than 20 μM, a correspondingly lower haemin concentration should be used during micrococcal nuclease treatment. Creatine phosphokinase can also be added during nuclease treatment although it is not essential at this stage. One report (4) suggests that 120 μM thymidine $3',5'$-bisphosphate is better than EGTA for inactivating the nuclease in a cell extract.

The optimal final concentration of nuclease to be used is ~ 100 units/ml, but this may vary between lysates and should be determined for each new preparation. Too low a concentration will not inactivate all endogenous mRNA and will leave a high background of protein synthesis. Too much enzyme may damage ribosomes and tRNAs and will result in less stimulation of amino acid incorporation by added mRNA.

2.4.2 *Incubation Conditions for Translation of Exogenous Messenger RNA*

With very few modifications, nuclease-treated reticulocyte lysate is used with exogenous mRNAs under exactly the same conditions as for untreated lysate (Section 2.3) except that extra tRNA (calf liver) is usually added to 50 μg/ml. This is required not because of destruction of endogenous tRNA, but rather because the composition of the reticulocyte tRNA population is related to the amino acid content of rabbit globins and may not be optimal for translation of other mRNAs such as tobacco mosaic virus (TMV) RNA (3). It may also be necessary to re-optimise the Mg^{2+} requirement of the system, to compensate for the weak chelating ability of excess EGTA towards this ion.

Nuclease-treated reticulocyte lysate is exquisitely sensitive to low concentrations of mRNAs and will detectably translate as little as 0.3 μg/ml. At the other extreme, up to 200 μg/ml of some mRNAs are translated well. A curve of amino acid incorporation versus mRNA concentration should be constructed with each new preparation of RNA, including a control with no added mRNA and one with a known amount of a standard mRNA, the translation characteristics of which are established. Incubations should be allowed to con-

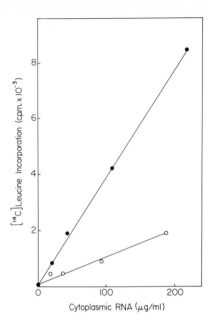

Cytoplasmic RNA (μg/ml)

Figure 2. Translation of exogenous total cytoplasmic RNA preparations in the nuclease-treated reticulocyte lysate system. Reticulocyte total cytoplasmic RNA was extracted from 6-fold diluted lysates by the method described in Chapter 8, *Table 3*. Various amounts of the RNA were translated *in vitro* under the conditions described in *Table 4* for 60 min at 30°C and then 10 μl aliquots were taken for determination of the radioactivity incorporated into protein as described in *Table 6*. The data for two different RNA preparations are illustrated: ●——●, RNA from a control lysate (previously incubated for 60 min); ○——○, RNA from a lysate incubated for 60 min with 2′,5′-oligoadenylate (see *Figure 6*). The latter RNA had been partially degraded by endonuclease activated by the 2′,5′-oligoadenylate.

tinue for up to 2 h at 30°C, to ensure complete synthesis of large products. Different RNA preparations vary considerably in their ability to stimulate amino acid incorporation. In general, viral RNAs and globin mRNA are good templates for the reticulocyte lysate system, giving several hundred-fold stimulations over background when added at 5 – 20 μg/ml. Cellular mRNAs may be less efficient. Both total cytoplasmic RNA and selected poly(A)$^+$ RNA stimulate translation. However, much higher amounts of total cytoplasmic RNA (100 – 200 μg/ml; see *Figure 2*) are required since most of it is rRNA. Provided the RNA solutions do not contain too high a concentration of salts (especially Mg^{2+}), quite large volumes can be added to assays. Alternatively, very dilute solutions (in water) can be lyophilised in the incubation tubes and the RNA dissolved directly in the assay mixture. Some RNA preparations may contain inhibitors of translation such as sulphated polysaccharides or double-stranded RNA. The presence of inhibitors may be checked by adding the preparations to assays containing a known amount of a standard translatable mRNA. However, even in cases where little stimulation of protein synthesis by an RNA is observed, it is still worth looking for specific products by gel electrophoresis or other methods (Section 5).

2.5 Advantages and Disadvantages

The reticulocyte lysate system, either in its native form or after micrococcal nuclease treatment, is probably the most widely used eukaryotic cell-free protein synthesising system at present. There are many reasons for this, notably the high translational activity with endogenous or exogenous mRNAs, the relative ease of preparation and the stability of the system on storage. Multiple rounds of protein synthesis occur on mRNAs, each active mRNA being translated 40 – 70 times in a 90 min incubation, with a polypeptide chain elongation rate of almost 1 amino acid per sec at 26°C (5).

To be set against these valuable characteristics, it should be noted that reticulocyte lysates are initially somewhat unpredictable in their activity and the success rate in preparing good ones is variable. In addition, they are uniquely sensitive to certain inhibitors, notably low concentrations of double-stranded RNA and oxidised thiol compounds which may sometimes be present in added components. Thus, for example, some RNA preparations which are translated well in the wheat germ system (Section 3) are poorly utilised, and even inhibitory, in the reticulocyte lysate system (unpublished observations of S. Trevor-Roper in the author's laboratory). The reticulocyte lysate may also exhibit unique regulatory features, such as the stringent requirement for haemin, which make it atypical in studies of translational control mechanisms.

Micrococcal nuclease-treated reticulocyte lysate is the most common cell-free system for the translation of exogenous eukaryotic cell and viral mRNAs. Not surprisingly, therefore, such preparations are now commercially available (see *Table 1*). Not only can these be used to examine specific translation products, but they are also valuable in studies on mRNA competition and inactivation, and in analysis of regulatory aspects of the protein synthetic process itself. Messenger-dependent reticulocyte lysates have been used in coupled transcription-translation systems with viral core particles containing endogenous RNA polymerases, often allowing the detection of early viral products not easily seen in infected cells (6). These systems also allow the requirements for post-transcriptional modifications to produce active RNA templates to be determined. Similar translational assays can be employed to detect coding sequences in the products of the whole cell extract transcription system of Manley (Chapter 3) which uses specific cloned or restriction enzyme-cleaved DNA molecules as templates.

With all these advantages it is important to bear in mind one possible problem associated with nuclease-treated lysates. It has been shown that significant amounts of mRNA fragments remain after enzymatic digestion, including sequences which contain initiation sites (8). Thus the formation of 80S initiation complexes does *not* become dependent on exogenous mRNA, even though little translation of the initiating fragments occurs. Added mRNAs have to compete with these fragments for ribosome binding, and weakly initiating species may not do so efficiently. Finally, it should be stressed that there is nothing magical about nuclease treatment. The cell-free translation system will only be as good as the original extract allows, and a poorly initiating extract will remain a poor system even after micrococcal nuclease digestion.

Table 8. Preparation of Wheat Germ Extract.

The following procedures should be carried out at 4°C.

1. Grind the wheat germ for 3 min with an equal weight of washed, autoclaved sand in a cold mortar in the presence of 4.5 ml of extraction buffer per gram of wheat germ. The extraction buffer contains:

    0.1 M KCl
    1 mM magnesium acetate
    2 mM CaCl$_2$
    6 mM 2-mercaptoethanol
    40 μM spermidine
    20 mM Hepes (pH 7.6 with KOH)

2. Centrifuge the resulting paste at 30 000 *g* for 10 min.

3. Recover the *central* portion of the supernatant, avoiding the floating layer of lipid.

4. *Either* dialyse this fraction of supernatant (use autoclaved dialysis tubing) *or* pass it through a column (at least 5 ml bed volume per ml of extract) of Sephadex G-25 (coarse) to equilibrate the extract with:

    0.12 M KCl
    5 mM magnesium acetate
    6 mM 2-mercaptoethanol
    40 μM spermidine
    20 mM Hepes (pH 7.6 with KOH)

5. Store the equilibrated extract in aliquots at -70°C or under liquid nitrogen.

3. THE WHEAT GERM SYSTEM

Wheat germ extracts are commercially available, with or without micrococcal nuclease treatment (*Table 1*). As with reticulocyte lysates, the cost is high (£90 per ml), in spite of the fact that the wheat germ system is extremely easy to prepare (see ref. 9 and Section 3.1).

3.1 Preparation and Storage of Extracts

One of the main advantages of the wheat germ cell-free system is the ready availability of the starting material; suitable commercial wheat germ (i.e., un-toasted and with no preservatives added) is available from most 'health food' shops. However, it may be necessary to prepare extracts from several different sources of wheat germ and to compare their activities (Section 3.2), since both the level of endogenous amino acid incorporation and the extent of stimulation by added mRNAs vary considerably between batches (10). The detailed preparation procedure is described in *Table 8*.

3.2 Incubation Conditions for Protein Synthesis

The optimal conditions for protein synthesis in the wheat germ system are broadly similar to those for the reticulocyte lysate (Section 2.3). However, unlike the reticulocyte lysate, the endogenous translational activity of the wheat germ system is low. Therefore, translation in this system is largely dependent on added mRNAs and so the precise requirements may vary with

Table 9. Incubation Conditions for Wheat Germ Extracts.

Stock Solutions

Salts-amino acids-energy mixture (5x stock)[a]:

0.17 – 0.37 M KCl[b]
5 – 10 mM magnesium acetate
0.5 mM amino acid minus methionine (*or* minus another amino acid depending on which radiolabelled amino acid is being used)
Up to 2.5 mCi/ml [^{35}S]methionine (~ 1200 Ci/mmol) (*or* another radiolabelled amino acid)
6.5 mM ATP
1.25 mM GTP
12.5 mM DTT
0.19 mM spermidine
70 mM Hepes (pH 7.6 with KOH)

Creatine phosphate – creatinine phosphokinase (10x stock):

0.16 M creatine phosphate
7 mg/ml creatine phosphokinase

Reaction Mixture

For each incubation (final volume 50 μl), mix the following components carefully on ice in a sterile microcentrifuge tube:

15 μl of wheat germ extract (thawed immediately before use)
10 μl of salts – amino acid – energy mixture (5x stock)
5 μl of creatine phosphate – creatine phosphokinase (10x stock)
20 μl of mRNA (to give a final concentration of 20 – 80 μg/ml)
Incubate at 25°C

[a]The exact requirements for K$^+$ and Mg^{2+} ions will depend on the RNA being translated and must be determined experimentally.
[b]Substitution of potassium acetate for potassium chloride will allow a higher K$^+$ ion concentration to be used for translation of certain exogenous mRNAs since high chloride ion concentrations inhibit initiation.

Table 10. Concentration of Components in Incubations with Wheat Germ Extract.

70 – 110 mM KCl[a]

2.5 – 3.5 mM magnesium acetate

100 μM amino acids minus methionine (*or* minus another amino acid depending on which radiolabelled amino acid is being used).

Up to 500 μCi/ml [^{35}S]methionine (*or* another radiolabelled amino acid)

1.3 mM ATP

0.25 mM GTP

16 mM creatine phosphate

0.7 mg/ml creatine phosphokinase

2.5 mM DTT

50 μM spermidine

20 mM Hepes (pH 7.6 with KOH).

Up to 80 μg/ml mRNA

[a]See the last footnote to *Table 9*.

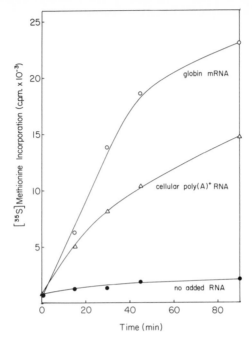

Figure 3. Translation of poly(A)$^+$ mRNA from rabbit reticulocytes and erythroleukaemia cells in the wheat germ cell-free system. Incubations were carried out at 25°C in the absence of added RNA or with mouse erythroleukaemia cell poly(A)$^+$ RNA (13 μg/ml) or globin mRNA (20 μg/ml). [^{35}S]Methionine was present at 130 μCi/ml. Aliquots (5 μl) were taken at the times indicated for determination of the radioactivity incorporated into proteins as described in *Table 6*.

different mRNAs. *Table 9* describes the preparation of a typical reaction mixture (50 μl final volume) and *Table 10* lists the final concentrations of each component. The final concentration of added mRNA is usually in the range of 20 – 80 μg/ml. A concentration curve should be run for each mRNA preparation at an incubation temperature of 25°C. Amino acid incorporation should be linear for at least 1 h. Addition of the human placental ribonuclease inhibitor (see Chapter 8, Section 2.7) may allow translation to continue up to 3 h. *Figure 3* illustrates the use of the wheat germ system in the translation of globin mRNA and RNA from Friend virus-transformed mouse erythroleukaemia cells.

3.3 Micrococcal Nuclease Treatment

The low, if somewhat variable, background incorporation of wheat germ extracts can be reduced even further by micrococcal nuclease treatment. The nuclease treatment protocol described earlier for reticulocyte lysate (Section 2.4.1) can be used without modification (except for the omission of haemin) to prepare a nuclease-treated wheat germ extract. The incubation conditions for use of nuclease-treated wheat germ extract are as described earlier for untreated extracts (Section 3.2).

3.4 **Advantages and Disadvantages**

The most obvious benefits of the wheat germ system are its ease of preparation and low cost. Unlike the reticulocyte lysate system, no animal handling facilities are required and little time need be invested in preparing many different extracts for comparison of their activities. Furthermore, wheat germ extracts are sensitive to stimulation of overall protein synthesis by exogenous mRNAs without prior micrococcal nuclease treatment, unlike the untreated reticulocyte lysate. If necessary, the already low background incorporation can be reduced even more by micrococcal nuclease treatment (Section 3.3). Under optimal conditions, an efficiently utilised mRNA will stimulate amino acid incorporation as much as 400-fold. The wheat germ extract has also been used in conjunction with *in vitro* transcription systems to synthesise polypeptide products coded by viral DNAs and complementary DNAs (11).

Potential disadvantages of this system are that the ionic optima for translation are quite sensitive to the nature and concentration of the mRNA, and may need to be determined for each template. There is also some evidence (12) that, at low levels of stimulation, added RNA can enhance endogenous protein synthesis so that the products observed may not be entirely coded by the RNA which is being characterised. However, perhaps the biggest criticism of the wheat germ system is its tendency to produce incomplete products due to premature termination and the release of peptidyl-tRNA. This can be a problem with large mRNAs, coding for polypeptides in excess of 60 000 daltons. However, the polyamines spermidine and spermine help to overcome this difficulty by stimulating the rate of chain elongation (to a value comparable with that in the reticulocyte lysate) and lowering the Mg^{2+} optimum. Since there is some endonucleolytic cleavage of large mRNAs in the wheat germ system, this may increase the probability of ribosomes completing the synthesis of full-length products before the template is degraded.

4. OTHER EUKARYOTIC CELL-FREE SYSTEMS

In addition to the widely used reticulocyte lysate and wheat germ systems, extracts from a variety of eukaryotic cell types have been prepared for the translation of exogenous mRNAs or in order to study aspects of the regulation of protein synthesis. Most of these systems are $10\,000 - 30\,000$ *g* supernatants which have been pre-incubated or nuclease-treated to eliminate endogenous mRNA and then dialysed or treated with Sephadex G-25 to standardise ionic conditions. The cells from which the supernatants are prepared are usually either grown as large-scale suspension cultures or as ascites tumours in mice. The latter yield larger amounts of material more easily but cannot be readily manipulated to examine translational responses to changes in cellular physiology.

Krebs II mouse ascites cells have been widely used as a source of a cell-free system which will translate added cellular and viral mRNAs (13). Extracts with similar properties can be prepared from mouse L-cells, Ehrlich ascites tumour cells, HeLa cells and Chinese hamster ovary (CHO) cells, all of which can be

245

Table 11. Preparation of Ehrlich Ascites Tumour Cell Extract.

The following procedures should be carried out at 4°C.

1. Harvest the cells (grown to a density of $\sim 10^6$/ml) by centrifugation at 1000 g for 5 min at 4°C[a].

2. Decant the supernatant, resuspend the cells in 250 ml of 0.15 M NaCl, 20 mM Hepes (pH 7.2 with KOH) and centrifuge as in step 1.

3. Repeat step 2.

4. Measure the packed cell volume by centrifugation in graduated conical tubes.

5. Swell the cells by suspension in 2 volumes of 10 mM KCl, 1.5 mM magnesium acetate, 7 mM 2-mercaptoethanol, 10 mM Hepes (pH 7.5 with KOH) (hypotonic buffer). Stand the suspension on ice for 10 min.

6. Homogenise with 20 strokes of a tight-fitting Dounce homogeniser or lyse the cells by addition of 0.25% Nonidet P-40.

7. Add 0.4x the packed cell volume of:

    0.55 M KCl
    2.5 mM spermidine
    3.5 mM 2-mercaptoethanol
    50% glycerol
    0.1 M Hepes (pH 7.5 with KOH)

8. Centrifuge at 20 000 g for 10 min at 4°C.

9. Avoiding any fat, collect the (post-mitochondrial) supernatant for storage, preincubation or micrococcal nuclease treatment (see Section 4.1).

---

[a]If the preservation of endogenous polysomes is important, the cells must first be rapidly chilled by pouring the culture onto an equal volume of crushed ice in the centrifuge bottles.

grown in tissue culture in multi-litre quantities. The procedures described here have been used successfully with Ehrlich cells (14,15) and should also be readily applicable to the other cell types mentioned above.

4.1 Preparation of Ehrlich Ascites Tumour Cell Extracts

The method is summarised in *Table 11*. The extract can either be stored directly in 200 μl aliquots in liquid nitrogen or subjected to preincubation (Section 4.2) or micrococcal nuclease treatment (Section 4.4) to lower the background of endogenous protein synthesis. Cytoplasmic extracts prepared from Ehrlich cells by this procedure contain $60-100$ A_{260} units/ml and the yield is about 4 ml/10^9 cells.

If desired, the cell extracts can be dialysed (four buffer changes over 4 h) or passed over Sephadex G-25 (coarse) (at least 5 ml bed volume per ml of extract) at 4°C to lower the concentrations of amino acids and to standardise the ionic conditions before storage. The equilibration buffer in either case is 90 mM KCl, 2 mM magnesium acetate, 7 mM 2-mercaptoethanol, 10 mM Hepes (pH 7.6 with KOH). Sephadex treatment usually results in a 2- to 3-fold dilution of the extracts.

Table 12. Concentration of Components Required for Protein Synthesis in Extracts of Cultured Mammalian Cells.

---

75 – 110 mM KCl (or up to 0.15 M potassium acetate)[a]

1.5 – 2.5 mM magnesium acetate[a]

50 μM amino acids minus methionine (*or* minus another amino acid depending on which radiolabelled amino acid is used)

Up to 500 μCi/ml [^{35}S]methionine (*or* another radiolabelled amino acid)

1 mM ATP

5 mM creatine phosphate

0.25 mM GTP

0.18 mg/ml creatine phosphokinase

0.5 mM DTT (*or* 6 mM 2-mercaptoethanol)

0.4 mM spermidine

30 mM Hepes (pH 7.5 with KOH)

Up to 40 μg/ml mRNA

---

[a]The optimal K^+ ion and Mg^{2+} ion concentrations vary with the nature of the mRNA to be translated.

4.2 Pre-incubation of Cell Extract

When cell extracts are incubated at 37°C under conditions optimal for protein synthesis (Section 4.3), endogenous mRNA translation occurs. This usually results in the completion of pre-existing nascent polypeptide chains with relatively little re-initiation of synthesis (see Section 7.2). Therefore, polysomes disaggregate and single ribosomes accumulate, a process often referred to as 'run-off'. The endogenous mRNA may either be destroyed by nuclease activity or sequestered in the form of untranslated RNP complexes, depending on the nature of the cell-free system. In either case the level of background protein synthesis falls to virtually zero, allowing the translation of added mRNAs to be examined.

Pre-incubation of cell extracts is performed as follows. To 1 volume of freshly prepared extract add 0.1 volume of a solution containing 10 mM ATP, 2 mM GTP, 0.1 M creatine phosphate and 1.6 mg/ml of creatine phosphokinase, and incubate at 37°C for 45 min. Centrifuge at 15 000 *g* for 5 min at 4°C and then dialyse or treat the supernatant with Sephadex G-25 as described above (Section 4.1) before freezing it in small aliquots. Pre-incubated systems prepared in this way are stable in liquid nitrogen for up to 2 years but should be thawed only once before use.

4.3. Incubation Conditions for Protein Synthesis

Table 12 describes the final concentrations of salts, amino acids and high energy compounds needed for optimal translation of both endogenous and added mRNAs in Ehrlich cell extracts and similar systems. Incubations are usually carried out in volumes of 25 – 100 μl in which the cell extract comprises

50 — 60% of the total. The standard additions are most conveniently achieved by pipetting in 0.1 volume of a 10-fold concentrated stock solution, the composition of which allows for the components already present in the cell extract. The stock solution can be stored at $-20°C$ for a few weeks without harm. Incubations are carried out at 30°C or 37°C. The time-course of amino acid incorporation depends not only on the temperature but also on the concentration of the extract, and on whether endogenous or exogenous mRNA translation is being assayed. In the case of endogenous mRNA, a substantial proportion of the amino acid incorporation is due to completion of pre-existing polypeptide chains rather than new initiation *in vitro*.

The optimal concentrations of Mg^{2+} and K^+ ions are particularly dependent on the nature of the mRNA being translated and range between the limits shown in *Table 12*. Some large viral RNAs, such as the genomes of polio or encephalomyocarditis (EMC) viruses, have unusually high optima. For these reasons, and because the precise conditions for maximum protein synthesis may depend on the previous treatment of the cell extracts (whether pre-incubated, dialysed, fractionated by gel filtration, etc.), optima should ideally be established for each new preparation and with each mRNA which is to be used. Addition of spermidine to 0.4 mM final concentration stimulates translation and lowers the Mg^{2+} ion optimum, particularly in systems which have been dialysed or fractionated by chromatography on Sephadex; therefore the optimum for this compound should also be determined. As in cell-free systems from reticulocytes (*Table 4*) or wheat germ (*Table 9*), substitution of potassium acetate for potassium chloride will allow a higher K^+ ion concentration to be used for protein synthesis since high chloride ion concentrations inhibit initiation. This gives some advantage when mRNAs are used which have a high K^+ ion requirement for maximum rates of initiation, although a concentration of acetate ions of up to 0.15 M can hardly be said to be physiological!

There are two alternative energy-generating systems which can be employed in eukaryotic cell-free systems. Pyruvate kinase and phosphoenol pyruvate have the advantage of converting both ADP to ATP and GDP to GTP directly, whereas creatine phosphokinase and creatine phosphate rely on the presence of nucleoside diphosphate kinase (endogenous to the cell extract) and ATP to regenerate GTP. There have been reports of contamination of creatine phosphokinase preparations with ribonuclease, and of creatine phosphate with pyrophosphate (an inhibitor of tRNA charging). Nevertheless this combination works well as an energy generating system in most circumstances. It may also be possible to utilise endogenous enzymes of glycolysis to generate ATP if fructose 1,6-bisphosphate is added, provided this pathway is sufficiently active *in vitro* to maintain the ATP level.

Of the other components, Hepes is preferable to Tris as a buffer because of its lower pK and lack of sensitivity to temperature changes between 0°C and 37°C. DTT or 2-mercaptoethanol is necessary for maximal activity in extracts which have been dialysed or fractionated by chromatography on Sephadex and may also protect endogenous ribonuclease inhibitor from inactivation if it is present in all buffers. The relatively low nuclease level in many eukaryotic cell-free systems may be attributable to this endogenous inhibitor being present in excess. Amino acid concentrations of ~40 μM each (less for the radiolabelled amino acid) are generally adequate in extracts from cultured cells. Since the activity of these systems is considerably less than that of the reticulocyte lysate (Section 2.3) there is little danger of running out of an amino acid during cell-free protein synthesis.

Amino acid incorporation is assayed in exactly the same way as described for the reticulocyte lysate (*Table 6*, Section 2.3), except that the bleaching step with hydrogen peroxide is omitted. Usually, $10-20$ μl samples are taken for determination of radioactivity.

4.4 Micrococcal Nuclease Treatment

Nuclease-treated cell-free extracts can be prepared by the protocol described for reticulocyte lysate (Section 2.4) except for the omission of haemin. As with reticulocyte lysate, this treatment degrades endogenous mRNA and therefore results in a cell-free extract which is mRNA-dependent and exhibits low background incorporation of amino acids. In some procedures (15), after nuclease digestion and addition of EGTA, extracts are fractionated by gel filtration to remove endogenous amino acids. However, this is by no means essential. The incubation conditions for use of nuclease-treated extracts are as described for untreated extracts (Section 4.3).

4.5 Other Cell-free Systems

Sections 4.1 to 4.4 have described the preparation and use of cell-free systems from Ehrlich ascites tumour cells and similar cell types, such as HeLa or mouse L-cells. In addition, active extracts have been successfully prepared from the yeast *Saccharomyces cerevisiae* (16). It has also proved possible to create fractionated systems containing purified or semi-purified components which together give rise to full-length polypeptide products (17,18). The latter systems require the isolation of ribosomal subunits, initiation, elongation and termination factors, tRNAs and aminoacyl-tRNA synthetases, as well as mRNAs. This involves considerable effort but such studies have given valuable information about the roles of the various components of the protein synthesising machinery. The reader is referred to the original literature (17,18) for details of the techniques involved in this more specialised field.

4.6 Advantages and Disadvantages of Cell-free Systems from Cultured Cells

Unfractionated post-mitochondrial extracts are readily prepared from mammalian cells grown either as ascites tumours or in culture. In the case of the former, extra work is involved in passaging the cells and maintaining their animal hosts. With cultured cells there is the need to grow sufficiently large numbers of cells. This is straightforward in the case of rapidly dividing cells in suspension culture, such as Ehrlich ascites or HeLa cells, and $1-5$ litres of culture will give workable amounts of cell extracts. However, there is more of a problem with cells which grow as monolayers, although even in these cases preparation of small amounts of active extracts is easily accomplished. Overall, the effort required is probably slightly less than for the preparation of reticulocyte lysates but more than for the preparation of the wheat germ system.

The major disadvantage of mammalian cell-free systems from non-erythroid cells is their comparatively low protein synthetic activity. Extracts from some cell types barely initiate at all on exogenous mRNAs, although most will elongate pre-existing nascent polypeptides. In those systems which are reasonably active in initiation (Ehrlich ascites cells, mouse L-cells and HeLa cells), the rate of chain elongation is often slow compared with *in vivo* and, in the case of translation of long mRNAs, there may be premature termination at certain sites along the sequence. Nevertheless, translation is usually faithful, with initiation occurring only at the correct start codon. Even with large templates such as polio or EMC virus RNAs, some full-size product can be found (up to 250 000 daltons), particularly if the K^+ ion concentration is high (0.155 M is optimal for elongation, although rather too high for maximal initiation). Furthermore the cell extracts can often carry out correct cleavages of large viral polyproteins to give authentic smaller products (19). Mammalian cell-free systems have also been used to translate faithfully the products of viral transcription formed *in situ* by virion cores added to the incubations. In this respect these systems have a unique advantage in that they can be used to study the changes in translational and post-translational activity which occur when the cells from which they are derived are subjected to physiological manipulation, such as viral infection or starvation (see Section 7.2).

It is not clear why initiation and elongation rates are depressed in non-reticulocyte mammalian systems. It may be partly a dilution problem, as a considerable proportion of the ribosomes (perhaps with associated factors and tRNAs) is lost during preparation of the extracts. In some cases initiation factor eIF-2 activity may be impaired by phosphorylation by endogenous protein kinase(s). Addition of haemin during preparation (20) or incubation (21) of extracts may prevent this and thus stimulate initiation. However, haemin is not always effective and often the ability to form an initiation complex between the 40S ribosomal subunit and Met-tRNA$_f$ (which requires active eIF-2) is unimpaired in systems which are nevertheless unable to initiate overall polypeptide synthesis. The main defect may well be at the stage of mRNA binding to ribosomes. It is possible that both endogenous and exogenous mRNA molecules become sequestered in an untranslatable form (as RNP complexes), rather than being degraded to any great extent. Indeed, there appears to be relatively little mRNA degradation during preincubation of some cell extracts since translation can be reactivated by subsequent gel filtration of the systems (15). It seems likely that inhibitory small molecules such as GDP may accumulate during incubation of mammalian cell-free systems, and this may lead to early failure of initiation. Rapid loss of ATP and GTP, even in the presence of an energy regenerating system, has been observed in a liver cell-free system (S. Morley, University of Cambridge, personal communication). The maintenance of adequate levels of ATP and GTP, and the prevention of the accumulation of their breakdown products, are clearly very important.

5. CHARACTERISATION OF TRANSLATION PRODUCTS

The successful preparation of an active cell-free extract, and of mRNA which is translated by it, is only the beginning of a full characterisation of the mRNA. In almost all cases it is necessary to investigate the nature of the polypeptide products which are synthesised, either to check that a known mRNA is being translated correctly or to determine what proteins the mRNA is coding for. Quantitation of specific products can also be used as an assay for the relative amounts of mRNA species in a mixture or between different preparations. There are many ways in which translation products can be characterised, most of which require the proteins to be labelled by incorporation of a radioactive amino acid during their synthesis. The most widely used techniques are considered briefly here.

5.1 Stimulation of Amino Acid Incorporation

As described in Sections 2 – 4, several types of translation system are available which are mRNA dependent, that is they show little or no incorporation of amino acids into protein unless exogenous mRNA is added. In such systems, if a single species of pure mRNA is added, the extent of incorporation should be a direct measure of the amount of product made. However, it is not safe to rely on this criterion alone, since it is important to know that the mRNA is translated faithfully and completely. Furthermore, there is the possibility that addition of exogenous RNA may stimulate the translation of residual endogenous mRNA (e.g., by protecting it from nucleases). Finally, when mixtures of mRNAs are translated, amino acid incorporation only indicates that protein is being synthesised but says nothing about its nature. For each of these reasons, one or more of the following techniques must be used when information on the identity of the product being synthesised is essential.

5.2 SDS-PAGE

The most widely applicable and versatile method for analysis of cell-free translation products is polyacrylamide slab gel electrophoresis under denaturing conditions in the presence of SDS. Combined with autoradiography or fluorography of the dried gels, this method can resolve a large number of radioactive proteins of different molecular weight in the same sample. It is therefore well suited to the characterisation of products synthesised in a cell-free system programmed with mixtures of mRNAs. Extension of the analysis to two-dimensional fractionation on polyacrylamide gels, using isoelectric focussing in the first dimension and SDS-PAGE in the second, resolves even more polypeptide products. The electrophoretic techniques themselves have recently been described in an earlier volume in this Series (22) and therefore the discussion here is confined to sample preparation before the gels are run.

It is important to ensure that the polypeptide products are sufficiently radiolabelled to allow their detection within a reasonably short time of exposure to X-ray film. The use of [^{35}S]methionine in incubations, at up to 500 μCi/ml, produces highly radioactive protein for relatively low cost, although it should be noted that this amino acid is rare in some polypeptides or even absent from others. For SDS-PAGE, whole translation assays should be diluted into gel sample buffer containing SDS, 2-mercaptoethanol, glycerol, Tris-HCl and Bromophenol blue to give the following final concentrations:

2% SDS
20% glycerol
70 mM 2-mercaptoethanol
0.001% Bromophenol blue
60 mM Tris-HCl (pH 6.8).

Samples are then boiled for 3 min before loading on the gel, or can be stored at $-20°C$ until required. The slab gels must not be overloaded with total protein as this will result in a loss of resolution. This is especially true with reticulocyte lysates, which contain massive amounts of haemoglobin. Usually, a maximum of 5 μl of translation mix (diluted 5- to 10-fold with sample buffer) can be run in one lane of a typical slab gel (0.15 cm thick). Molecular weight markers should be run in a parallel lane to calibrate the gel. If radiolabelled markers are available they should be added to a sample of unlabelled cell extract to ensure uniform behaviour of markers and products on the gel. It is important to run a control incubation with no added mRNA to check for the presence of labelled proteins not coded for by the exogenous template. In the case of the micrococcal nuclease-treated reticulocyte lysate, [^{35}S]methionine often labels a protein of 42 000 daltons by a process which does not require protein synthesis! If necessary, small labelled nascent peptides, still attached to tRNA, can be eliminated from the samples by treating incubation mixtures with 100 μg/ml pancreatic ribonuclease for 10 min at 30°C before adding the gel sample buffer.

The interpretation of autoradiographs of labelled cell-free translation products should also be considered. As illustrated in *Figure 4*, single mRNA species sometimes give rise to complex patterns of products to an extent which depends on the system being used. This is particularly true for translation of large viral RNAs and may be a consequence both of premature termination of synthesis at specific points along the mRNA and of some post-translational cleavage of large polypeptide products (see Section 6.2).

5.3 Immunoprecipitation of Translation Products

If specific antibodies are available against products of cell-free protein synthesis, they provide a powerful means of identifying and quantifying these products, especially when mixtures of many mRNA species are translated and yield a large number of different polypeptides. Numerous examples of immunoprecipitation procedures for cell-free translation products are to be found in the literature. The most suitable method will depend on the charac-

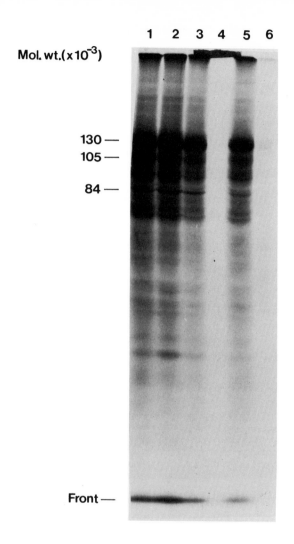

Mol. wt.(×10⁻³)

1 2 3 4 5 6

130 —
105 —

84 —

Front —

Figure 4. SDS-PAGE and autoradiography of translation products. EMC virus RNA (40 μg/ml) was translated in a preincubated and dialysed extract prepared from cultured mouse L-cells (see Section 4) in the presence of [³⁵S]methionine (100 μCi/ml) for 150 min at 30°C. Aliquots (35 μl) were processed for analysis of labelled products by SDS-PAGE using a 7.5% polyacrylamide slab gel followed by autoradiography of the dried gel. The relative molecular masses of the major EMC-specific polypeptides are shown on the left. Note the multiple species resulting from translation of the single pure mRNA. **Lane 1,** control conditions; **lane 2,** plus post-ribosomal supernatant from interferon-treated L-cells; **lane 3,** plus double-stranded RNA (1 μg/ml); **lane 4,** plus the extract from interferon-treated cells and double-stranded RNA; **lane 5,** as **lane 1** but EMC-RNA added after 30 min of incubation; **lane 6,** as **lane 4** but EMC-RNA added after 30 min of incubation. Note the ability of the system to translate RNA added as late as 30 min, and the 80–90% inhibition of translation in the samples examined in **lanes 4** and **6.** The inhibition is due to the double-stranded RNA-dependent activation of interferon-induced 2′,5′-oligoadenylate synthetase and protein kinase activities, both of which impair protein synthesis (see Section 7).

Table 13. Immunoprecipitation of Translation Products[a].

1. At the end of the incubation for cell-free translation, dilute the reaction mixture 5- to 10-fold with immunoprecipitation buffer containing:

   0.1 M NaCl
   1 mM EDTA
   1% Nonidet P-40
   10 mM Tris-HCl (pH 7.5)

2. Add an appropriate volume of antiserum (determined by prior experimentation) and allow the mixture to stand at room temperature for 3 h for antigen-antibody complexes to form.

3. *Either* (a) add carrier antigen and excess antibody to immunoprecipitate the labelled antigen-antibody complexes

   or (b) add Protein A-Sepharose to adsorb the complexes.
   The amounts of these components required will depend on the antiserum used and must be determined empirically.

4. Incubate the mixture, with shaking, in a small, round-bottomed plastic tube at 20°C for 40 min.

5. Transfer the sample to a conical microcentrifuge tube (1.5 ml capacity). Centrifuge at 14 000 g for 1 min.

6. Remove and discard the supernatant. Wash the pellet three times by resuspending it in 1 ml of immunoprecipitation buffer and recentrifugation.

7. Resuspend the pellet in 0.1 ml of immunoprecipitation buffer without detergent and transfer it to a fresh microcentrifuge tube (0.4 ml capacity)[b]. Centrifuge as in step 5.

8. Remove and discard the supernatant. Resuspend the pellet in 20 μl of SDS-containing sample buffer (Section 5.2) and heat at 90°C for 3 min.

9. Remove the Protein A-Sepharose by centrifugation (if step 3b was used) and analyse the solubilised antigen by SDS-PAGE (Section 5.2).

[a]This procedure is best carried out immediately after the translation incubation since freezing the labelled translation products before immunoprecipitation may cause some non-specific precipitation of denatured protein. However, if freezing of the sample is unavoidable, any insoluble material must be removed by centrifugation prior to the addition of antiserum.
[b]The transfer to fresh tubes minimises contamination by protein adsorbed to the walls of the previous tube.

teristics of the particular antibody and antigen in question. A prime consideration is the need to avoid contamination by non-specific precipitation of other labelled proteins or by adsorption of proteins to surfaces of tubes during the washing of the insoluble antigen-antibody complexes. It is essential to check for such contamination of the final product by subjecting it to SDS-PAGE and autoradiography or fluorography.

A generally applicable immunoprecipitation procedure is described in *Table 13*. The inclusion of a non-ionic or an ionic detergent during the immunoprecipitation and washing procedures helps to reduce non-specific contamination of the antigen-antibody-Protein A-Sepharose complexes but it should be noted that some antigen-antibody combinations may dissociate in the presence of detergents. Therefore, again it is important to perform preliminary experiments to optimise the conditions necessary to give the maximum yield of immunoprecipitate consistent with the minimum acceptable level of contamination. Detergent should be omitted from the last wash as a precaution

Table 14. Trypsin Digestion of Translation Products.

1. Add pancreatic ribonuclease to 25 μg/ml and EDTA to 50 mM final concentration. Incubate at 37°C for 15 min (to hydrolyse the tRNA part of peptidyl-tRNA).
2. Precipitate the proteins by adding TCA (5% final concentration) and incubating on ice for 30 min.
3. Recover the precipitate by centrifugation (2000 g for 5 min) and wash it successively with 5% TCA, ethanol:ether (2:3, v/v) and ether.
4. *Either* subject the protein to performate oxidation *or* aminoethylate the protein as follows:

   Performate oxidation

   First mix 950 μl of formic acid with 50 μl of 100 volume hydrogen peroxide and incubate for 2 h at room temperature. Meanwhile allow the protein to dry and then dissolve it in formic acid: methanol (5:1, v/v) using 60 μl per mg protein. Now add the formic acid-hydrogen peroxide mixture to the redissolved protein (100 μl per mg protein). Incubate for 2.5 h at 0°C. Add distilled water (2 ml per mg protein) and freeze-dry overnight.

   Aminoethylation:

   Dissolve the precipitate in 8 M urea (deionised)[a], 5 mM EDTA, 0.19 M 2-mercapto-ethanol, 0.5 M Tris-HCl (pH 8.6) (solution fluxed with argon for 5 min before use) to give a concentration of ~15 mg protein/ml. Incubate for 2 h at room temperature. Then add 25 μl of ethylenimine per ml and incubate for a further 2 h. Precipitate and wash the proteins as in steps 2 and 3 and lyophilise the protein pellet.
5. Dissolve the pellet from step 4 in 0.1 M ammonium bicarbonate, pH 8.9, add trypsin [enzyme: substrate (weight ratio) = 1:100] and incubate at 37°C for 5 h. Add a further equal weight of trypsin after 2 h.
6. Lyophilise the sample and then dissolve it in a minimal volume of water for subsequent analysis of the tryptic peptides (see Section 5.4).

[a]Deionise the urea by passage through a column of Bio-Rad AG 501-X8 mixed bed resin. Alternatively, purchase ultra-pure urea and use this without further treatment.

against causing abnormal migration of translation products during SDS-PAGE.

5.4 Tryptic Peptide Analysis

Another very specific assay for synthesis of the correct translation product is to examine the pattern of radioactive peptides obtained on digestion of the protein with trypsin. Such an approach can be used with a total incubation mixture if a purified mRNA is the template, or it can be applied in the characterisation of immunoprecipitated material or of individual bands from a gel if sufficient radioactivity is present. It is particularly useful if tryptic peptides labelled *in vitro* with one kind of isotope can be compared with those derived from the authentic protein labelled with another isotope.

Table 14 describes a procedure for trypsin digestion of labelled products immediately after the translation incubation. When it is necessary to analyse the tryptic peptide patterns of polypeptides after their resolution by electrophoresis, gel slices or crushed fragments containing the relevant bands can be incubated in 10 mM sodium phosphate (pH 7.2) containing 0.1% SDS for 24 h at 37°C to elute the proteins. Alternatively, dried gel fragments can be treated with 6.0 M urea, 0.5% SDS. The solubilised proteins are then precipitated with TCA in the presence of 100 μg of carrier BSA and treated as described in *Table 14*.

Tryptic peptides may be analysed by two-dimensional separation on 20 cm x 20 cm thin layer silica plates (23) or Whatman 3MM chromatography paper (24), using high voltage electrophoresis (400 V, 3 – 4 h) in an aqueous pyridine/acetic acid buffer at either pH 6.5 (474 ml water:25 ml pyridine:1 ml acetic acid) or pH 3.5 (289 ml water:1 ml pyridine:10 ml acetic acid) followed by chromatography in butanol/acetic acid/water (3:1:1, by vol.). The radioactive spots are identified by autoradiography. Alternatively, peptides can be resolved by ion-exchange chromatography on a column (0.9 cm x 16.0 cm) of Aminex A-5 (Bio-Rad Laboratories) and the radioactivity in the fractions then determined (25). The latter is a particularly useful technique when combined with a double isotope method. For example, an *in vitro* translation product labelled with ^{3}H can be mixed with a sample of the authentic protein labelled with ^{35}S (or *vice versa*) and the isotope ratio in the column fractions will indicate whether the tryptic peptides co-fractionate, thus supporting or disproving identity.

5.5 Amino Acid Sequence Analysis

It is also possible to characterise and quantitate specific translation products of purified mRNAs by determining their N-terminal amino acid sequences (26). However, this is an expensive procedure, requiring large amounts of radioactive amino acids and a protein sequencer of high sensitivity. A similar approach, which is also particularly useful for characterising the initiation sites on mRNAs, is to measure the synthesis of short initiation peptides labelled with N-terminal [^{35}S]methionine. In order to limit protein synthesis to the first few amino acids, an inhibitor of chain elongation such as sparsomycin is added to the cell-free system and incubation times are kept relatively short (27). The peptidyl-tRNA is phenol extracted in the presence of carrier RNA and ethanol precipitated. Samples are then exhaustively digested with ribonuclease, lyophilised and the labelled peptides separated by high voltage paper electrophoresis at pH 3.5 (28). The spots are detected by autoradiography and identified by comparison of their mobility with that of marker methionyl peptides.

5.6 Analysis of Biological Activity

Finally, translation products may be identified by their biological activity. However, this is only feasible in rare cases because the amounts of proteins synthesised in cell-free systems are minute. Nevertheless, if an enzyme or other protein has a sufficiently high specific biological activity it may be possible to quantify its synthesis by direct assay of this activity. Success in this approach also requires that all post-translational modifications necessary for its function will have been carried out *in vitro* (see Section 6). Examples where an assay of this kind has proved possible are the translation of mRNAs coding for β-interferon (29) and for vaccinia virus thymidine kinase (30), both in reticulocyte lysate.

6. *IN VITRO* PROCESSING OF TRANSLATION PRODUCTS

The use of cell-free protein synthesising systems is not confined to the characterisation of the initial products of translation. Under appropriate conditions, a number of post-translational modifications of primary products can occur, including proteolytic cleavages, glycosylation, acetylation and phosphorylation at specific sites. Some of the procedures used to study such events *in vitro* are described in this section.

6.1 **Signal Sequences**

Cell-free translation studies have provided much of the evidence for the existence of N-terminal sequences which act as signals for the insertion of certain proteins into or through membranes of the endoplasmic reticulum (31,32). Such a process is characteristic of the biosynthesis of many secreted and membrane proteins and occurs as the nascent polypeptide chain emerges from the ribosome in the early stages of its synthesis. Indeed, transmembrane insertion of nascent polypeptides is an important factor in the binding of polysomes to the membranes of the rough endoplasmic reticulum. Signal sequences are commonly 20−30 amino acids long, usually have highly hydrophobic regions within them and, in most cases, are cleaved off by a membrane-located enzyme before the polypeptide chain is completed.

In order to study these processes *in vitro* it is necessary to prepare not only the mRNAs for the appropriate proteins and the systems in which to translate them, but also the membranes which recognise the signal sequences. Choice of suitable incubation conditions, and the development of assays for the signal sequences and for the passage of nascent chains into membrane-bounded vesicles, must also be considered.

6.1.1 *Preparation of Microsomal Membranes*

Table 15 summarises a method for preparing microsomal membranes from dog pancreas, the tissue which has been most widely used in the past in studies on processing of nascent secretory proteins *in vitro*. The procedure results in nuclease-free preparations with high processing activity. The minced pancreas is homogenised and a rough microsome fraction obtained by differential centrifugation through sucrose. If necessary, the membranes are stripped of their endogenous polysomes by treatment with EDTA and re-isolated by sucrose gradient centrifugation (*Table 15*). Treatment with puromycin in 0.5 M KCl has also been used (33). It is possible to obtain similar preparations of microsomal membranes, with signal sequence recognition and signal peptidase activities, from rat liver or ascites tumour cells (34). Since there is no specificity in the recognition of signal sequences by microsomal membranes, these sources may be equally satisfactory. However, the relative activities of such preparations and the extent of their contamination with ribonuclease may vary with the tissue source and should be checked before adopting these membranes for routine use.

Table 15. Preparation of Stripped Microsomal Membranes from Dog Pancreas[a].

Preparation of 'Rough' Microsomes

1. Surgically remove the pancreas from a dog and place it in cold TKM buffer (50 mM triethanolamine (pH 7.5), 50 mM KCl, 5 mM $MgCl_2$) containing 0.25 M sucrose.

2. Remove connective tissue, fat and large blood vessels. Chop the pancreas and pass the pieces through a stainless steel tissue press (1 mm diameter).

3. Add 2 volumes of TKM buffer containing 0.25 M sucrose and homogenise the tissue in a motor-driven Teflon homogeniser (5 – 10 strokes at 400 r.p.m.).

4. Centrifuge at 13 000 g for 10 min at 4°C.

5. Collect the supernatant and layer it onto a discontinuous gradient of 1.5 M, 1.75 M and 2.2 M sucrose (5 ml each) in TKM buffer.

6. Centrifuge at 140 000 g for 24 h at 4°C.

7. Recover the 1.75 M sucrose layer ('rough' microsomes). Dilute this with an equal volume of TKM buffer and layer over 2 ml of 1.3 M sucrose in this buffer.

8. Centrifuge at 100 000 g for 30 min at 4°C.

9. Decant the supernatant. The pellet ('rough' microsomes) may be stored at $-70°C$[b].

Preparation of 'Stripped' Microsomes

10. Resuspend the 'rough' microsome pellet in TK buffer [50 mM triethanolamine-HCl (pH 7.5), 50 mM KCl] to a concentration of 100 A_{260} units/ml. Add 0.2 M EDTA (pH 7.0) to give 3 μmol of EDTA per 10 A_{260} units of 'rough' microsomes.

11. Layer 0.5 ml aliquots onto 12.5 ml gradients of 10 – 55% sucrose in TKM buffer and centrifuge at 190 000 g for 2 h at 4°C.

12. Collect the turbid band ('stripped' microsomes) located at ~40 – 45% sucrose and dilute this with 2 volumes of TK buffer.

13. Centrifuge at 100 000 g for 30 min at 4°C.

14. Decant the supernatant and store the pellet at $-70°C$.

15. Before use, resuspend the pellet by brief sonication in:
    0.25 M sucrose
    0.1 M KCl
    3 mM $MgCl_2$
    2 mM DTT
    20 mM Hepes (pH 7.3 with KOH).

[a]Adapted from ref. 35.
[b]This preparation of 'rough' microsomes can be purified further by passage through Sepharose CL-2B (ref. 33).

6.1.2 *Incubation Conditions for Processing*

The conditions under which mRNAs are translated, and their products processed by added preparations of microsomal membranes, are essentially the same as those described earlier (Sections 2 – 4). In some experiments, protease inhibitors such as pepstatin, chymostatin, antipain, leupeptin and Trasylol (35) have been included in incubations when analysis of membrane-mediated proteolytic cleavages is to be carried out. The membranes themselves are added at concentrations of up to 5 A_{260} units/ml. It has frequently been noted that high concentrations of membranes (necessary to drive processing reactions to completion) inhibit overall protein synthesis, although this is more of a problem in the wheat germ system than in the micrococcal nuclease-treated reticulocyte lysate. Choice of incubation times will depend on the kinetics of protein synthesis and the post-translational processing events. The latter can be synchronised, and their temporal relationship to initiation of polypeptide synthesis determined, by deliberately inhibiting further initiation shortly after the beginning of the incubation. This may be achieved with specific agents such as m^7GMP or edeine (36). Similarly, the fate of pre-existing nascent chains on polysomes bound to rough endoplasmic reticulum membranes (not stripped by EDTA treatment) can be studied *in vitro* under conditions where initiation is completely blocked and only elongation of polypeptides occurs, that is, 'run-off' systems (37).

6.1.3 *Assays of Protein Translocation across Membranes and Cleavage of Signal Sequences*

The existence of additional N-terminal signal sequences was first revealed by the observation that primary translation products of mRNAs coding for secretory proteins, synthesised in the absence of microsomal membranes, are slightly larger than the authentic proteins (31,32). The presence of an extra 20 – 30 amino acids on a protein is just sufficient for it to be resolved from the normal-size product by SDS-PAGE, particularly in the case of small proteins such as preproinsulin or pretrypsinogen. Electrophoretic mobility can therefore be used as a simple assay for the presence or absence of a signal sequence (33). However, it should be noted that additional modifications to a primary translation product, especially extensive glycosylations, can also change its behaviour on gels (sometimes in an unpredictable way). The most rigorous criterion for the presence of an additional N-terminal sequence, or for its loss during translation *in vitro*, is therefore direct amino acid sequencing of the radiolabelled protein using an automated Edman degradation procedure. The presence or absence of an N-terminal methionine residue does not provide any indication of whether a signal sequence has been cleaved off, since this initiating amino acid is usually removed anyway, shortly after elongation of the polypeptide chain has begun. It should also be noted that many *in vitro* systems are capable of acetylating the N-terminal residue of a nascent polypeptide chain, thus causing problems in sequencing studies. Such acetylations can be prevented by depleting incubation mixtures of endogenous acetyl CoA by inclusion of 1 mM oxaloacetate and 25 units/ml citrate synthase.

Cleavage of signal sequences is normally a co-translational process which occurs concomitantly with the translocation of the nascent polypeptides across the microsomal membranes. Since the proteins become sequestered within the lumen of the microsomal vesicles they are protected from attack by proteases. This has provided a method for assaying such translocation events. Addition of a mixture of trypsin and chymotrypsin (50 μg/ml each) at the end of a cell-free translation reaction, followed by incubation for 3 h at $0-2°C$, will destroy any products not protected within the membrane vesicles. Precipitation with TCA, followed by SDS-PAGE and autoradiography or fluorography will then indicate which proteins are resistant to proteolysis under these conditions. Both negative controls (no protease treatment) and positive controls (protease treatment of a detergent-solubilised incubation) should be included in such an analysis.

In recent years, several developments in the field of membrane-mediated protein translocation *in vitro* have occurred. For example, it has proved possible to solubilise the signal peptidase activity from rough endoplasmic reticulum membranes by use of detergents. This enzyme can be inhibited by chymostatin. It has also been shown that a 'signal recognition protein' is associated with microsomal membranes and this mediates the selective binding of nascent secretory proteins to the membranes during translation *in vitro* (35). This protein, in its soluble form, has the ability to arrest the chain elongation of polypeptides with signal sequences; the effect is alleviated in the presence of the membranes themselves. This illustrates the major role which cell-free translation systems have played in the characterisation of the mechanism of protein translocation across membranes. Fortunately, progress has been aided by the apparently ubiquitous nature of the signals and components involved, as exemplified by the successful use of dog pancreas membranes in a wheat germ system translating mRNA derived from as exotic a source as islet tissue of the angler fish (38).

6.2 Proteolysis of Primary Products

In addition to the removal of N-terminal signal sequences in the presence of microsomal membranes, primary translation products can undergo other proteolytic processing events in cell-free systems. This is particularly common in the case of the 'polyproteins' of several animal and plant viruses, which are synthesised by complete translation of large mRNAs. Polypeptide chain cleavages can occur at various specific sites along the primary translation products (19,39). The proteases responsible are soluble enzymes, distinct from the membrane-bound signal peptidase activity, and can be present as endogenous components of the translation system used (e.g., reticulocyte lysate; ref. 39). Alternatively, they may themselves be products of viral mRNA translation which are formed during the course of the *in vitro* incubations (39).

Co-translational and post-translational processing of newly-synthesised products in cell-free systems are best studied by kinetic analysis of the appearance and disappearance of specific polypeptides, using SDS-PAGE and autoradiography or fluorography of [^{35}S]methionine-labelled material. Many experimental approaches are possible for pulse-labelling and pulse-chasing (using excess unlabelled methionine in the chase), employing various incubation times. Initiation of protein synthesis may be synchronised by the addition of initiation inhibitors such as 'cap' analogues or edeine shortly after the mRNA is added (36). Peptide sequences nearer the C-terminal end of the translation product will be preferentially labelled if the radioactive amino acid is added after the initiation inhibitor. N-terminal peptides can be exclusively labelled if N-formyl[^{35}S]methionine-tRNA$_f$ (see Sections 7.1.1 and 7.2.1) is used in place of methionine. It is also possible to use specific protease inhibitors to block the processing of primary translation products. Conversely, proteolysis may be enhanced if cell extracts which contain elevated levels of particular enzymes are added to translation systems. This is often the case, for example, when extracts from virus-infected cells are used to carry out processing of homologous virus proteins synthesised *in vitro* (19).

A cautionary note is necessary concerning the interpretation of labelled polypeptide patterns after SDS-PAGE. Because of the tendency of some cell-free systems to terminate polypeptide synthesis prematurely (see Section 4.6), a complex set of products often arises from the translation of a single large mRNA (e.g., *Figure 4*). The appearance of these incomplete proteins may obscure, or complicate the analysis of, the proteolytic processing of large polyproteins to give smaller species. This problem is least acute in the reticulocyte lysate system and this is therefore the system of choice for most studies of this type.

7. ANALYSIS OF TRANSLATIONAL REGULATION USING CELL-FREE SYSTEMS

Most of this chapter has concentrated on the use of cell-free protein synthesising systems for studying the translation of exogenous mRNAs and for analysing the products formed. It is also possible to use these systems to investigate the mechanism and regulation of the process of protein synthesis itself. A vast body of literature has accumulated as a result of such work. Examples of the types of study which are possible are described here.

7.1 The Reticulocyte Lysate System

Because of its high activity, the reticulocyte lysate system has been the most popular choice for investigation of the mechanism of eukaryotic cell protein synthesis and a number of aspects of translational regulation. Its low endogenous ribonuclease activity, high content of ribosomes and initiation factors and its well-characterised major products of translation have all contributed to this popularity. Fortunately the reticulocyte lysate also responds to a variety of controls, such as haemin deficiency and low concentrations of double-stranded RNA, which can be analysed in detail. Whether this specialised cell type is sufficiently typical of eukaryotic cells for the conclusions reached to be universally applicable, remains to be seen.

7.1.1 *Formation of Initiation Complexes*

The assembly of the various complexes involved in the initiation of protein synthesis can be studied relatively easily in the reticulocyte lysate system. The binding of initiator Met-tRNA$_f$ to the initiation factor eIF-2 (in the presence of GTP), the subsequent association of this ternary complex with the native 40S ribosomal subunit and the mRNA-dependent joining of the 60S subunit to form an 80S initiation complex can all be assayed. In each case it is first necessary to prepare [^{35}S]Met-tRNA$_f$. This can be achieved using a crude preparation of aminoacyl-tRNA synthetases from *Escherichia coli* and deacylated rabbit liver tRNA. *Table 16* summarises the procedure for preparing these enzymes and *Table 17* describes their use for the preparation of [^{35}S]-Met-tRNA$_f$. Only the mammalian tRNA$_f^{Met}$ species is charged by the bacterial synthetase. A brief phenol extraction and overnight dialysis gives a preparation of Met-tRNA$_f$ containing $2 \times 10^4 - 5 \times 10^4$ c.p.m./μl. If it is necessary to purify the Met-tRNA$_f$ further (for example, to separate it from tRNA$_m^{Met}$ and from inhibitory contaminants such as sulphated polysaccharides), chromatography on a small column of DEAE-cellulose or benzoylated DEAE-cellulose is a convenient means of achieving this. For the latter, the Met-tRNA preparation is loaded in a buffer containing 0.4 M NaCl at pH 5.0 and the purified Met-tRNA$_f$ is eluted by raising the salt concentration to 0.5 M. Alternatively, elution with a gradient ($0.4 - 1.0$ M NaCl) can be used.

When N-formyl-Met-tRNA$_f$ is required (for example to label N termini of newly synthesised proteins without subsequent cleavage of the initiating amino acid), 10 μM Ca-leucovorin (folinic acid, made up in 0.25 M HCl) is included in the tRNA charging mixture described in step 1 of *Table 17*. This acts as a formyl-group donor in a reaction catalyzed by another enzyme present in the *E. coli* synthetase preparation. About $70 - 80\%$ of the Met-tRNA$_f$ is formylated under these conditions.

The binding of [^{35}S]Met-tRNA$_f$ in a ternary complex with eIF-2 and GTP can be assayed in ribosome-free supernatants from reticulocyte lysates by retention of radioactivity on cellulose nitrate filters (40). Certain characteristics are diagnostic for eIF-2 and should be checked to determine the specificity of the assay. These are:

(i)   the binding should be specific for Met-tRNA$_f$,
(ii) complex formation should be GTP-dependent and strongly inhibited by GDP,
(iii) the rate and extent of the reaction should be stimulated by phosphoenol pyruvate and pyruvate kinase (which regenerate GTP from GDP).

In order to examine the next stage in the initiation process, namely 40S initiation complex formation, it is necessary to fractionate the lysate, after incubation, on a sucrose gradient. The legend to *Figure 5* describes the conditions for this. Because of the rapid turnover of [40S.eIF-2.Met-tRNA$_f$] complexes during protein synthesis, a brief (2 min) incubation with [^{35}S]Met-tRNA$_f$ is sufficient to label the complexes maximally. Under normal conditions of translation, the steady-state level of these complexes is, however, quite low and greater labelling can be achieved by blocking polypeptide elongation with, for example, cycloheximide (*Figure 5* legend). Under these conditions, 40S complexes accumulate, because they cannot be utilised for protein synthesis, to an extent which probably depends on the amount of active eIF-2 in the system.

Under conditions of protein synthesis, 40S initiation complexes are rapidly converted to larger complexes, each by associating with an mRNA molecule and a 60S ribosomal subunit. In a typical lysate, most of the mRNA is in polysomes and so it is difficult to monitor the formation of [80S.Met-

Table 16. Preparation of *E. coli* Aminoacyl-tRNA Synthetases[a].

---

1. Grind 5 g of lyophilised *E. coli* B with 15 g of alumina (type 305) in the presence of sufficient buffer to create a paste. The buffer contains 10 mM $MgCl_2$, 10% glycerol, 10 mM Tris-HCl (pH 8.0).

2. Add a further 1 volume of buffer and centrifuge at 700 *g* for 5 min to remove the alumina.

3. Centrifuge the supernatant at 100 000 *g* for 2 h.

4. Load the supernatant onto a column (1 cm x 9 cm) of DEAE-cellulose equilibrated with running buffer:

    20 mM 2-mercaptoethanol
    1 mM $MgCl_2$
    10% glycerol
    20 mM KH_2PO_4 (pH 7.5 with KOH).

5. Wash the column with running buffer to remove non-bound proteins.

6. Elute the aminoacyl-tRNA synthetases with running buffer in which the KH_2PO_4 concentration is increased to 0.25 M (pH 6.5 with KOH).

7. Pool the protein peak and dialyse against:

    40 mM 2-mercaptoethanol
    10% glycerol
    15% polyethylene glycol 6000
    2 mM KH_2PO_4 (pH 7.0 with KOH).

8. Add an equal volume of glycerol to the sample and store at $-20°C$. Met-tRNA$_f$ synthetase activity is stable for prolonged periods under these conditions.

---

[a]Adapted from ref. 61.

Table 17. Preparation of [^{35}S]Met-tRNA$_f$.

---

1. Incubate the following components in a final volume of 1 ml:

    50 mM KCl
    8 mM magnesium acetate
    2 mM DTT
    4 mM ATP
    1 mM CTP
    100 μCi [^{35}S]methionine (1100 – 1300 Ci/mmol)
    50 mM Hepes (pH 7.5 with KOH).
    1 mg deacylated calf liver tRNA (Boehringer)
    80 μl *E. coli* aminoacyl-tRNA synthetases[a]

2. After 30 min at 30°C, cool on ice, add 0.1 ml of 2 M sodium acetate (pH 4.4) and 2 ml of water-saturated phenol. Shake vigorously for 5 min at 4°C.

3. Centrifuge at 400 *g* for 10 min and collect the upper (aqueous) layer.

4. Re-extract the phenol layer by shaking with 0.8 ml of 50 mM sodium acetate (pH 5.0), 5 mM magnesium acetate.

5. Centrifuge as in step 3. Recover the aqueous layer and pool it with the previous one from step 3.

6. *Either* dialyse the pooled aqueous sample for 6 h against 0.5 M NaCl, 50 mM sodium acetate (pH 5.0) and then for 17 h against 20 mM sodium acetate (pH 5.0),

    or precipitate the tRNA by adding 2.5 volumes of ethanol at $-20°C$; centrifuge (10 000 *g*, 10 min), wash the pellet in ethanol, dry and dissolve in 20 mM sodium acetate.

7. Store the [^{35}S]Met-tRNA$_f$ preparation in 0.1 ml aliquots at $-70°C$.

---

[a]Prepared according to the procedure described in *Table 16*.

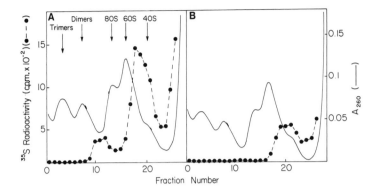

Figure 5. Formation of [40S ribosomal subunit.Met-tRNA$_f$] initiation complexes in the reticulocyte lysate. Incubations (150 μl) were carried out under the conditions used for protein synthesis assays (*Table 4*), except that cycloheximide (100 μg/ml) was present. The labelled precursor was [^{35}S]Met-tRNA$_f$ (2.22 x 10^5 c.p.m.), added after preincubation of the lysate for 35 min at 30°C. The reaction was stopped 2 min later by adding 250 μl of cold buffer (25 mM KCl, 10 mM NaCl, 1 mM magnesium acetate, 0.25 mM DTT, 10 mM Tris-HCl, pH 7.6). The samples were centrifuged through 12.5 ml gradients of 20−40% sucrose in the above buffer for 4.5 h at 205 000 *g*. Fractions were collected, precipitated with 1 ml of 2% (w/v) cetyl trimethyl ammonium bromide in the presence of 0.5 mg carrier yeast RNA, filtered on glass fibre (GF/C) discs and the radioactivity determined by scintillation counting. The sedimentation positions of 40S and 60S ribosomal subunits and of 80S ribosomes, dimeric and trimeric polysomes are indicated by arrows. **A**, control lysate. **B**, lysate preincubated with a preparation of the haem-controlled repressor. Note the presence of 40S and 80S Met-tRNA$_f$-containing initiation complexes in **A**, which sediment on the heavy side of the corresponding A_{260} peaks. Binding of Met-tRNA$_f$ to these complexes is strongly inhibited by the haem-controlled repressor (acting through phosphorylation of initiation factor eIF-2).

tRNA$_f$.mRNA] complexes. However, when elongation is inhibited, the conversion of 40S to 80S complexes can be achieved by adding exogenous mRNA. This is the basis of the 'shift assay' described by Darnbrough *et al.* (41), the beauty of which is that it enables mRNA binding to be assayed using [^{35}S]Met-tRNA$_f$ as the radiolabelled component. Radioactive mRNA may, of course, be used if it is available or, alternatively, one can assay for poly(A)$^+$ RNA by hybridisation with [^{3}H]poly(U) (ref. 42). Again, sucrose gradient centrifugation and fractionation (*Figure 5*) are used to separate and monitor the levels of the complexes formed.

7.1.2 *Mode of Action of Inhibitors of Protein Synthesis*

The reticulocyte lysate system has been extensively used to study the mechanisms by which inhibitors block protein synthesis *in vitro*. In particular, a great deal of attention has been paid to the effects of specific protein kinases such as the haem-controlled repressor and the inhibitor activated by double-stranded RNA (43). These enzymes inhibit polypeptide chain initiation by phosphorylating the 38 000 dalton α-subunit of eIF-2. They also undergo autophosphorylation. These phosphorylations can be shown by incubating lysates in the presence of [γ-^{32}P]ATP and analysing the labelled proteins by one- or two-dimensional polyacrylamide gel electrophoresis and/or isoelectric

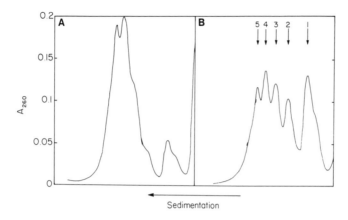

Figure 6. Polysome degradation by the 2′,5′-oligoadenylate-activated endonuclease in the reticulocyte lysate. **A**, lysates incubated for 25 min at 30°C in a reaction volume of 100 μl in the presence of all components necessary for protein synthesis, but with cycloheximide added at 100 μg/ml. **B**, reaction mixture as in **A** but with 2′,5′-oligoadenylate (15 nM) also present. Incubations were stopped by adding 250 μl of cold buffer (0.1 M KCl, 5 mM magnesium acetate, 20 mM Tris-HCl, pH 7.6) and the samples were analysed by centrifugation through 5 ml gradients of 20−50% sucrose in this buffer (35 min at 233 000 g). Optical density at 260 nm was monitored by pumping the gradients through a flow-cell (10 mm path length, 80 μl chamber volume). The sedimentation positions of single ribosomes and of polysomes containing 2−5 ribosomes are indicated by arrows in **B**. Note the 2′,5′-oligoadenylate-dependent cleavage of polysomes containing predominantly four or five ribosomes to fragments with one to three ribosomes. For further details see ref. 47.

focussing, followed by autoradiography (44). Whole lysates (44), pH 5 precipitates (45) or eIF-2 partially purified by phosphocellulose chromatography (46) can be examined by these techniques. The effects of these eIF-2 kinases on initiation have been studied by the methods described in Section 7.1.1.

Another class of inhibitors with a totally different mode of action is the 2′,5′-oligoadenylate series of nucleotides. The mechanism of inhibition [i.e., activation of a specific endonuclease (47)] can be shown by carrying out sucrose gradient analyses of polysomes from lysates incubated in the presence of high levels of cycloheximide or emetine (to prevent polysome breakdown caused by ribosome 'run-off'). *Figure 6* illustrates the conversion of polysomes to smaller size classes by 2′,5′-oligoadenylate under these conditions. The presence of some endogenous 2′,5′-oligoadenylate synthetase in the reticulocyte lysate, together with the double-stranded RNA-activated eIF-2 kinase (48), accounts for the sensitivity of protein synthesis to inhibition by double-stranded RNA in this system.

Reticulocyte lysates are remarkably sensitive to certain classes of compounds, in addition to the haemin and double-stranded RNA already mentioned. Notable amongst these are oxidized thiol compounds, which inhibit translation by a mechanism similar to that of haemin deficiency. Further categories of inhibitors of reticulocyte protein synthesis include various low molecular weight RNA species of uncertain function (49), compounds which are analogues of the 'cap' structure at the 5' end of most mRNAs (50), and DNA sequences which are complementary to the mRNAs present and prevent the utilisation of the latter by 'hybrid-arrested translation' (51).

The detailed mode of action of all of these agents, and any other inhibitor of protein synthesis, can be analysed by the approaches described above. The important general points which guide the interpretation of such analyses are:

(i) initiation inhibitors cause polysome breakdown only under conditions of protein synthesis,

(ii) elongation or termination inhibitors impair protein synthesis without causing polysomal breakdown, and may even cause formation of larger than normal polysomes (ribosome 'loading'),

(iii) endonucleases cause polysome breakdown whether or not protein synthesis is allowed to take place.

7.2 Extracts from Cultured Cells and Normal Tissues

There are many types of investigations of translation for which the reticulocyte lysate system is not suitable. As implied above (Section 7.1), it may not be safe to assume that reticulocytes are representative of eukaryotic cells in general in the way they regulate translation. They are, after all, a highly specialised cell type. Furthermore, they cannot or do not respond to various physiological stimuli (e.g., hormones) which are important regulators of cellular functions in other cells, and they cannot be used for studies on viral infection. For these reasons, much work on translational control has been carried out using extracts from many other cell types, in spite of the disadvantages and difficulties inherent in some of these systems (see Section 7.2.3).

7.2.1 *Formation of Initiation Complexes*

Cell-free systems prepared from a wide range of cells and tissues have been successfully used to study the assembly and regulation of initiation complexes on native 40S ribosomal subunits and 80S ribosomes. To date, such experiments have been carried out with extracts from Krebs II ascites cells (52), Ehrlich ascites cells (14,21), mouse L-cells (47), HeLa cells (53), CHO cells (54), myeloma cells (53), normal and mitogen-stimulated lymphocytes (55), liver and muscle (C. Harmon and V.M. Pain, University of Sussex, unpublished data),and yeast (56). It appears that initiation factor eIF-2 remains active and able to bind Met-tRNA$_f$ to 40S subunits in all of these systems, in some cases producing a level of 40S complexes comparable with that seen *in vivo* (14,57). Moreover, in many cases the initiation activity in the extracts reflects the physiological changes occurring in the intact cells. For example, initiation of protein synthesis in cultured cells is regulated by factors such as starvation for essential amino acids, virus infection and interferon treatment and these effects can be studied in the appropriate extracts (58). In many cases, the addition of eIF-2 to a cell-free extract reverses the physiologically-induced inhibition of initiation. Similar observations have been made with extracts from rat muscle, in which diabetes or starvation regulate initiation (C. Har-

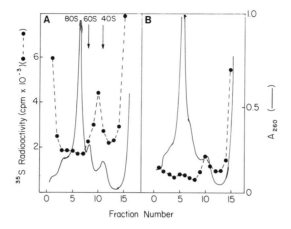

Figure 7. Formation of [40S ribosomal subunit.Met-tRNA_f] initiation complexes in extracts from fed and amino acid-starved Ehrlich ascites tumour cells. Incubations (100 μl) were carried out under protein synthesis conditions with endogenous mRNA (Section 4.3). [^{35}S]Methionine was present at 250 μCi/ml. After 2 min at 30°C the reaction was stopped by rapid cooling and the samples were centrifuged through 5 ml gradients of 20−50% sucrose for 3 h at 233 000 *g*. Fractions were collected and the radioactivity determined as in *Figure 5*. The sedimentation positions of native 40S and 60S ribosomal subunits and of 80S ribosomes are indicated by arrows. **A**, extract from exponentially growing, fully-fed cells; **B**, extract from cells starved for lysine for 30 min before harvesting. Note the labelled material at the top (free Met-tRNA_f) and bottom (nascent polypeptide chains) of the gradients, as well as in the 40S complexes. Amino acid starvation inhibits formation of the latter through inactivation of initiation factor eIF-2. For further details see ref. 21.

mon and V.M. Pain, University of Sussex, unpublished data).

These events can be monitored using techniques essentially identical to those described for the reticulocyte lysate (Section 7.1.1). Either [^{35}S]Met-tRNA_f or free [^{35}S]methionine can be used to label 40S (and, in some cases, 80S) initiation complexes *in vitro*. In the case of [^{35}S]methionine, endogenous aminoacyl-tRNA synthetases will charge both tRNA_f^{Met} and tRNA_m^{Met} so that nascent polypeptide chains also become labelled under conditions of protein synthesis. However, these are well separated from 40S complexes by sucrose gradient centrifugation; *Figure 7* illustrates a typical experiment using such a technique to compare initiation in extracts from fully-fed and amino acid-starved Ehrlich ascites cells. It should be noted that when extracts are used which have not been dialysed or fractionated by Sephadex chromatography, the specific activity of the labelled methionine is lowered by the presence of endogenous amino acid. This should be corrected for by carrying out isotope dilution experiments or by direct amino acid analysis of the extracts.

7.2.2 *Messenger RNA Translation Studies*

Cell-free systems from various sources may also be used to study the complete translation process. In the case of endogenous mRNA, for example, it is often important to determine the extent to which protein synthesis represents 'run-off' of pre-existing polysomes and how much is attributable to initiation *in*

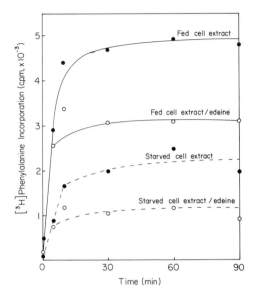

Figure 8. Effect of an initiation inhibitor on the time-course of endogenous mRNA-directed protein synthesis in extracts from fed and amino acid-starved Ehrich ascites tumour cells. Incubations (50 μl) were carried out as in *Figure 7* with [^{3}H]phenylalanine (50 μCi/ml) as labelled precursor in the presence or absence of edeine (4 μM). Aliquots (5 μl) were taken at the times indicated for measurement of the radioactivity incorporated into protein. Note the early cessation of protein synthesis, even in the absence of edeine, and the proportion of amino acid incorporation which is due to initiation of new polypeptide chains (edeine-sensitive counts).

vitro. The kinetics of amino acid incorporation in the presence and absence of an initiation inhibitor can be used to distinguish these alternatives. *Figure 8* shows such an experiment. It is important that the inhibitor should not also affect elongation or termination; a good choice is the antibiotic edeine, which is commercially available. Other possibilities are pactamycin (used at 10 μM) or a non-translatable homopolynucleotide such as polyinosinic acid (at 400 μg/ml).

When exogenous mRNAs are used, particularly in preincubated and/or micrococcal nuclease-treated systems, mRNA competition studies can be performed. This has proved especially useful in analysing the relationships between viral mRNAs and host cell mRNAs, for example, and for determining the relative efficiencies of translation of different mRNAs under different ionic conditions (59). In such studies it is, of course, essential to have a specific assay for the translation of each template. Product analysis by SDS-PAGE (Section 5.2) is widely used for this purpose.

7.2.3 *Problems and Limitations*

The most serious pitfall in using cell extracts to analyse physiological changes in cells or tissues is that one finishes up studying the properties of the extracts rather than the cells. Clearly the preparation of cell-free systems itself introduces some artifacts and these must be weighed against the advantages of being able to perform experiments which are not possible with intact cells or whole animals. The low translational activity in many systems (see Section 4.6) and the short time during which initiation on endogenous mRNA is maintained (*Figure 8*) are worrying problems which need to be tackled. The apparent sequestration of mRNA in a non-translatable form, and the marked contrast between the ability of some non-erythroid mammalian cell-free systems to form [40S.Met-tRNA$_f$] initiation complexes and to bind mRNA to such complexes are also technical disadvantages which merit attention.

Another difficult problem is why some cell extracts fail to respond to controls which act rapidly and efficiently on intact cells. A particularly puzzling case is the different responses of amino acid-starved cells and their extracts to restoration of the missing amino acid (60). In whole cells, protein synthesis is controlled at the initiation level and responds within minutes to added amino acids. In contrast, addition of amino acids to extracts from starved cells is completely ineffective although exogenous eIF-2 does restore initiation (14).

It seems likely that by understanding the causes of these defects in the behaviour of eukaryotic cell-free systems, we will not only overcome technical barriers but will also improve our knowledge of the mechanism and control of protein synthesis in eukaryotes.

8. ACKNOWLEDGEMENTS

I am very grateful to several colleagues who have communicated helpful information about practical aspects of mRNA preparation and translation. I particularly wish to thank Tim Hunt, Richard Jackson and Simon Morley (University of Cambridge) for unpublished data and for a copy of their laboratory manual 'The Rabbit Reticulocyte Lysate Translation System'. I also thank Drs Edgar Henshaw and Richard Panniers (University of Rochester Cancer Center) for the manuscript of their paper (*Methods in Enzymology*, in press). Parts of the data presented here were obtained in collaboration with Drs Sara Austin, Margaret McNurlan, Joachim Kruppa, Bryan Williams and Ian Kerr. I am grateful to Vivienne Tilleray and Sigrid Burridge for technical assistance, Barbara Bashford for the drawings and Vivienne Marvell for preparation of the manuscript. This work has been supported by a Career Development Award from the Cancer Research Campaign and by grants from the Medical Research Council and the Anna Fuller Fund.

9. REFERENCES

1. Clemens,M.J., ed. (1980) *The Biochemistry of Cellular Regulation,* Vol. **1**, CRC Press, Boca Raton, p.159.
2. Jackson,R.J., Campbell,E.A., Herbert,P. and Hunt,T. (1983) *Eur. J. Biochem.,* **131**, 289.
3. Pelham,H.R.B. and Jackson,R.J. (1976) *Eur. J. Biochem.,* **67**, 247.
4. Skup,D. and Millward,S. (1977) *Nucleic Acids Res.,* **4**, 3581.
5. Palmiter,R.D. (1973) *J. Biol. Chem.,* **248**, 2095.
6. Pelham,H.R.B., Sykes,J.M.M. and Hunt,T. (1978) *Eur. J. Biochem.,* **82**, 199.
7. Cepko,C.L., Hansen,U., Handa,H. and Sharp,P.A. (1981) *Mol. Cell Biol.,* **1**, 919.
8. Kay,J.E. and Benzie,C.R. (1982) *Biochim. Biophys. Acta,* **698**, 218.
9. Roberts,B.E. and Paterson,B.M. (1973) *Proc. Natl. Acad. Sci. USA,* **70**, 2330.
10. Marcu,K. and Dudock,B. (1974) *Nucleic Acids Res.,* **1**, 1385.
11. Kronenberg,H.M., Roberts,B.E., Habener,J.F., Potts,J.T. and Rich,A. (1977) *Nature,* **267**, 804.
12. Senger,D.R. and Gross,P.R. (1976) *Dev. Biol.,* **53**, 128.

13. Mathews,M.B. (1972) *Biochim. Biophys. Acta,* **272**, 108.
14. Pain,V.M., Lewis,J.A., Huvos,P., Henshaw,E.C. and Clemens,M.J. (1980) *J. Biol. Chem.,* **255**, 1486.
15. Henshaw,E.C. and Panniers,L.R.V. (1983) in *Methods in Enzymology,* Vol. **101**, Wu,R., Grossman,L. and Moldave,K. (eds.), p. 616.
16. Tuite,M.F., Plesset,J., Moldave,K. and McLaughlin,C.S. (1980) *J. Biol. Chem.,* **255**, 8761.
17. Schreier,M.H. and Staehelin,T. (1973) *J. Mol. Biol,* **73**, 329.
18. Trachsel,H., Erni,B., Schreier,M.H. and Staehelin,T. (1977) *J. Mol. Biol.,* **116**, 755.
19. Esteban,M. and Kerr,I.M. (1974) *Eur. J. Biochem.,* **45**, 567.
20. Weber,L.A., Feman,E.R. and Baglioni,C. (1975) *Biochemistry (Wash.),* **14**, 5315.
21. Austin,S.A. and Clemens,M.J. (1981) *Eur. J. Biochem.,* **117**, 601.
22. Hames,B.D. and Rickwood,D., eds. (1981) *Gel Electrophoresis of Proteins,* IRL Press Ltd., Oxford and Washington, DC.
23. Dobos,P., Kerr,I.M. and Martin,E.M. (1971) *J. Virol.,* **8**, 491.
24. Heywood,S.M. and Rourke,A.W. (1974) in *Methods in Enzymology,* Vol.**30**, Moldave,K. and Grossman,L. (eds.), Academic Press, New York and London, p. 669.
25. Woodward,W.R., Wilairat,P. and Herbert,E. (1974) in *Methods in Enzymology,* Vol. **30**, Moldave,K. and Grossman,L. (eds.), Academic Press, New York and London, p. 740.
26. Devillers-Thiery,A., Kindt,T., Scheele,G. and Blobel,G. (1975) *Proc. Natl. Acad. Sci. USA,* **72**, 5016.
27. Smith,A.E. (1973) *Eur. J. Biochem.,* **33**, 301.
28. Wigle,D.T. and Smith,A.E. (1973) *Nature New Biol.,* **242**, 136.
29. Raj,N.B.K. and Pitha,P.M. (1977) *Proc. Natl. Acad. Sci. USA,* **74**, 1483.
30. Hruby,D.E. and Ball,P.M. (1977) *J. Virol.,* **40**, 456.
31. Davis,B.D. and Tai,P-C. (1980) *Nature,* **282**, 433.
32. Sabatini,D.D., Kreibich,G., Morimoto,T. and Adesnik,M. (1982) *J. Cell Biol.,* **92**, 1.
33. Blobel,G. and Dobberstein,B. (1975) *J. Cell Biol.,* **67**, 852.
34. Strauss,A.W., Zimmerman,M., Boime,I., Ashe,B., Mumford,R.A. and Alberts,A.W. (1979) *Proc. Natl. Acad. Sci. USA,* **76**, 4225.
35. Walter,P., Ibrahimi,I. and Blobel,G. (1981) *J. Cell Biol.,* **91**, 545.
36. Rothman,J.E. and Lodish,H.F. (1977) *Nature,* **269**, 775.
37. Scheele,G., Dobberstein,B. and Blobel,G. (1978) *Eur. J. Biochem.,* **82**, 593.
38. Shields,D. and Blobel,G. (1977) *Proc. Natl. Acad. Sci. USA,* **74**, 2059.
39. Pelham,H.R.B. (1978) *Eur. J. Biochem.,* **85**, 457.
40. Proud,C.G., Clemens,M.J. and Pain,V.M. (1982) *FEBS Lett.,* **148**, 214.
41. Darnbrough,C., Legon,S., Hunt,T. and Jackson,R.J. (1973) *J. Mol. Biol.,* **76**, 379.
42. Jagus,R. and Safer,B. (1979) *J. Biol. Chem.,* **254**, 6865.
43. Austin,S.A. and Clemens,M.J. (1980) *FEBS Lett.,* **110**, 1.
44. Farrell,P.J., Balkow,K., Hunt,T., Jackson,R.J. and Trachsel,H. (1977) *Cell,* **11**, 187.
45. Leroux,A. and London,I.M. (1982) *Proc. Natl. Acad. Sci. USA,* **79**, 2147.
46. Wong,S.-T., Mastropaolo,W. and Henshaw,E.C. (1982) *J. Biol. Chem.,* **257**, 5231.
47. Clemens,M.J. and Williams,B.R.G. (1978) *Cell,* **13**, 565.
48. Williams,B.R.G., Gilbert,C.S. and Kerr,I.M. (1979) *Nucleic Acids Res.,* **6**, 1335.
49. Sarkar,S., Mukherjee,A.K. and Guha,C. (1981) *J. Biol. Chem.,* **256**, 5077.
50. Fresno,M. and Vazquez,D. (1980) *Eur. J. Biochem.,* **103**, 125.
51. Hastie,N.D. and Held,W.A. (1978) *Proc. Natl. Acad. Sci. USA,* **75**, 1217.
52. Austin,S.A. and Kay,J.E. (1975) *Biochim. Biophys. Acta,* **395**, 468.
53. Lenz,J.R., Chatterjee,G.E., Maroney,P.A. and Baglioni,C. (1978) *Biochemistry (Wash.),* **17**, 80.
54. Fischer,I. and Moldave,K. (1981) *Anal. Biochem.,* **113**, 13.
55. Ahern,T., Sampson,J. and Kay,J.E. (1974) *Nature,* **248**, 519.
56. Gasior,E., Herrera,F., McLaughlin,C.S. and Moldave,K. (1979) *J. Biol. Chem.,* **254**, 3970.
57. Pain,V.M. and Henshaw,E.C. (1975) *Eur. J. Biochem.,* **57**, 335.
58. Austin,S.A. and Kay,J.E. (1982) in *Essays in Biochemistry,* Vol. **18**, Campbell,P.N. and Marshall,R.D. (eds.), The Biochemical Society London, p. 79.
59. Brendler,Y., Godefroy-Colburn,T., Carlill,R.D. and Thach,R.E. (1981) *J. Biol. Chem.,* **256**, 11747.
60. Austin,S.A., Pain,V.M., Lewis,J.A. and Clemens,M.J. (1982) *Eur. J. Biochem.,* **122**, 519.
61. RajBhandary,U.L. and Ghosh,H.P. (1968) *J. Biol. Chem.,* **244**, 1104.

CHAPTER 10

Translation of Eukaryotic Messenger RNA in Xenopus Oocytes

ALAN COLMAN

1. INTRODUCTION

During oogenesis the oocytes of *Xenopus laevis* accumulate large quantities of enzymes, storage proteins and organelles, which form a maternal reserve for use during early embryonic development. A striking aspect of this early development is the large, post-fertilisation increase in protein synthesis which results from the increased translation of maternally-produced mRNA. Not surprisingly, a significant proportion of the maternal reserve is represented by components required for this translational burst, for example, ribosomes, transfer RNAs etc. (1,2). However, the bulk of the maternal reserve consists of the yolk proteins, phosvitin and lipovitellin. It is the acquisition of large amounts of these proteins (~ 200 μg/oocyte) from the maternal blood supply which largely accounts for the impressive diameter (~ 1.2 mm) of the full-grown *X. laevis* oocyte. These two facets of the oocyte — the maternal reserve and the large size — have made the oocyte an attractive test system for studies on the transcription (3), replication (4), assembly (5), partition (6) and translation (7 − 10) of injected macromolecules. This chapter will concentrate on the ability of the oocyte to translate injected mRNAs. The transcription of injected DNA by oocytes is described in Chapter 2.

Injected mRNAs translate very efficiently in *Xenopus* oocytes. This was first demonstrated by John Gurdon and his colleagues (11) when rabbit globins were synthesised after the microinjection of rabbit reticulocyte 9S RNA. These authors estimated that the mRNAs were translated about 30 times more efficiently in the oocyte than in reticulocyte lysate over the same incubation period. Moreover, this discrepancy between oocyte and lysate efficiency could be increased to 1000-fold simply by culturing injected oocytes for prolonged periods. Since then, a large variety of mRNAs have been translated after injection into oocytes *(Table 1)*. In the original globin work, the radioactive globin product was detected using column chromatography. Although these chromatographic procedures have now been largely superceded by immunological and electrophoretic techniques, direct radiolabelling of the foreign protein is still routine.

For the majority of routine assays of mRNA, cell-free extracts remain the simplest and most economical translation systems available. However, the *Xenopus* oocyte has clear advantages over cell-free systems in special circumstances. Thus, unlike the cell-free systems, the oocyte can correctly carry

Table 1. RNAs Translated after Microinjection into *Xenopus* Oocytes.[a]

Source of mRNA	Reference
Virus	
Adenovirus	12
Alfalfa mosaic	13
Avian myeloblastosis	14
Avian sarcoma	15
Barley mosaic	16
Bovine leukaemia	17
Brome mosaic	18
Citrus exocortis	19
Cucumber mosaic	20
Encephalomyocarditis	21
Influenza	unpublished[b]
Mengovirus	22
Moloney	23
Mouse mammary tumour	24
Polyoma	25
Rabies	26
Reovirus	27
Simian Virus 40	25
Tobacco mosaic	28
Plant	
Barley secretory proteins	25
Castor bean lectin	unpublished[c]
Common bean *(Phaseolus vulgaris)* storage protein	29
Maize storage proteins	30
Invertebrate	
Honey bee promelittin	31
Locust vitellins	32
Sea urchin histones	33
Vertebrate	
Bovine lens α and β crystallins	34, 35
Bovine prochymosin	unpublished[d]
Bovine thyroglobulin	36
Carp proinsulin	37
Cat acetylcholine receptor	38
Chick conalbumin	39
Chick lysozyme	39
Chick ovalbumin	39, 40
Chick ovomucoid	39
Duck globin	41
Frog albumin	25
Frog vitellogenin	25,32
Frog collagen	42
Guinea pig caseins	43
Guinea pig α-lactalbumin	43
Human immunoglobulin	unpublished[e]
Human interferon	44
Human chorionic gonadotropin	45
Mouse globin	41
Mouse interferon	46

Mouse collagen	7
Mouse immunoglobulin	47 – 50
Mouse β-glucuronidase	51
Mouse thyrotropin	52
Pig immunoglobulins	53
Rabbit globin	11
Rabbit uteroglobulin	54
Rat immunoglobulin	55
Rat cytochrome	56
Rat epoxide hydratase	56
Rat seminal vesicle proteins	57
Rat prostatic steroid binding protein	58
Torpedo acetylcholine receptor	59
Trout protamine	60

[a]For a more complete set of references see references 7 – 10.
[b]Colman,A., Dimmock,N. and Davey,J.
[c]Colman,A. and Lord,J.M.
[d]Colman,A., Doel,M. and Harris,T.
[e]Colman,A. and Williamson,A.

out the post-translational modification of many foreign proteins such as precursor processing, phosphorylation and glycosylation *(Table 2)*. It can also direct proteins into the correct intracellular membrane and will even export secretory proteins *(Table 3)*. Finally, biologically-active, foreign proteins synthesised in oocytes can acquire their native activity even when this involves complex subunit interactions. In these cases a suitable bioassay, offering higher sensitivity than conventional radiochemical analysis, can be used to detect the translation product *(Table 4)*.

2. OBTAINING AND PREPARING OOCYTES FOR MICROINJECTION

2.1 Purchase and Maintenance of Xenopus laevis

2.1.1 *Purchase*

The natural habitat for *X. laevis* frogs is South Africa. The author has traditionally purchased frogs from South African sources, having found that the 'wild' animals are larger and often in better condition than laboratory-reared animals. However, when only small numbers of frogs are required (<20) the handling charges, airfreight, etc., make importation uneconomical. Fortunately, several countries have commercial outlets which offer the buyer the option of laboratory-reared, laboratory-conditioned or 'wild' animals, for example, South African Snake Farm, Xenopus Ltd. and Nasco. The relevant addresses are given in Appendix II. Make sure, when ordering, to specify the sex of animal required!

2.1.2 *Accommodation*

Keep the frogs in water in large tanks (one animal per 3 – 4 litres). The depth of water should be no more than 15 cm since the frogs need to come to the surface to breathe. Ensure that the tanks have firmly-attached covers (soft plastic

Table 2. Post-translational Modification of Foreign Proteins in *Xenopus* Oocytes.

Modification	Protein modified[a]	References
αNH_2-acetylation	Calf lens αA2 crystallin	34
ϵNH_2-acetylation	Sea urchin histones	33
N-glycosylation	Chicken ovalbumin	39
	Mouse immunoglobulins	47
Hydroxylation	Mouse fibroblast collagen	7
Phosphorylation	Trout testis protamine	60
Polyprotein cleavage	*Xenopus* vitellogenin	61
Covalent subunit assembly	Rabbit uteroglobulin	54
	Immunoglobulins (various)	48, 49, 55
Non-covalent subunits	*Torpedo* acetylcholine receptor	59
Signal sequence removal	Various proteins	see *Table 3* footnote

[a]Selected examples of each modification are shown.

Table 3. Correct Segregation of Foreign Proteins made in *Xenopus* Oocytes after mRNA Injection.

Protein	Final location	References
Retained within the oocyte		
Rat liver epoxide hydratase	oocyte membranes	56
cytochrome P-450	oocyte membranes	56
Maize storage proteins	oocyte membranes	30
Common bean storage proteins	oocyte membranes	29
Non-secretory mouse immunoglobulin[a]	oocyte membranes	48
Castor bean lectin[a]	oocyte membranes	unpublished[b]
Acetylcholine receptor	oocyte membranes	38, 59
Polyoma large tumour antigen	oocyte nucleus	25
Influenza nuclear protein	oocyte nucleus	unpublished[c]
Xenopus liver phosvitin	oocyte yolk platelet	61
lipovitellin	oocyte yolk platelet	61
Exported from the oocyte		
Barley seed low mol. wt. proteins		25
Bovine prochymosin[a]		unpublished[d]
Chick ovalbumin		39
conalbumin		39
ovomucoid		39
lysozyme[a]		39
Guinea pig caseins		62
Human interferons		62
Locust vitellin polypeptides		25
Mouse interferons		46
thyroid-stimulating hormone		52
immunoglobulins[a]		48 – 50
Rat albumin		25
seminal vesicle proteins[a]		57

[a]Signal sequence shown to be removed by the oocyte.
[b]Colman,A. and Lord,J.M.
[c]Colman,A. and Dimmock,N.
[d]Colman,A., Harris,T. and Doel,M.

Table 4. Foreign Proteins made in Oocytes and detected by Bioassay.

Protein	Bioassay	References
Interferons (various)	Inhibition of viral activity	44, 46, 62
Immunoglobulins (various)	Antigen binding	50, 55
Mouse kidney β-glucuronidase	Enzyme activity	51
Torpedo electric organ and cat muscle acetylcholine receptors	α-Bungarotoxin binding	38, 59
Xenopus vitellogenin	Uptake by oocytes	32
Rat liver cytochrome P-450	Enzyme assay	56
Mouse antigen-specific suppressor factors	Suppressor activity	63

mesh is ideal) to prevent escape. Out of water the frogs can die from dehydration within 6 h. The optimum water temperature is approximately $19-21\,^{\circ}$C. Tap water can be used but it is important to remove any high concentrations of chlorine or heavy metals which may be present. This can be achieved by storing the water in large open tanks for 24 h (to allow chlorine to evaporate) or by ion-exchange chromatography, respectively.

2.1.3 *Feeding*

The frogs require feeding only twice per week. In the author's experience, the best diet is one feed of coarsely-minced beef and one feed of live blowfly larvae. Clean the tanks out after each meal and more often if the water becomes murky. A more controlled diet ('frog pellets') can be obtained from Xenopus Ltd. but the author has found that frogs usually have to be weaned onto these pellets over a long period, ideally during their growth. This regime is often not feasible for frogs obtained from the wild. Further details on the upkeep and rearing of *Xenopus* can be found in reference 64.

2.2 **Removal of Ovary Tissue from Xenopus**

A fully mature *Xenopus* female contains about 30 000 large (>1 mm diameter) oocytes. Since a typical experiment rarely requires more than $500-1000$ oocytes, it is unnecessarily wasteful to kill a frog for one experiment. Indeed, in some cases, it has proved highly desirable to re-use the same female intermittently due to differences in the levels of certain oocyte components from frog to frog (65). However, where several experiments are contemplated or other *Xenopus* tissues are required, it is necessary to kill the frog.

2.2.1 *Limited Removal of Oocytes Under Anaesthesia*

(i) Anaesthetise a large *Xenopus* female by immersion for $15-30$ min in a 0.1% solution of ethyl *m*-aminobenzoate (also called Tricaine or MS 222) in water. Wear gloves throughout these procedures since this compound is a potential carcinogen.

(ii) When the frog is anaesthetised, place it on a flat surface and make a small (~ 1 cm) incision through the loose skin and body wall in the posterior, ventral side as indicated in *Figure 1a*. The ovary, which consists of 16

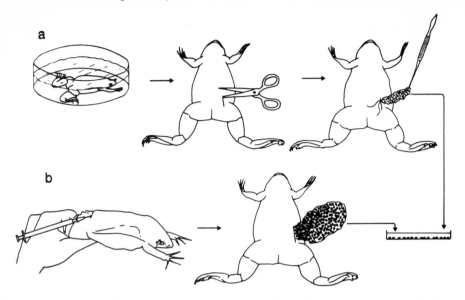

Figure 1. Removal of oocytes from *Xenopus* frogs. Schematic representations of **(a)** the anaesthesia of a frog and subsequent removal of a single lobe of ovary through an incision in the ventral side of the animal, **(b)** barbiturate injection to kill a frog and subsequent removal of the complete ovary. Details of the techniques are given in the text. (Drawings by Dr.J.Brown).

lobes and completely fills the abdomen, is exposed by this surgery.

(iii) Using ordinary forceps, carefully tease out one lobe of the ovary and tie a ligature at its base.

(iv) Excise the lobe using scissors and transfer it to modified Barths' saline (see *Table 5* for composition and preparation).

(v) Suture the incision in the body wall separately from that in the skin.

(vi) Leave the frog to recover for 24 h on a slope partially-immersed in shallow water supplemented by 0.5% NaCl, ensuring that its head is above the water to prevent the frog drowning.

2.2.2 *Removal of Complete Ovary*

(i) Kill the frog by injection of a suitable veterinary preparation of concentrated barbiturate (e.g., 'Euthetal', May and Baker). To do this, grip the frog firmly by the legs, using a towel or tissue to prevent escape, and inject 0.3 ml of barbiturate per 100 g body weight under the skin in the lower back just above the legs (see *Figure 1b*).

(ii) Place the frog in shallow water until it is anaesthetised.

(iii) Remove the complete ovary, preferably via an incision made as described in Section 2.2.1 since more radical surgery results in the rupture of major abdominal blood vessels and consequent mess.

2.3 **Preparation of Oocytes for Injection**

It is imperative to their long-term survival that *Xenopus* oocytes are thoroughly washed in modified Barths' saline (see *Table 5*) soon after their

Table 5. Composition and Preparation of Modified Barths' Saline.

Composition[a]

    88.0 mM NaCl
    1.0 mM KCl
    2.4 mM NaHCO$_3$
    15.0 mM Hepes-NaOH *or* Tris-HCl (pH 7.6)
    0.30 mM CaNO$_3$·4H$_2$O
    0.41 mM CaCl$_2$·6H$_2$O
    0.82 mM MgSO$_4$·7H$_2$O
    10 μg/ml sodium penicillin
    10 μg/ml streptomycin sulphate

Preparation[b]

  A. *High salt stock*

NaCl	128 g
KCl	2 g
NaHCO$_3$	5 g
Hepes	89 g (*or* Tris base 45 g)

Dissolve these reagents in distilled water and adjust to pH 7.6 with conc. HCl (when using Tris) or 1 M NaOH (when using Hepes). Make up to 1 litre. Store as 40 ml aliquots at −20°C.

  B. *Divalent cation stock*

CaNO$_3$·4H$_2$O	1.90 g
CaCl$_2$·6H$_2$O	2.25 g
MgSO$_4$·7H$_2$O	5.00 g

Dissolve these reagents in 1 litre of distilled water. Store as 40 ml aliquots at −20°C.

  C. *Antibiotic stock*

Freshly prepare a solution of sodium penicillin and streptomycin sulphate, each at 10 mg/ml.

To prepare modified Barths' saline, mix in the following order:-
    919 ml of distilled water
    40 ml of high salt stock solution A
    40 ml of divalent cation stock solution B
    1 ml of antibiotic stock solution C

[a]Optional additions: for studies of secretion, include gentamicin (100 μg/ml), nystatin (20 U/ml) and, in some circumstances (see Section 4.1), 5% *dialysed* calf or horse serum.
[b]All reagents should be analytical grade.

removal from the frog. Using watchmaker's forceps, tear apart the ovarian lobe(s) into small clumps of approximately 10−30 large oocytes. There will also be smaller, less mature oocytes present in these clumps *(Figure 2)* but, for convenience, only the large oocytes are used for microinjection. Store the clumps in Petri dishes at a density of about 30 clumps per 20 ml Barths' saline. For long term storage (several days), keep the clumps at 18−21°C in a cooled incubator (e.g., Gallenkamp). The smaller clumps ($\sim$10 oocytes) may be used directly for injection. However, it is often more desirable to inject oocytes singly rather than in clumps. Although more damage can be sustained by the further dissection needed to obtain these, a single oocyte is more manoeuvrable. This is especially important when injection of the oocyte nucleus is performed (see Chapter 2). In addition, many smaller oocytes and

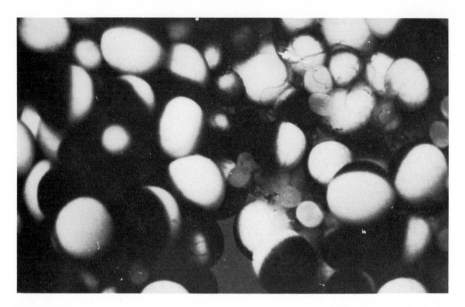

Figure 2. Part of an ovarian lobe is shown soon after its removal from the frog. Oocytes of various sizes representing different stages of growth are seen.

other ovarian tissues are present in clumps and, although these will remain uninjected, they can complicate the experimental analysis in certain situations. The production of large numbers of single oocytes can be accomplished either by manual dissection or by enzymatic 'stripping'.

2.3.1 *Manual Dissection*

This is the recommended method unless several thousand single oocytes are required. Grip a large clump (>30 oocytes) with one pair of watchmaker's forceps whilst stripping the oocytes off the clump with another pair of forceps in a manner similar to removing grapes from a bunch. The success of this method varies from batch to batch of oocytes. Inevitably some damage is incurred. However, the dissected oocytes are routinely incubated overnight before injection, after which time the damaged oocytes can be easily identified. In the event of this technique being too disruptive, individual dissection of oocytes from clumps can be accomplished under a stereomicroscope. The single oocytes are most easily handled using a wide-mouthed Pasteur pipette.

2.3.2 *Enzymatic 'Stripping'*

Gently swirl the oocyte clumps for 2 – 4 h at 18 – 24°C in 2 mg/ml collagenase (Sigma, Type II) dissolved in Barths' saline. This procedure liberates the oocytes from their follicles. However, the released oocytes are still surrounded by three layers of somatic cells.

3. MICROINJECTION OF OOCYTES WITH MESSENGER RNA

3.1. **Essential Equipment**

A stereomicroscope, micromanipulator, syringe, light source and micropipette-making apparatus are considered essential for efficient microinjection. Some of this equipment is displayed in the position used for microinjection in *Figure 3a* and *3b*.

3.1.1 *The Stereomicroscope*

Any good stereomicroscope will suffice. However, a microscope offering a lens with adjustable magnification and a good depth of field is preferable. The latter facility is useful since it is desirable to have both the meniscus of liquid in the micropipette and the oocyte in reasonable focus simultaneously.

3.1.2 *The Micromanipulator*

A low power micromanipulator modelled on the original Singer Micromanipulator Mark 1 is more practical than any higher power alternatives, since it allows three-dimensional movement which will follow the movement of the hand with a reduction ratio of 4:1. Manipulators of this type can be obtained from the Singer Instrument Company or Oxford Microinstruments.

3.1.3 *The Syringe*

The principle of the mRNA delivery system is that the sample in the micropipette can be placed under positive pressure via flexible polythene tubing (~ 1 mm diameter) to a syringe. Both the syringe and tubing are filled with medicinal grade paraffin oil (from a local pharmacist). The author uses an 'Agla' hand-operated micrometer syringe (Wellcome Reagents Ltd.) whereas Birnstiel and his colleagues (66) use a 10 μl Hamilton syringe which is connected to a motorised micrometer screw. A similar design, but incorporating the Agla syringe, is shown in *Figure 3b*. This 'automated' delivery system has the advantages of freeing one hand and allowing more reproducible and quantitative sample delivery into oocytes. Using either method, air should be flushed from the system regularly.

3.1.4 *The Light Source*

Any available microscope lamp can be used. However, fibre-optic lamps (e.g., Schott fibre optic lamp from Micro Instruments) offer the advantage that the intense illumination they provide is cold and hence the drying out of oocytes during injection is not such a problem.

3.1.5 *Micropipette-making Apparatus*

Micropipette construction requires a reliable micropipette puller which essentially consists of a tungsten heating coil with a stabilised and adjustable power supply. Although a 'home-made' version *(Figure 4a)* will suffice, a rugged

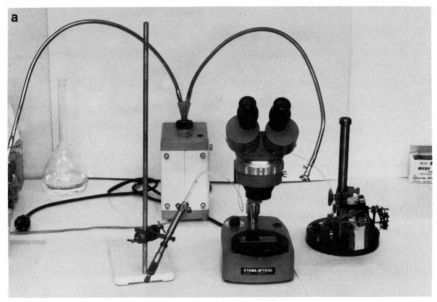

Figure 3. Microinjection equipment. (**a**) Syringe, micromanipulator and stereomicroscope are displayed in their normal working positions. (**b**) Motorised syringe featuring a small motor-driven micrometer syringe connected to a foot-operated pedal via a transformer (to vary the speed fo the micrometer screw between 0 and 30 r.p.m.).

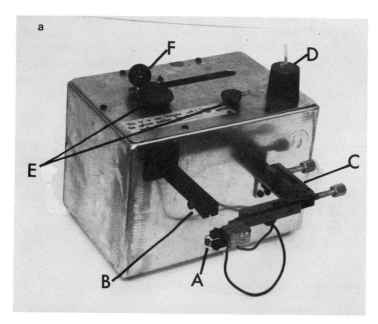

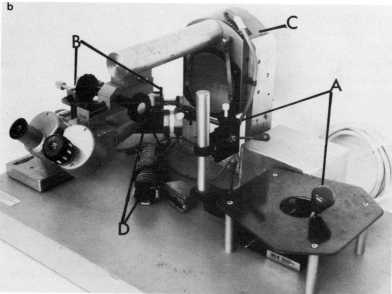

Figure 4. Apparatus for making micropipettes. **(a)** A 'home-made' micropipette puller consisting of a tungsten heating coil (A), pipette clamp (B), stage providing XY movement of the heating coil (C), pipette weight (D), spare heating coils (E) and rheostat control (F). **(b)** A microforge. This particular instrument incorporates two 3-dimensional manipulators, one (A) to hold the glass micropipette and the other (B) to hold the platinum heating wire. These manipulators are supplied by Research Instruments Ltd. A stereomicroscope (C) is shown mounted horizontally. A rheostat (D) with a 1 ohm resistance in series is incorporated into a 'home-made' power circuit. A complete microforge unit incorporating all of the above features can also be purchased from Research Instruments Ltd.

281

commercial instrument is the microelectrode puller marketed by Searle Bioscience. The use of a microforge, to produce a sharp tip to micropipettes, is unnecessary for mRNA injections into oocytes. It is, however, important for the injection of fertilised eggs and facilitates injections into the oocyte nucleus (see Chapter 2). A suitable instrument *(Figure 4b)* can be purchased from Research Instruments Ltd.

3.1.6. *Ancilliary Equipment*

Xenopus oocytes are best cultured at about 20°C. At 25°C, protein synthesis is higher but the oocytes do not survive so long. Above 30°C, oocytes will only synthesise one major protein, a 70 K 'heat-shock' protein (67) and injected mRNAs are not translated (A.Colman, unpublished data). In order to maintain a constant temperature of 20°C, a cooled incubator is desirable though not essential; an alternative strategy is to use an ordinary incubator in a cold-room.

3.2 **Construction, Calibration and Storage of Micropipettes**

The construction and calibration of micropipettes is, in the author's experience, the most problematic technical aspect of oocyte microinjection. Whilst the tip of the micropipette should have an external diameter no greater than 30 μm, a sufficient part of the main barrel of the pipette (external diameter ~ 300 μm) should remain within the field of view during injection to ensure visual monitoring and control of the progressive delivery of sample to several oocytes. In addition, this barrel should be parallel sided for at least 15 mm to facilitate calibration.

3.2.1 *Construction*

The procedures used in making micropipettes are shown diagrammatically in *Figure 5*. Hard glass capillaries, 10 cm long (e.g., from BDH Chemicals Ltd.) are used.

(i) Manually draw out each capillary over a Bunsen micro-burner flame to give an external diameter of approximately 200 μm *(Figure 5a)*. Standardising this initial manual pull is crucial to the reproducible and rapid construction of micropipettes.

(ii) Place the capillary into position on the micropipette puller so that the tungsten heating coil surrounds part of the drawn-out section *(Figure 5b)*. Attach a suitable weight to the end of the capillary.

(iii) Allow current to flow to the coil. As the glass heats up, the weight will pull the capillary towards the ground, leaving a finely drawn-out central section.

(iv) Break the central section obliquely using watchmaker's forceps to give a micropipette with the dimensions shown in *Figure 5b*. The final diameter of the micropipette tip depends on the size of the heating coil, the current flow, the weight, etc. Exact conditions must be established empirically.

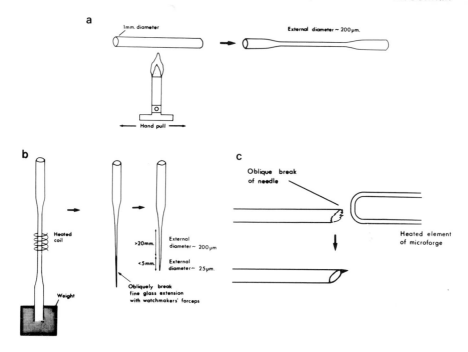

Figure 5. Construction of micropipettes. The procedures used in making micropipettes are shown diagrammatically and described in detail in the text. This diagram has been modified from that in reference 68.

(v) If desired, the tip of the micropipette can be smoothed and sharpened by touching it briefly to the heated element of a microforge *(Figure 5c)*.

3.2.2 *Calibration*

After penetration of the oocyte during microinjection, injection of the sample via the hydraulic system is monitored by watching the meniscus in the micropipette barrel. Routinely about 50 nl should be injected. Pipette calibration, described below, serves for training purposes alone. It rapidly becomes unnecessary; after two or three sessions, the researcher will be able to deliver 50 nl ± 20 nl, without recourse to prior calibration or even marking the pipette.

To calibrate a micropipette.

(i) Mark it at several equally-spaced positions away from the tip as shown in *Figure 6a*.

(ii) Fill the micropipette with paraffin oil then with water, leaving an air gap between the two liquids.

(iii) Slowly expel the water to create a spherical droplet at the pipette tip by allowing the movement of the meniscus between two adjacent marks *(Figure 6b)*.

(iv) Determine the volume of the droplet by measuring its diameter using an eyepiece micrometer and referring this value to a calibration curve *(Figure 6c)* derived mathematically.

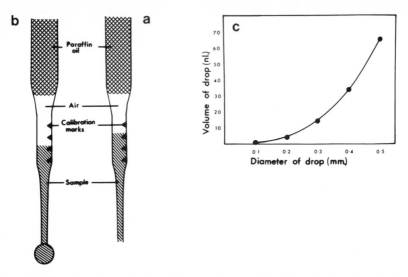

Figure 6. Calibration of micropipettes. The procedures used in **(a)** loading the micropipette and **(b)** expelling a droplet of sample for calibration purposes are shown diagrammatically and describ-ed in detail in the text. **(c)** Calibration curve derived by plotting the volume $(4/3\ \pi\ r^3)$ against the diameter $(2r)$ of a sphere where r is the radius.

3.2.3 *Storage of Micropipettes*

Keep the micropipettes upright in long ($>$10 cm), slender glass tubes, one micropipette per tube. Place the tubes upright in a beaker and cover with foil. Before mRNA injection, sterilise the tubes by heating ($>$1 h, 200°C).

3.3 **Microinjection of Messenger RNA**

3.3.1 *Choice of mRNA Concentration*

Dissolve the mRNA in distilled water and store in small ($\sim$5 μl) aliquots at -70°C. Purified mRNA [e.g., by oligo(dT)-cellulose chromatography; see Chapter 8] is preferable to unpurified cellular RNA since the amount of pro-tein made will be greater per ng of injected RNA. However, the presence of ribosomal RNA in the sample has no detrimental effect on translation. As shown in *Figure 7*, for mRNAs specifying non-secretory proteins the amount of protein synthesised increases with the amount of mRNA injected, at least up to 100 ng per oocyte (i.e., injection of 50 nl of a 2 mg/ml solution). However, because of the significant mRNA degradation which occurs with high concentrations of injected mRNA *(Figure 8)*, this increase is not linear, that is, doubling the amount of mRNA injected does not lead to a proportional increase in protein synthesis. When injected at high concentrations, mRNAs specifying secretory proteins are similarly degraded, although saturation of the translational apparatus occurs after 20 ng mRNA has been injected per oocyte *(Figure 7)*. The implication of these findings (69) is that, unless detection of the protein is difficult, it is more economical to inject less than 10 ng of mRNA per oocyte, i.e., mRNA concentrations less than 0.2 mg/ml.

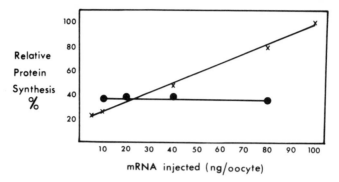

Figure 7. Relationship between protein synthesis and the amount of mRNA injected into *Xenopus* oocytes. These data indicate the relative incorporation of [³H]leucine into a non-secretory protein, rabbit globin ($\times$ —— $\times$) and a group of secretory proteins, zein ($\bullet$ —— $\bullet$) made in oocytes after injection of the indicated amounts of mRNA. Values are expressed relative to the maximum (i.e., 100%) incorporation seen after globin mRNA injection. (Based on the data of ref. 69.)

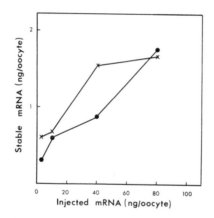

Figure 8. Degradation of mRNA in oocytes. The stabilities of injected globin ($\times$ —— $\times$) and zein mRNAs ($\bullet$ —— $\bullet$) in oocytes 10 h after injection were examined using hybridisation analysis. (Based on the data in ref. 69.)

3.3.2 *Loading Micropipettes with Sample Messenger RNA*

Transfer a small volume $(1-2\ \mu l)$ of the mRNA solution to a strip of 'Parafilm' or 'Nescofilm' and load the pipette by sucking up mRNA until the meniscus is at the edge of the field of view. Although this can be done keeping an air interface between the sample and the paraffin oil *(Figure 6)*, more control is possible if the micropipette is completely filled with paraffin oil prior to loading with sample. The paraffin oil has no effect on the mRNA, but the aqueous sample/paraffin oil interface is difficult to see clearly. This difficulty can be remedied by the inclusion of the fat-soluble dye, Fast Red, in the paraffin oil phase. Once the micropipette has been correctly loaded, store the unused mRNA plus its support on a bed of ice.

285

3.3.3 *Microinjection*

(i)   Transfer 4−5 oocytes (or 9−10 once the operator is skilled) to a clean microscope slide using a wide-mouthed Pasteur pipette.

(ii) Suck off excess liquid and position the slide on the stereomicroscope stage.

(iii) Various methods can be used to immobilise the oocytes during injection. *Either* gently hold the oocyte with a pair of watchmaker's forceps, *or* align the oocytes along a ridge made by placing a second slide in an overlapping position on the first slide.

(iv) Inject 50 nl (±20 nl) of RNA into the vegetal (yellow) region of the oocyte *(Figure 9a)*; injection into the animal (pigmented) region could hit the oocyte nucleus which can cause greater oocyte mortality[1]. Avoid allowing the oocytes to dry out during the injection procedure.

(v)   After injection, transfer the oocytes back to modified Barths' saline.

Oocytes can be re-injected with further mRNA samples at later times if necessary (69,70). With practice, 200−300 oocytes can be injected per hour.

4. CULTURE AND RADIOACTIVE LABELLING OF INJECTED OOCYTES

4.1 Culture Medium

The most common medium used for culturing oocytes is modified Barths' saline containing either Tris or Hepes as buffer *(Table 5)*. Although Hepes is a better buffer at physiological pH, a considerable latitude in pH (pH 7.5 ± 0.4) is tolerated by oocytes. The saline should be supplemented with penicillin and streptomycin and, for studies on protein secretion, the inclusion of gentamicin and the fungicide, nystatin, is advisable. In addition, the inclusion of 5% foetal calf or horse serum (dialysed against Barths' saline in each case) in the medium prevents reduction of sulphydryl bonds in the secreted products (49). This may also prevent proteolytic digestion of secretory proteins by proteases secreted by the follicle layers surrounding the oocytes (71). This additional serum is necessary only in medium which is to be analysed for secreted proteins.

4.2 Choice of Radioactive Amino Acids

Since oocytes are permeable to amino acids, simply adding radioactive amino acids to the medium results in their incorporation into oocyte protein. However, the uptake and equilibration of the exogenous radioactive amino acid with the internal pool takes 30−60 min. Therefore, if short pulses of incorporation into protein are desired, a concentrated solution of labelled amino acid should be injected instead (72). Even then, diffusion within the oocytes takes time. For amino acids with small pool sizes, the radioactive amino acid may be completely utilised before it equilibrates throughout the oocyte.

Whilst the requirements of the particular experiment will influence the

---

[1]The distribution of mRNA throughout the oocyte also occurs much more rapidly if an equatorial or vegetal injection is·performed rather than an injection into the pigmented half (Colman and Drummond, unpublished observations).

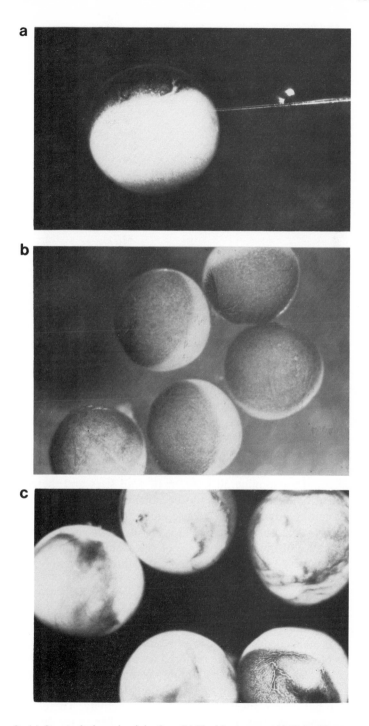

Figure 9. (a) Oocyte during microinjection. (b) Healthy oocytes. (c) Unhealthy oocytes; contrast the uneven pigmentation with the more uniform pigmentation of the healthy oocytes shown in (b).

Table 6. Amino Acid Pools in Large *Xenopus* Oocytes.[a]

Amino Acid	pmol/oocyte	
	Batch 1	Batch 2
Lysine	182	158
Histidine	67	84
Arginine	96	95
Aspartic acid	265	470
Threonine	283	240
Serine	374	495
Glutamic acid	857	741
Proline[b]	67	78
Glycine	194	261
Alanine	273	153
Cysteine	36	—
Valine	84	125
Methionine[b]	29	31
Isoleucine	37	56
Leucine[b]	45	84
Tyrosine	73	32
Phenylalanine	54	45

[a]This table contains unpublished data kindly provided by Drs.H.R.Woodland and E.D.Adamson.
[b]As evidenced by the efficiency of protein labelling using methionine, leucine and proline, the total amino acid pool is probably synonymous with the precursor pool.

choice of amino acid used, [^{35}S]methionine is particularly useful because of its relatively low cost, high specific activity, and high-energy β-particle emission (cf. tritium). The efficiency of a chosen amino acid in labelling newly synthesised oocyte proteins will depend to a large extent on the size of that amino acid pool. As a guide to experimental design, *Table 6* indicates the relative pool sizes of the different amino acids within the *Xenopus* oocyte. Histidine, proline, leucine and methionine all have small pool sizes and therefore frequent use of their radioactive isomers has been made. However, in special circumstances, it can be advantageous to use labelled amino acids which have large oocyte pool sizes. For example, [^{3}H]lysine is useful in the labelling of histones (73) because of the predominance of this residue in these proteins.

Oocytes tolerate very high concentrations of radioactive amino acid, and concentrations in the incubation medium of $0.1 - 5$ mCi/ml of radioactive amino acid are routine. Five oocytes incubated for 10 h in [^{35}S]methionine (30 μl, 1 mCi/ml, 350 Ci/mmol) will each incorporate 2 x 10^6 d.p.m. into protein. Considering that up to 10% of newly-synthesized protein can be represented by protein encoded by the injected mRNA, it is clear that only very small amounts of oocyte material are required for analysis. It should be noted, however, that the combination of small amino acid pool sizes and media containing labelled amino acids of high specific activity leads to newly-synthesized proteins of high but progressively changing specific activity. Therefore, when quantitative data are required, great care should be taken in the design of experiments. For example, in order to have linear incorporation of [^{35}S]-

methionine into protein over a 24 h labelling period (using the regime indicated above), the specific activity of the amino acid should be lowered to 50 Ci/mmol by addition of unlabelled methionine to a final concentration in the medium of 0.06 mM.

If the experimental design requires a 'chase' with non-radioactive medium, the author has found the inclusion of unlabelled amino acid to a final concentration of 10 mM to be completely effective within 1 h in the case of methionine. In fact, when high specific activity [^{35}S]methionine (>350 Ci/mmol) is used, incubation beyond about 6 h itself constitutes a chase. Therefore, if the possibility of an unstable protein product exists, incubation beyond this time is not recommended.

4.3 Other Radioactive Precursors

Several radioactive precursors other than labelled amino acids are useful in particular experimental protocols.

[^{3}H]Acetate. Use at $10-100$ mCi/ml in the medium to demonstrate post-translational acetylation of proteins. However, since [^{3}H]acetate is metabolised to [^{3}H]amino acids to a limited extent (72), newly synthesised proteins in general also become labelled.

[^{32}P]Phosphate. Use at $10-20$ mCi/ml in the medium to demonstrate phosphorylation of proteins. However, even at this concentration of isotope, there is little incorporation (73) due to the large inorganic phosphate pool within the oocyte.

[^{3}H]- or [^{14}C]mannose. Use 10 μCi/ml [^{14}C]mannose (47) or 1 mCi/ml [^{3}H]-mannose (45) in the medium to demonstrate glycosylation.

4.4 Choice of Labelling Period

A period of several hours can elapse before maximal translation of mRNAs injected into oocytes is seen (61). For many purposes this period of 'recruitment' can be ignored and oocytes can be incubated soon after injection. This procedure should also be utilised when poly(A)$^-$ mRNA is injected since some poly(A)$^-$ mRNAs (33), but not all (27), are unstable in oocytes. In contrast, poly(A)$^+$ mRNAs are very stable in oocytes (9) (but see Section 3.3.1). For secretion studies, leave the injected oocytes overnight before labelling. This delay has the advantage that unhealthy oocytes (see *Figure 9c*) can be removed prior to labelling. This selection ensures economical use of labelled precursor and minimises non-specific leakage of oocyte proteins into the culture medium. Such leakage can complicate subsequent analyses as well as providing a substrate for antibiotic-resistant micro-organisms in the medium.

4.5 Procedure for Labelling Oocytes

As described in Section 4.2, the labelling of oocytes can be achieved either by their incubation in radioactive precursor added to the medium or by injection of the isotope. The former method is described here whilst the injection procedure is the same as that described earlier for mRNA (Section 3.3).

Label oocytes in the wells of disposable microtitre plates. Incubate in batches of five oocytes per well in 30 μl of medium, checking that the oocytes are submerged. For secretion studies, pre-coat the wells with protein by incubating with a solution of bovine serum albumin (0.5 mg/ml in saline) at 37°C for 20 min before rinsing with saline.

When oocytes are being cultured for bioassay, larger volumes of incubation media may be used. The plastic tops of scintillation vial inserts (virtually any commercial source) are ideal for this, allowing the culture of approximately 25 oocytes per 300 μl of medium per top.

5. COLLECTION, FRACTIONATION AND ANALYSIS OF CULTURED OOCYTES AND THE SURROUNDING MEDIUM

As stated in the Introduction, *Xenopus* oocytes will not only translate and modify foreign proteins but will also direct proteins to the correct intracellular site as well as exporting secretory proteins. The type of fractionation used will depend on the objectives of the experiment. Homogenisation of oocytes in the presence of detergents will solubilise any membrane-associated protein whereas homogenisation in the absence of detergent followed by sucrose gradient fractionation gives an indication of the intracellular location of that protein. In some cases, for example, mammalian interferons, export of the protein is so rapid that analysis of the incubation medium alone is sufficient as an assay (62).

The type of analysis used will also be determined by the particular protein under study, and by the fact that the oocyte system, unlike some cell-free systems, has high endogenous translational activity. In many situations, especially when the specific mRNA concentration is low, a suitable bioassay, immunoprecipitation, two-dimensional gel electrophoresis, or selective solvent extraction techniques have to be considered. Only the most common fractionation and analytical techniques will be described below.

5.1 Collection of Oocytes and Medium after Incubation

At the end of the incubation period, inspect the microtitre plate wells. Collect medium only from wells containing no unhealthy oocytes (see *Figure 9*). Flush out and wash the oocytes with modified Barths' saline. If the oocytes are to be fractionated, they must not be frozen since the intracellular compartmentation of proteins is altered by the freezing and thawing process. If the oocytes are to be analysed without fractionation, they can be frozen quickly on dry ice and then stored at -20°C.

5.2 Homogenisation of Oocytes

When no fractionation is required:
(i) Homogenise the fresh or frozen oocytes in homogenisation buffer (*Table 7*), allowing $20-50$ μl of buffer per oocyte.
(ii) Centrifuge at 10 000 g in a microcentrifuge.

Table 7. Solutions for Oocyte Homogenisation, Fractionation, Immunoprecipitation and Electrophoresis.

Homogenisation Buffer
    0.1 M NaCl
    1% Triton X-100
    1 mM PMSF[a]
    20 mM Tris-HCl (pH 7.6)

T Buffer
    50 mM NaCl
    10 mM magnesium acetate
    1 mM PMSF[a] [*Omit* if protease digestion to follow]
    20 mM Tris-HCl (pH 7.6)

Immunoprecipitation Buffer
    0.1 M KCl
    5 mM $MgCl_2$
    1% Triton X-100
    0.5% SDS
    1% sodium deoxycholate
    1 mM PMSF[a]
    0.1 M Tris-HCl, (pH 8.2)

For the correct preparation of this buffer, mix the reagents in the following order:-

1 M Tris-HCl pH 7.6	5.0 ml
1 M KCl	5.0 ml
0.5 M $MgCl_2$	0.5 ml
100% Triton X-100	0.5 ml
10% SDS	2.5 ml
10% sodium deoxycholate in water (freshly made)	5.0 ml
100 mM PMSF[a]	0.5 ml

Titrate the buffer to pH 8.2 with NaOH and adjust to 50 ml final volume.

Electrophoresis Sample Buffer
    1 M sucrose
    0.01% Bromophenol blue
    5 mM EDTA
    2% SDS
    0.2 M Tris-HCl, pH 8.8

---

[a]PMSF stock solution: 0.1 M in ethanol. Store at $-20°C$. Dispense just prior to use.

(iii) Remove the supernatant, avoiding the lipid pellicle if possible. This supernatant can be prepared directly for electrophoresis by addition of electrophoresis sample buffer *(Table 7)* or it can be immunoprecipitated first (see Section 5.4).

5.3 Fractionation of Oocytes by Sucrose Gradient Centrifugation

Membrane and secretory proteins can be resolved from cytosolic and nuclear proteins by fractionation of oocytes using sucrose step gradients. Fresh, not frozen, oocytes must be used (see Section 5.1). During this procedure, over 90% of the secretory proteins can be recovered in membranous elements which represent fragments of the endoplasmic reticulum, Golgi apparatus and plasma membranes of the oocyte. Contamination of these membranes by the nuclear membrane (though not the nuclear contents) does occur but can be avoided by prior enucleation of oocytes (see Section 6.1). Two methods of sucrose gradient fractionation are described below (modified from reference 56).

Using either procedure, it is important to work rapidly and to keep solutions ice-cold.

5.3.1 *'Step' Method*

(i) Homogenise about 20 oocytes in 0.5 ml of 10% (w/v) sucrose in T buffer *(Table 7)* but supplemented with additional NaCl to a final concentration of 0.15 M.

(ii) Layer the homogenate onto a step gradient consisting of 1 ml of 20% (w/v) sucrose above 1 ml of 50% (w/v) sucrose, both in T buffer. Centrifuge at 15 000 *g* for 30 min, in a swing-out rotor.

(iii) After centrifugation, there will be a lipid pellicle on top of the gradient. The 10% sucrose layer contains the cytosolic components and the 10−20% interface contains mainly mitochondria. The 20−50% interface contains membraneous elements and is where all the membrane and secretory proteins are to be found. The heavy yolk granules form a pellet. Remove the desired fraction with a Pasteur pipette. This allows the integrity of the membranes to be maintained should further handling be necessary, for example, protease-protection assays (see Section 6.4).

5.3.2 *'Cushion' Method*

For many purposes, it is more convenient to pellet the membranes. Proceed as in Section 5.3.1 but omit the 50% sucrose layer in the gradient. Under these conditions the membranes pellet with the yolk. Resuspend the pellet with homogenisation buffer to solubilise the membrane and secretory proteins and then spin for 1 min in a microcentrifuge to pellet the yolk protein.

5.4 **Treatment of Incubation Medium**

If the incubation medium is to be examined for the presence of secreted proteins using immunoprecipitation and subsequent gel electrophoresis (see below), clarify the medium by centrifugation for 5 min in a microcentrifuge. This will remove any contaminating bacteria which could otherwise co-sediment with immunoprecipitated material.

5.5 **Immunoprecipitation**

(i) To 100 μl aliquots of oocyte sample (homogenate, sucrose fraction or clarified culture medium), add 400 μl of immunoprecipitation buffer *(Table 7)*. Leave at 4°C for 30 min.

(ii) Add antibody and then leave for an additional 30 min at 4°C.

(iii) Add 20 μl of a 10% (v/v) suspension of formaldehyde-treated envelopes of *Staphylococcus aureus* (Pharmacia) and leave shaking for at least 2 h at 4°C.

(iv) Pellet the envelopes in a microcentrifuge (20 sec) and then wash them three times with fresh immunoprecipitation buffer.

(v) Resuspend the washed pellet in 30 μl of electrophoresis buffer *(Table 7)*. Add 1 μl of 0.1 M DTT (stored at −20°C), mix and boil for 3 min.

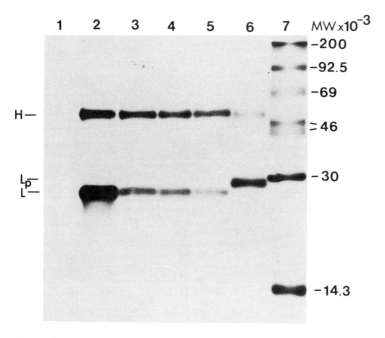

Figure 10. Fractionation of microinjected oocytes. Oocytes were microinjected with immunoglobulin light (L) and heavy (H) chain mRNA and labelled with [³⁵S]methionine for 24 h. The labelled oocytes were fractionated by the 'step' method (Section 5.3.1) followed by immunoprecipitation and electrophoresis in a 12.5% reducing polyacrylamide gel. **Lanes 1, 2** and **3** show immunoprecipitated cytosol, membrane, and medium samples, respectively. **Lanes 4** and **5** show immunoprecipitates of the labelled parental myeloma cells and the surrounding medium, respectively. Immunoglobulin mRNA translated in a wheat germ cell-free translation system and showing the precursor light chain (Lp) is shown in **lane 6** whilst ¹⁴C-labelled marker proteins are in **lane 7**, (reproduced from ref. 49 with permission).

5.6 Alkylation and Analysis by Reducing SDS-PAGE

Alkylation is recommended since it is very simple and can prevent intrachain S-S bond formation occurring during electrophoresis which may otherwise result in anomalous gel profiles.

(i) To the 30 μl sample in electrophoresis buffer *(Table 7)*, add 5 μl of 0.5 M iodoacetamide (sodium salt; stored in aliquots at −70°C). Leave at room temperature for at least 20 min.

(ii) Centrifuge for 5 min in a microcentrifuge to pellet any bacterial debris.

(iii) Load the sample (supernatant) onto a polyacrylamide slab gel for SDS-PAGE under reducing conditions (74).

An experiment involving analysis of the type discussed in Sections 5.3 – 5.6 is shown in *Figure 10*.

5.7 Partial Alkylation and Analysis by Non-reducing SDS-PAGE

Occasionally, for example, studies of tetrameric immunoglobulin (49), it is useful to alkylate oocyte proteins immediately upon homogenisation in order

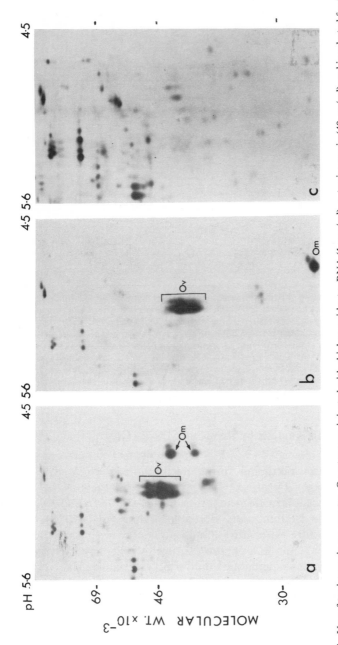

Figure 11. Use of tunicamycin on oocytes. Oocytes were injected with chicken oviduct mRNA (1 mg/ml) ± tunicamycin (40 μg/ml) and incubated for 24 h in [35S]methionine ± tunicamycin (2 μg/ml). Membrane fractions from (**a**) injected (minus tynicamycin), (**b**) injected (plus tunicamycin) and (**c**) uninjected oocytes were subjected to two-dimensional electrophoresis followed by fluorography. Abbreviations: Ov, chick ovalbumin polypeptides; Om, chicken ovomucoid. (Reproduced from ref. 39 with permission.)

to examine multimeric proteins under conditions which prevent their formation from monomers. To do this:

(i) Homogenise the oocytes in homogenisation buffer containing 0.1 M iodo-acetamide (freshly added). Leave for 20 min at room temperature.

(ii) Precipitate the proteins by addition of 5 volumes of ice-cold acetone and incubate on ice for 10 min.

(iii) Centrifuge for 5 min in a microcentrifuge and air dry the pellets.

(iv) The pellets can then be resuspended in immunoprecipitation buffer or electrophoresis buffer and treated as in Sections 5.4 and 5.5, respectively, except that no DTT should be added prior to electrophoresis on SDS-PAGE so that this occurs under non-reducing conditions.

5.8 Two-dimensional Gel Electrophoresis

For two-dimensional electrophoresis:

(i) Mix 20 μl of homogenate, membrane, cytosol or clarified incubation medium (secreted samples) with 0.5 ml of lysis buffer containing 9.5 M urea (ultrapure), 2% (v/v) Nonidet P-40, 2% (v/v) Ampholines (one volume pH range $3.5-10$ + 2 volumes pH range $5-7$), 5% 2-mercapto-ethanol and 0.1% SDS. When concentration of fractions is required before electrophoresis, lyophilise the sample before dissolving it in lysis buffer as above. Immunoprecipitates can be dissolved directly in lysis buffer.

(ii) Centrifuge in a microcentrifuge for 10 min.

(iii) Remove the clear supernatant by puncturing the tubes (to avoid the floating pellicle of lipid).

(iv) This supernatant can be run on isoelectric focussing gels without further treatment.

An analysis of this type is shown in *Figure 11*.

6. SPECIALISED PROCEDURES

6.1 Enucleation of Oocytes

Frog oocytes are rare amongst cells in that their nucleus can be removed mannually. Enucleation is often required in experiments where the contribution of the oocyte's transcriptional apparatus might complicate interpretation of the data. In addition, enucleation also provides a highly efficient method of oocyte fractionation in situations where the destination of a protein encoded by the injected mRNA is thought to be the nucleus. The methods described below are fast and allow the collection of both the nucleus and enucleated oocyte for analysis. They are not suitable for experiments where injection of the enucleated oocyte is planned; the methods of Ford and Gurdon (75) can be used in these situations.

6.1.1 *Method 1*

(i) Using watchmaker's forceps, hold the oocyte gently under modified Barths' saline keeping its pigmented half uppermost.

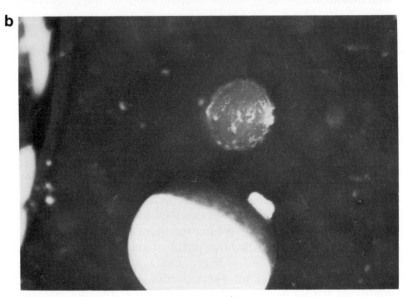

Figure 12. Oocyte enucleation. Stages in the enucleation of an oocyte by method 1 (Section 6.1.1) are shown. A superficial slit is made in the top of the pigmented half of the oocyte with a 23-gauge needle. Squeezing the oocyte gently around the equatorial region causes gradual extrusion of the nucleus **(a)** which eventually becomes free **(b)**.

(ii) Make a slit in the apex of the pigmented half with the tip of a 23-gauge syringe needle. Ensure that the slit is superficial.

(iii) Gently squeeze the oocyte round its middle. The spherical, transparent nucleus, which is about 0.3 mm in diameter, will be extruded (see *Figure 12*).

The nucleoplasm is rather gelatinous and, if the procedure is performed too vigorously, this jelly-like mass will be seen to be released from the nuclear envelope. Do not collect such damaged nuclei.

(iv) The extruded nucleus should be denuded of adhering cytoplasmic material by pipetting it up and down in a wide-mouthed plastic micropipettor tip before transfer to a collecting tube kept on ice. With practice, a nucleus can be collected within 30 sec of puncturing the oocyte. The enucleated oocytes can then be subjected to further analysis as described for intact oocytes in Section 5.

6.1.2 *Method 2*

(i)   Place the oocyte in cold $5-20\%$ TCA for 15 min.
(ii) Tear each oocyte open using watchmaker's forceps and then remove the acid-fixed nucleus.
(iii) Pool the nuclei from the desired number of oocytes and pellet these by centrifugation in a microcentrifuge for 30 sec.
(iv) Discard the supernatant and remove the acid present in the nuclei by two successive washes of the nuclear pellet with ice-cold acetone.
(v)   The proteins in the precipitated nuclei can be solubilised directly in electrophoresis sample buffer *(Table 7)* ready for analysis by SDS-PAGE.

6.2 Inhibition of Translation

Cycloheximide is effective in blocking translation when used in the incubation medium at $100\ \mu g/ml$. However, oocytes will not survive for more than 36 h in the presence of this drug (62).

6.3 Inhibition of Post-translational Modification

Tunicamycin should be dissolved at 1 mg/ml in dimethyl sulphoxide to provide a stock solution which is stored at $-20°C$. The drug is best co-injected at a concentration of $40\ \mu g/ml$ with the mRNA (39). Also include tunicamycin $(2\ \mu g/ml)$ in the incubation medium. Allow 24 h for the drug to be effective before labelling the oocytes with radioactive precursor. Tunicamycin is only partially effective if used in the incubation medium alone, even at $20\ \mu g/ml$. The use of tunicamycin is illustrated in *Figure 11*.

6.4 Protease-protection Experiments

It is often necessary to determine the disposition of proteins relative to membranes with which they co-sediment. Often during tissue fractionation, proteins will adhere adventitiously to membranous debris. However, such proteins are accessible to exogenous proteases so that the resistance of proteins to proteolysis in the absence of detergent is a reasonable indication that the proteins are sequestered within membranes. The following method works well with oocytes (e.g., see *Figure 13)*.

TRITON — + —
TRACK 1 2 3

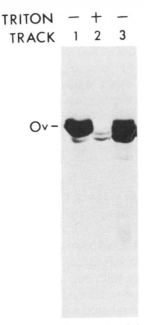

Ov —

Figure 13. A protease-protection experiment. Oocytes were injected with chick oviduct mRNA and labelled in media containing [³⁵S]methionine. A membrane fraction was prepared from homogenised oocytes by the 'cushion' method (Section 5.3.2), and exposed to Proteinase K in the presence or absence of Triton X-100 as indicated in Section 6.4. The treated samples were immunoprecipitated with anti-ovalbumin antibody and electrophoresed on a 12.5% polyacrylamide gel (Sections 5.4, 5.5). **Lane 1**, membranes kept at 0°C in the absence of Proteinase K. **Lanes 2** and **3**, membranes incubated in the presence of Proteinase K for 90 min at 20°C. Abbreviation: Ov, ovalbumin.

(i)   Prepare oocyte membranes by the sucrose gradient 'step' method (Section 5.3.1).

(ii) Dilute the membranes with three volumes of T buffer *(Table 7)* and divide them into three aliquots.

(iii) To the first aliquot add Proteinase K (Calbiochem) to a final concentration of 100 μg/ml. To the second aliquot add both Proteinase K (to 100 μg/ml) and Triton X-100 (to 1% v/v). No additions are made to the third aliquot.

(iv) Incubate all three aliquots at 20°C for 90 min.

(v)   Finally add PMSF to a final concentration of 1 mM and analyse the protein products as indicated in Sections 5.4 – 5.7.

7. APPLICATIONS OF THE OOCYTE SYSTEM

The *Xenopus* oocyte can be programmed with a wide variety of eukaryotic mRNAs obtained from sources as diverse as plants and mammals. For the most part, with the qualification mentioned in Section 3.3.1, these mRNAs are translated efficiently. However, it appears that the oocyte will not translate prokaryotic or organellar (e.g., mitochondrial) mRNAs (10). It is also possible that mRNAs specifying proteins whose normal cellular compartment is absent from oocytes will yield products which are highly unstable and therefore

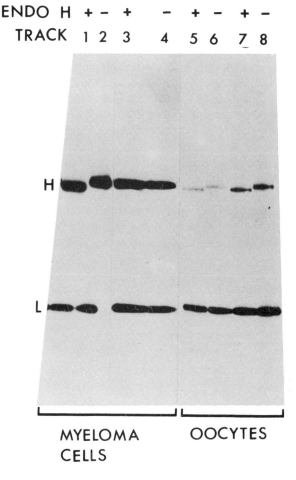

ENDO H + - +   - + - + -
TRACK 1 2 3   4 5 6 7 8

H

L

MYELOMA CELLS   OOCYTES

Figure 14. Comparison of oligosaccharide side chains added to immunoglobulin heavy chains in oocytes and in myeloma (MOPC 21) cells. Shortly before secretion from myeloma cells, the oligosaccharide side chains of immunoglobulin heavy chains are converted from 'simple' to complex forms. Only the complex form is resistant to cleavage by endoglycosidase H (Endo H). The effects of Endo H digestion on the immunoglobulin heavy chain (H), either extracted from myeloma cells (**lanes 1** and **2**) or the surrounding medium (**lanes 3** and **4**), were analysed by reducing SDS-PAGE using 12.5% polyacrylamide gel. The secreted heavy chain is resistant to the enzyme. This contrasts with the same heavy chain made and secreted from oocytes after mRNA injection when both the intracellular (**lanes 5** and **6**) and secreted (**lanes 7** and **8**) heavy chains are sensitive. The light chain (L) is not glycosylated in either cell type.

undetectable in oocyte cytoplasm. For example, the author has never been able to detect any translation products after the injection of nucleus-encoded mRNAs which specify pea chloroplast proteins, nor castor bean malate dehydrogenase, a protein destined for glyoxysomes, a plant organelle absent from oocytes. In contrast, the latter mRNA preparation did yield the castor bean lectin which partitioned into the endoplasmic reticulum of the oocyte as expected. This mRNA preparation programmed both proteins in cell-free translation systems.

A major advantage of the oocyte system is its ability to modify many proteins correctly after translation. However, for some proteins these modifications are not completed correctly (31,37). Fortunately, in certain cases this may not matter, even when a bioassay is contemplated. Thus, the author has shown that mouse immunoglobulin chains are incorrectly glycosylated in oocytes *(Figure 14)* but these chains are still assembled into biologically active antibodies (50). Another example of this phenomenon is given in reference 51.

In summary, the oocyte is a highly efficient translational assay system. However, its use in this way can only be justified, in preference to the cell-free systems, under special circumstances. These include situations where the formation of a biologically active product offers the only practical detection method, or when very small amounts of the mRNA are available. Finally, where the main objective is to study post-translational events the oocyte is clearly an excellent system, especially now that mutant proteins can be synthesized in oocytes after injection of manipulated, cloned DNAs (reference 76 and see Chapter 2).

8. REFERENCES

1. Davidson,E. (1976) *Gene Activity in Early Development* (2nd Edn.), published by Academic Press Inc., New York and London.
2. Woodland,H. (1982) *Biosci. Rep.,* **2**, 471.
3. Gurdon,J.B. and Melton,D.A. (1981) *Annu. Rev. Genet.,* **15**, 189.
4. Harland,R.M. and Laskey,R.A. (1980) *Cell,* **21**, 761.
5. Laskey,R.A., Honda,B.M., Mills,A.D., Morris,N.R., Wyllie,A.H., Mertz,J.E., De Robertis,E.M. and Gurdon,J.B. (1978) *Cold Spring Harbor Symp. Quant. Biol.,* **18**, 643.
6. Zehavi-Willner,T. and Lane,C. (1977) *Cell,* **11**, 683.
7. Lane,C.D. and Knowland,J.S. (1975) in *The Biochemistry of Animal Development,* Vol. 3, Weber,R.A. (ed.), Academic Press Inc., New York and London, p. 145.
8. Asselbergs,F.M. (1979) *Mol. Biol. Rep.,* **5**, 199.
9. Marbaix,G. and Huez,G. (1980) in *The Transfer of Cell Constituents into Eukaryotic Cells,* Vol. **31**, Celis,J.E., Graessmann,A. and Loyter,A. (eds.),NATO Advanced Study Series A, Plenum Press, New York and London, p. 347.
10. Lane,C.D. (1983) *Curr. Top. Dev. Biol.,* **18**, 89.
11. Gurdon,J.B., Lane,C.D., Woodland,H.R. and Marbaix,G. (1971) *Nature,* **233**, 177.
12. De Robertis,E.M., Partington,G.A., Longthorne,R.F. and Gurdon,J.B. (1977) *J. Embryol. Exp. Morphol.,* **40**, 199.
13. Rutgers,T., Neeleman,L., Van Vloten-Doting,L., Cleuter,Y., Hubert,E., Huez,G. and Marbaix,G. (1976) *Arch. Int. Physiol. Biochem.,* **84**, 654.
14. Salden,M., Asselbergs,F. and Bloemendal,H. (1976) *Nature,* **259**, 696.
15. Katz,R., Maniatis,E. and Guntaka,R.V. (1979) *Biochem. Biophys. Res. Commun.,* **86**, 447.
16. Rutgers,T., Molen-Doting,L., Marbaix,G., Huez,G., Hubert,E. and Cleuter,Y. (1977) *11th FEBS Meeting,* Copenhagen, Abstracts A2-5/205/3.
17. Ghysdael,J., Kettmann,R. and Burny,A. (1979) *J. Virol.,* **29**, 1087.
18. Kondo,M., Marbaix,G., Moens,L., Huez,G., Cleuter,Y. and Hubert,E. (1975) *10th FEBS Meeting,* Paris, Abstract No. 352.
19. Semancik,J.S., Conejero,V. and Gerhart,J. (1977) *Virology,* **80**, 218.
20. Schwinghamer,M. and Symons,R. (1977) *Virology,* **79**, 88.
21. Laskey,R., Mills,H., Gurdon,J. and Partington,G. (1977) *Cell,* **11**, 335.
22. Littauer,U., Soreq.H. and Cornelis,P. (1980) in *Enzyme Regulation on Mechanism of Action,* Mildner,P. and Reis,B. (eds.),Pergamon, New York, p. 233.
23. Hesselink,W.G., Van der Kamp,A. and Bloemers,H. (1981) *Virology,* **110**, 375.
24. Nusse,R., Asselbergs,F., Salden,M., Michalides,R. and Bloemendal,H. (1978) *Virology,* **91**, 106.

25. Lane,C.D., Colman,A., Mohun,T., Morser,J., Champion,J., Kourides,I., Craig,R., Higgins,S., James,T.C., Appelbaum,S.W., Ohlsson,R.I., Paucha,E., Houghton,M., Mathews,J. and Miflin,B.J. (1981) *Eur. J. Biochem.,* **111**, 225.
26. Wunner,W., Curtis,P. and Wiktov,T.J. (1980) *J. Virol.,* **36**, 133.
27. McCrae,M. and Woodland,H. (1981) *Eur. J. Biochem.,* **116**, 467.
28. Knowland,J. (1974) *Genetics,* **78**, 383.
29. Matthews,J., Brown,J. and Hall,T. (1981) *Nature,* **294**, 175.
30. Hurkman,W., Smith,L.D., Richter,J. and Larkins,B.A. (1981) *J. Cell Biol.,* **89**, 292.
31. Lane,C., Champion,J., Haiml,L. and Kreil,G. (1981) *Eur. J. Biochem.,* **113**, 273.
32. Lane,C., Champion,J., Colman,A., James,T. and Applebaum,S. (1983) *Eur. J. Biochem.,* **130**, 529.
33. Woodland,H. and Wilt,F. (1980) *Dev. Biol.,* **75**, 214.
34. Asselbergs,F., Koopmans,M., Van Venrooij,W. and Bloemendal,H. (1978) *Eur. J. Biochem.,* **91**, 65.
35. Asselbergs,F., Koopmans,M., Van Venrooij,W. and Bloemendal,H. (1979) *Exp. Eye Res.,* **28**, 475.
36. Vassart,G., Refetoff,S., Brocas,H., Dinsart,C. and Dumont,J.E. (1977) *Proc. Natl. Acad. Sci. USA,* **74**, 4462.
37. Rapoport,T., Thiele,B., Prehn,S., Marbaix,G., Cleuter,Y., Hubert,E. and Huez,G. (1978) *Eur. J. Biochem.,* **87**, 229.
38. Miledi,R., Parker,I. and Sumikawa,K. (1982) *EMBO J.,* **1**, 1307.
39. Colman,A., Lane,C., Craig,R., Boulton,A., Mohun,T. and Morser,J. (1981) *Eur. J. Biochem.,* **113**, 339.
40. Chan,L., Kohler,P.O. and O'Malley,B.W. (1976) *J. Clin. Invest.,* **57**, 576.
41. Lane,C.D., Gregory,C.M. and Morel,C. (1973) *Eur. J. Biochem.,* **34**, 219.
42. Rollins,J.W. and Flickinger,R. (1972) *Science (Wash.),* **178**, 1204.
43. Lane,C., Shannon,S. and Craig,R. (1979) *Eur. J. Biochem.,* **101**, 485.
44. Reynolds,R., PremKumar,E. and Pitha,P. (1975) *Proc. Natl. Acad. Sci. USA,* **72**, 4881.
45. Mous,J., Peeters,B. and Rombauts,W. (1980) *FEBS Lett.,* **122**, 105.
46. Lebleu,B., Hubert,E., Content,J., De Wit,L., Braude,I. and de Clercq,E. (1978) *Biochem. Biophys. Res. Commun.,* **82**, 665.
47. Jilka,R., Calvalieri,R., Yaffe,L. and Pestka,S. (1977) *Biochem. Biophys. Res. Commun.,* **79**, 625.
48. Valle,G., Besley,J., Williamson,A., Mosmann,T. and Colman,A. (1983) *Eur. J. Biochem.,* **132**, 131.
49. Valle,G., Besley,J. and Colman,A. (1981) *Nature,* **291**, 338.
50. Valle,G., Jones,E. and Colman,A. (1982) *Nature,* **300**, 71.
51. Labarca,C. and Paigen,K. (1977) *Proc. Natl. Acad. Sci. USA,* **74**, 4462.
52. Kourides,I. and Weintraub,B. (1979) *Proc. Natl. Acad. Sci. USA,* **76**, 298.
53. Kortbeek-Jacobs,N. and van der Donk,H. (1978) *J. Immunol. Methods,* **24**, 195.
54. Beato,M. and Rungger,D. (1975) *FEBS Lett.,* **59**, 305.
55. Deacon,N. and Erbinger,A. (1977) *FEBS Lett.,* **79**, 191.
56. Olhsson,R., Lane,C. and Guengerich,F. (1981) *Eur. J. Biochem.,* **115**, 367.
57. Higgins,S., Colman,A., Fuller,F. and Jackson,P. (1981) *Mol. Cell. Endocrinol.,* **21**, 255.
58. Mous,J., Peeters,B., Rombauts,W. and Heyns,W. (1977) *Biochim. Biophys. Res. Commun.,* **79**, 1111.
59. Sumikawa,K. Houghton,M., Emtage,J., Richards,B. and Barnard,E. (1981) *Nature,* **292**, 862.
60. Gedamu,L., Dixon,G. and Gurdon,J. (1978) *Exp. Cell Res.,* **117**, 325.
61. Berridge,M. and Lane,C. (1976) *Cell,* **8**, 283.
62. Colman,A. and Morser,J. (1979) *Cell,* **17**, 517.
63. Taniguichi,M., Takuhisa,T., Kanno,M., Yaoita,Y., Shimizu,A. and Hanjo,T. (1982) *Nature,* **298**, 172.
64. Gurdon,J. and Woodland,H. (1967) in *Handbook of Genetics,* Vol. **4**, King,R. (ed.),Plenum Press, New York and London, p. 35.
65. Korn,L. and Gurdon,J. (1981) *Nature,* **289**, 461.
66. Kressmann,A., Clarkson,S., Pirotta,V. and Birnstiel,M. (1978) *Proc. Natl. Acad. Sci. USA,* **75**, 1176.
67. Bienz,M. and Gurdon,J. (1982) *Cell,* **29**, 811.
68. Gurdon,J. (1974) *Control of Gene Expression in Animal Development,* published by Oxford and Harvard University Press, London and New York.

69. Richter,J. and Smith,L. (1981) *Cell,* **27**, 183.
70. Colman,A., Besley,J. and Valle,G. (1982) *J. Mol. Biol.,* **160**, 459.
71. Soreq,H. and Miskin,R. (1981) *FEBS Lett.,* **128**, 305.
72. Shih,R., O'Connor,C., Keem,K. and Smith,L.D. (1978) *Dev. Biol.,* **66**, 172.
73. Woodland,H. (1979) *Dev. Biol.,* **68**, 360.
74. Hames,B.D. (1981) in *Gel Electrophoresis of Proteins — A Practical Approach,* Hames,B.D. and Rickwood,D. (eds.), IRL Press, Oxford and Washington DC, p. 1.
75. Ford,C. and Gurdon,J. (1977) *J. Embryol. Exp. Morphol.,* **37**, 203.
76. McKnight,S. and Gavin,E. (1980) *Nucleic Acids Res.,* **8**, 5931.

Nucleic Acid and Polypeptide Molecular Weight Markers[1]

STEPHEN J. MINTER and PAUL G. SEALEY

1. DNA SIZE MARKERS

1.1 Plasmid pBR322

Plasmid pBR322 is extremely convenient for use as a molecular weight marker. By using a relatively small range of restriction endonucleases a wide range of molecular weight markers can be generated. The linear 4362-bp molecule, useful as a marker for agarose gels, can be obtained using any one of the following restriction enzymes, each of which cleaves at a single site in the plasmid:

*Aat*II, *Ava*I, *Bal*I, *Bam*HI, *Cla*I, *Eco*RI, *Hind*III, *Nde*I, *Nru*I, *Pst*I, *Pvu*II, *Rru*I, *Sal*I, *Sph*I, *Tth*I, *Xma*III, *Xor*II.

Other enzymes which have multiple restriction sites, together with the sizes of the fragments obtained, are listed in *Table 1*.

---

[1]This Appendix is updated and extended from Appendix I of ref. 1.

Appendix I

Table 1. Sizes of Restriction Fragments of pBR322.[a,b,c]

BvuI	HincII	AccI	XmnI	EcoK	AhaIII	BglI	RsaI	NarI	NaeI	ThnII
4348	3256	2767	2430	2377	3651	2319	2117	3571	3481	3196
14	1106	1595	1932	1985	692	1809	1565	657	367	1127
					19	234	680	113	354	32
							21	21	160	7

MstI	HgiDI	BstNI	TaqI	AvaII	DdeI	HgiAI	HgiCI	HinfI	HaeII	HgaI
2134	2699	1857	1444	1746	1652	1161	2027	1631	1876	867
1095	657	1060	1307	1433	542	826	1123	517	622	731
1035	490	928	475	303	540	604	439	506	439	633
98	382	383	368	279	465	587	294	396	430	578
	113	121	315	249	426	498	218	344	370	415
	21	13	312	222	409	310	113	298	227	314
			141	88	166	291	84	221	181	245
				42	162	85	43	220	83	239
							21	154	60	158
								75	53	150
									21·	32

NciI	MboII	FokI	HphI	Sau96	AluI	ScrFI	Sau3A	SfaNI	HaeIII	TacI
724	790	1171	1106	1461	910	696	1374	1052	587	581
699	755	853	853	616	659	592	665	424	540	493
696	753	659	576	352	655	525	358	395	504	452
632	592	649	415	279	521	363	341	375	458	372
363	492	287	387	274	403	351	317	248	434	355
351	271	188	282	249	281	347	272	243	267	341
328	254	181	227	222	257	328	258	234	234	332
308	196	141	221	191	226	308	207	223	213	330
226	109	78	207	189	136	218	105	220	192	145
35	78	62	45	179	100	199	91	192	184	129
	72	48	34	124	63	184	78	164	124	129
		45	9	88	57	121	75	134	123	122
				79	49	42	46	96	104	115
				42	19	40	36	88	89	103
				17	15	35	31	78	80	97
					11	13	27	63	64	69
							18	39	57	66
							17	37	51	61
							15	25	21	27
							12	12	18	26
							11	11	11	10
							8	9	7	5
										2

[a]The enzymes are listed in order of increasing number of fragments generated.
[b]The fragment sizes (in base pairs) do not include any single-stranded extensions which may be generated by the restriction endonuclease.
[c]Data derived using the pBR322 sequence (2) on file in the EMBO DNA Sequence Library.

Table 1 continued.

HpaII	MnlI	HhaI	Fnu4H
622	591	393	480
527	400	348	328
404	334	337	299
309	262	332	277
242	247	270	229
238	218	259	206
217	206	206	189
201	206	190	165
190	206	174	155
180	204	153	150
160	185	152	143
160	179	151	129
147	166	141	123
147	156	132	119
122	150	131	118
110	96	109	116
90	88	103	107
76	81	100	97
67	77	93	95
34	61	83	85
34	60	75	83
26	58	67	81
26	38	62	79
15	36	60	71
9	30	53	65
9	27	40	57
		36	53
		33	51
		30	46
		28	45
		21	34
			27
			18
			14
			7
			3
			3
			3
			3
			3
			3
			3

1.2 Plasmid pAT153

pBR322 and pAT153 are closely-related plasmids. pAT153 was derived from pBR322 by removing the B (622 bp) and G (83 bp) *Hae*II restriction fragments of pBR322 (3). Linear molecules of pAT153 are 3657 bp long; they can be obtained by restriction with any one of the following enzymes:

*Aat*II, *Acc*I, *Ava*I, *Bal*I, *Bam*HI, *Cla*I, *Eco*K, *Eco*RV, *Hind*III, *Nru*I, *Pst*I, *Rru*I, *Sal*I, *Sph*I, *Xma*III, *Xmn*I, *Xor*II.

Other enzymes which have multiple restriction sites, together with the sizes of the fragments obtained, are listed in *Table 2*.

Table 2. Sizes of Restriction Fragments of pAT153[a].

*Bvu*I	*Rsa*I	*Hinc*II	*Tth*II	*Aha*III	*Bgl*I	*Nar*I	*Nae*I	*Mst*I	*Hgi*DI	*Bst*NI	*Dde*I
3643	3486	2551	3618	2946	1809	2866	2776	1429	1994	1857	1652
14	165	1106	32	692	1614	657	367	1095	657	928	540
			7	19	234	113	354	1035	490	383	464
						21	160	98	382	355	426
									113	121	409
									21	13	166

*Fok*I	*Taq*I	*Ava*II	*Hgi*AI	*Nci*I	*Hae*II	*Hinf*I	*Hgi*CI	*Hga*I	*Hph*I	*Mbo*II	*Alu*I
1584	1444	1433	1161	724	1876	1631	1322	867	986	790	659
853	602	1320	619	696	439	517	1117	731	853	755	655
659	475	303	604	665	430	396	439	578	415	592	622
287	368	249	587	632	370	298	294	501	387	492	521
181	315	222	310	363	227	221	218	314	282	271	403
48	312	88	291	351	181	220	113	245	227	254	257
45	141	42	85	226	60	154	84	239	221	196	226
					53	145	43	150	207	109	136
					21	75	21	32	45	78	100
									34	72	63
										48	15

[a]See footnotes to *Table 1*.

Table 2 continued.

Sau96	SfaNI	ScrFI	TacI	Sau3A	MnlI	HaeIII	HpaII	HhaI	Fnu4H
1224	1052	696	581	876	591	587	622	393	480
616	437	592	493	665	400	458	492	337	328
352	424	525	452	358	247	434	404	332	277
274	395	363	355	341	218	339	242	270	229
249	375	351	332	317	206	267	238	259	206
222	248	313	330	272	206	234	217	206	189
191	234	218	182	258	206	213	201	190	165
179	192	199	145	105	204	192	190	174	155
124	164	184	129	91	200	184	160	153	150
88	88	121	129	78	179	124	160	152	143
79	25	42	122	75	166	123	147	151	129
42	12	40	115	46	156	104	122	132	119
17	11	13	97	36	150	89	110	131	118
			66	31	96	80	90	109	113
			61	27	88	64	76	100	107
			27	18	81	57	67	93	95
			26	17	77	51	34	75	85
			10	15	61	21	26	67	83
			5	12	60	18	26	62	79
				11	38	11	15	60	71
				8	27	7	9	53	65
							9	40	57
								36	51
								33	45
								28	34
								21	27
									18
									14
									7
									3 x 6

1.3 **Bacteriophage lambda (λ)**

Bacteriophage λ is the most extensively-studied phage of *Escherichia coli*. The DNA of strain λ cI *indl ts* 857 Sam7 has been sequenced (4). The native molecule is circular and double-stranded. Heating at 68°C for 10 min, in a buffer containing a low concentration of salt, followed by rapid cooling causes separation of the cohesive ends at the 'cos' site. This generates a linear molecule with single-stranded 'sticky' ends 12 nucleotides long. *Table 3* lists the sizes of fragments generated by a variety of restriction endonucleases.

Table 3. Sizes of Restriction Fragments of Phage λ DNA (strain cI *ts* 857)[a,b].

*Xho*I	*Xba*I	*Sal*I	*Kpn*I	*Sst*I	*Avr*II	*Sma*I	*Pvu*I
33 498	24 508	32 745	29 942	24 776	24 322	19 399	14 321
15 004	23 994	15 258	17 057	22 621	24 106	12 220	12 712
		499	1503	1105	74	8612	11 936
						8271	9533

*Sst*II	*Eco*RI	*Bam*HI	*Bgl*II	*Hind*III	*Ava*I	*Hpa*I	*Pvu*II
20 323	21 226	16 841	22 010	23 130	14 677	8666	21 088
18 780	7421	7233	13 286	9416	8614	6911	4421
8113	5804	6770	9688	6557	6888	5414	4268
1076	5643	6527	2392	4361	4720	4535	4194
210	4878	5626	651	2322	4716	4491	3916
	3530	5505	415	2027	3730	4347	3638
			60	564	1881	3408	2296
				125	1674	3384	1708
					1602	3042	636
						2240	579
						734	532
						441	468
						410	343
						251	211
						228	141
							63

[a]See footnotes a and b to *Table 1*.
[b]Data derived from the sequence of λ cI *ts* 857 on file in the EMBO DNA Sequence Library.

1.4 **Simian Virus 40 (SV40)**

The SV40 genome is a double-stranded circular DNA 5243 bp long. It can be linearised using *Acc*I, *Bam*HI, *Eco*RI, *Eco*RV, *Hae*II, *Kpn*I, *Msp*I (*Hpa*II), *Nae*I or *Taq*I. *Table 4* lists the sizes of restriction fragments generated by other enzymes which have multiple cleavage sites.

Table 4. Sizes of Restriction Fragments from SV40[a,b,c].

*Hha*I	*Nde*I	*Pst*I	*Xho*II	*Pvu*II	*Hph*I	*Hpa*I	*Ava*II
4753	4225	4027	3007	2007	3091	2147	1580
490	1018	1216	1566	1719	1856	2009	1525
			760	1446	160	1067	995
					136	20	682
							430
							31

*Hind*III	*Hinc*II	*Mbo*II	*Hinf*I	*Hae*I	*Aha*III	*Rsa*I	*Mbo*II
1768	1980	1347	1845	1739	1753	1605	1350
1169	1538	1264	1085	1661	739	708	756
1116	1067	945	766	383	565	675	687
526	369	610	543	348	430	551	645
447	240	396	525	329	411	497	409
215	29	384	237	300	364	351	395
	20	234	109	227	318	294	383
		60	83	179	315	226	375
			24	33	141	153	69
			24	30	136	111	65
				14	71	57	31
						15	30
							14
							13
							11
							10

*Hae*III	*Alu*I	
1661	775	46
765	483	41
540	329	38
373	288	30
329	275	29
325	253	28
300	253	27
299	243	12
227	224	10
179	223	8
49	177	7
45	157	
41	154	
33	153	
30	146	
29	144	
14	123	
9	75	
6	54	
	53	
	50	
	49	

[a]See footnotes a and b to *Table 1*.
[b]From sequence data for SV40 DNA reported in references 5 and 6.
[c]There are no cleavage sites for *Ava*I, *Bgl*II, *Cla*I, *Fnu*DII, *Pvu*I, *Sal*I, *Sma*I, *Xba*I, *Xho*I or *Xma*II.

1.5 Coliphage M13 mp7

This is a widely-used cloning and sequencing vector which consists of a circular DNA molecule, 7238 bp long (RF form). It can be linearised by *Acy*I, *Ava*I, *Ava*II, *Avr*I, *Bgl*I, *Bgl*II, *Mst*II, *Nae*I, *Nar*I, *Pvu*I, *Pst*I, *Sau*I, *Sua*I. *Table 5* lists the sizes of fragments generated from M13 mp7 DNA by enzymes which have multiple restriction sites.

Table 5. Sizes of Restriction Fragments for Phage M13 mp7[a,b].

*Acc*I	*Sal*I	*Hinc*II	*Bam*HI	*Eco*RI	*Gdi*II	*Pvu*II	*Hgi*AI	*Xmn*I	*Cla*I
7226	7226	7226	7214	7196	6993	6835	6516	4949	4343
12	12	12	24	42	245	310	722	2289	2895

*Asu*I	*Cau*II	*Nci*I	*Xho*I	*Hae*I	*Hgi*CI	*Hae*II	*Eco*RII	*Sfa*NI	*Hga*I
6578	4313	4313	4020	2836	4428	3514	3975	2625	1638
446	4162	2336	2535	2813	2022	2520	1809	1684	1517
190	558	558	659	1325	324	434	952	966	1089
24	31	31	24	264	322	433	179	871	1075
					130	329	139	716	846
					12	8	127	363	758
							57	13	315

*Sau*3A	*Bbv*I	*Kpn*I	*Taq*I	*Hae*III	*Fnu*4HI	*Hph*I	*Rsa*I	*Fnu*DII	*Hpa*II
4020	1826	1694	1018	2527	1826	2212	1345	1290	1596
1696	1739	1241	971	1623	1739	1595	1334	889	829
507	1154	1226	927	849	891	1013	1004	772	818
434	665	792	703	341	629	589	742	681	651
332	611	666	639	311	611	489	604	646	545
129	436	332	612	309	436	271	522	541	543
96	420	318	579	245	420	241	322	496	472
24	387	301	564	214	387	194	258	495	454
		196	441	169	164	144	201	361	357
		196	381	158	45	135	190	353	183
		163	239	117	27	127	163	304	176
		113	152	106	24	76	143	190	156
			12	102	22	39	107	63	130
				98	14	39	102	57	123
				69	3	33	93	54	79
						26	65	24	60
						9	27	20	30
						6	16	2	18
									18

[a]See footnotes a and b of *Table 1*.
[b]There are no cleavage sites for *Bst*EII, *Hind*III, *Hpa*I, *Kpn*I, *Mst*I, *Sma*I, *Xho*I or *Xma*I.

Table 5 continued.

HhaI	AluI	HinfI	DdeI	EcoRI*	MnlI	
967	1446	1288	998	1402	423	31
732	1330	771	866	665	421	30
725	600	571	672	567	391	28
714	555	486	652	508	384	27
683	484	413	600	416	351	22
495	336	348	563	406	340	21
434	331	345	399	323	318	19
340	313	328	381	283	313	18
329	220	324	376	272	304	16
312	204	274	303	247	286	15
293	201	261	294	213	269	15
272	180	253	272	209	263	15
244	159	234	160	176	243	15
190	151	232	153	174	234	15
115	140	212	128	152	226	15
92	111	209	72	142	217	15
74	111	160	63	109	158	15
65	104	137	46	109	143	12
56	93	96	42	102	130	8
28	72	80	39	99	119	6
26	63	63	28	96	117	5
22	39	46	27	88	111	4
13	39	45	24	76	105	4
9	27	22	15	69	102	
8	26	21	15	63	98	
	24	19	15	55	90	
			15	53	86	
			15	41	76	
			14	40	74	
				36	69	
				33	66	
				12	63	
				1	51	
				1	49	
					48	
					48	
					43	
					38	

1.6 **Bacteriophage ΦX174**

This is a bacteriophage of *E. coli* with a single-stranded DNA molecule. The RF form is double-stranded and 5386 bp long. It can be linearised using *Ava*I, *Ava*II, *Mst*II, *Pst*I, *Sac*II or *Xho*I. The restriction fragment sizes for enzymes with multiple restriction sites are given in *Table 6*.

Table 6. Sizes of Restriction Fragments of Phage φX174[a,b,c].

*Aha*III	*Nar*I	*Acc*I	*Eco*RII	*Xmn*I	*Hpa*II	*Hae*I
4307	3429	3034	2767	4126	2748	2712
1079	1957	2352	2619	974	1697	872
				286	374	738
					348	603
					219	389
						72

*Hae*II	*Hph*I	*Taq*I	*Rsa*I	*Hae*III	*Mbo*II	*Hinc*II
2314	1638	2914	1560	1353	1103	1057
1560	1116	1175	964	1078	1066	770
786	791	404	645	872	857	612
269	777	327	525	605	812	495
185	777	321	472	310	396	397
125	110	141	392	281	394	392
93	77	87	247	271	324	345
54	43	54	197	232	224	341
	39	33	157	194	118	335
		20	138	118	89	291
			89	72	3	210
						162
						79

*Dde*I	*Fnu*DII	*Hha*I
1012	1050	1553
998	870	640
927	718	614
542	695	532
486	530	305
393	490	300
303	259	269
302	176	201
186	156	192
165	127	145
38	114	143
18	103	123
10	79	101
6	19	93
		84
		54
		35
		2

[a]See footnotes a and b to *Table 1*.
[b]Data from reference 7.
[c]*Hinf*I gives 21 fragments and *Alu*I gives 24 fragments. There are no cleavage sites for *Bam*HI, *Eco*RI, *Hind*III, *Nae*I, *Pvu*I, *Pvu*II, *Sal*I, *Sma*I, *Xba*I or *Xho*II.

1.7 **Coliphage fd**

This is a male-specific coliphage closely related to the phage M13. The virion is a single-stranded circular DNA molecule 6408 nucleotides long. The fd DNA can be linearised using either *Acc*I or *Hinc*II. The restriction fragment sizes for enzymes with multiple cleavage sites are given in *Table 7*.

Table 7. Sizes of Restriction Fragments for Phage fd DNA[a,b,c].

*Bam*HI	*Hae*II	*Mbo*II	*Hae*III	*Taq*I	*Msp*I(*Hpa*II)	*Alu*I
3425	3550	4349	2528	2019	1596	1446
2983	2033	666	1633	850	829	1330
	817	384	849	703	819	705
	8	332	352	652	652	554
		318	311	579	648	484
		196	309	441	501	366
		163	154	381	454	314
			106	357	381	257
			103	287	156	220
			69	139	129	204
					123	201
					60	166
					42	111
					12	29
					6	27
						24

[a]See footnotes a and b of *Table 1*.
[b]Data from reference 8.
[c]*Hinf*I gives 24 fragments and *Mnl*I gives 51 fragments. There are no cut sites for *Bcl*I, *Eco*RI, *Hind*III or *Kpn*I.

2. RNA SIZE MARKERS

RNA can be electrophoresed on non-denaturing agarose or polyacrylamide gels, or on denaturing gels, either agarose containing methyl mercuric hydroxide or polyacrylamide gels containing 98% formamide or 8 M urea (9). Hence the apparent molecular weight will depend upon the conditions of electrophoresis. In addition to the RNA molecular weight markers listed in *Table 8*, restriction endonuclease fragments of DNA (see *Tables 1 – 7*) may also be useful as size markers in certain situations.

Table 8. Molecular Weight Markers for Gel Electrophoresis of RNA.

RNA species	*Molecular weight*[a]	*Number of nucleotides*	*Reference*
Myosin heavy chain mRNA (chicken)	2.02×10^6	6500	13[c]
28S rRNA (HeLa)[d]	1.90×10^6	6333	10
25S rRNA *(Aspergillus)*[d]	1.24×10^6	4000	19
23S rRNA *(E. coli)*[d]	1.07×10^6	3566	11
18S rRNA (HeLa)[d]	0.71×10^6	2366	10
17S rRNA *(Aspergillus)*[d]	0.62×10^6	2000	19
16S rRNA *(E. coli)*[d]	0.53×10^6	1776	12
A2 crystallin mRNA (calf lens)	0.45×10^6	1460	14[c]
Immunoglobulin light chain mRNA (mouse)	0.39×10^6	1250	15[c]
β-globin mRNA (mouse)	0.24×10^6	783	16
β-globin mRNA (rabbit)	0.22×10^6	710	17[c]
α-globin mRNA (mouse)	0.22×10^6	696	16
α-globin mRNA (rabbit)	0.20×10^6	630	17[c]
Histone H4 mRNA (sea urchin)	0.13×10^6	410	18[b]
5.8S RNA *(Aspergillus)*	4.89×10^4	158	19
5S RNA *(E. coli)*	3.72×10^4	120	20
4S RNA *(Aspergillus)*	2.63×10^4	85	19

[a]Molecular weights are approximate only and based upon average 'molecular weight' of 310 for each nucleotide residue.
[b]Non-denaturing gel system.
[c]Denaturing (formamide) gel system.
[d]A more extensive list of sizes of rRNAs is given in ref. 21.

3. POLYPEPTIDE SIZE MARKERS

Selected polypeptides of known molecular weight which have proved useful as size markers in SDS-PAGE are listed in *Table 9*.

Table 9. Molecular Weights of Polypeptide Standards[a,b,c].

Polypeptide	Molecular weight
Myosin heavy chain (rabbit muscle)	212 000
RNA polymerase β'-subunit *(E. coli)*	165 000
RNA polymerase β subunit *(E. coli)*	155 000
β-Galactosidase *(E. coli)*	130 000
Phosphorylase a (rabbit muscle)	92 500
Bovine serum albumin	68 000
Catalase (bovine liver)	57 500
Pyruvate kinase (rabbit muscle)	57 200
Glutamate dehydrogenase (bovine liver)	53 000
Fumarase (pig liver)	48 500
Ovalbumin	43 000
Enolase (rabbit muscle)	42 000
Alcohol dehydrogenase (horse liver)	41 000
Aldolase (rabbit muscle)	40 000
RNA polymerase α-subunit *(E. coli)*	39 000
Glyceraldehyde-3-phosphate dehydrogenase (rabbit muscle)	36 000
Lactate dehydrogenase (pig heart)	36 000
Carbonic anhydrase	29 000
Chymotrypsinogen A	25 700
Trypsin inhibitor (soybean)	20 100
Myoglobin (horse heart)	16 950
α-Lactalbumin (bovine milk)	14 400
Lysozyme (egg white)	14 300
Cytochrome c	11 700

[a]This table is reproduced from ref. 22 which lists the original bibliography upon which the molecular weights are based.

[b]Several proteases, for example trypsin, chymotrypsin and papain, have been used as molecular weight standards by various workers but these may sometimes cause proteolysis of other polypeptide standards and so are omitted here.

[c]The molecular weight range $\sim 12\ 000 - 68\ 000$ is reasonably well covered but there are few suitable proteins with subunit molecular weights above this range. This can be overcome by using polypeptides which are cross-linked to form an oligomeric series (22). Kits of these are also commercially available.

4. REFERENCES

1. Rickwood,D. and Hames,B.D., eds. (1982) *Gel Electrophoresis of Nucleic Acids – A Practical Approach,* IRL Press Ltd., Oxford and Washington, DC.
2. Sutcliffe,J.G. (1979) *Cold Spring Harbor Symp. Quant. Biol.,* **43**, 77.
3. Twigg,A.J. and Sherrat,D. (1980) *Nature,* **283**, 216.
4. Sanger,F., Coulson,A.R., Hong,G.F., Hill,D.F. and Petersen,G.B. (1982) *J. Mol. Biol.,* **162**, 729.
5. Fiers,W., Contreras,R., Haegeman,G., Rogiers,R., Van de Voorde,A., Van Heuverswyn,H., Van Herreweghe,J., Volckaert,G. and Ysebaert,M. (1978) *Nature,* **273**, 113.
6. Van Henversuryn,H. and Fiers,W. (1979) *Eur. J. Biochem.,* **100**, 50.
7. Sanger,F., Coulson,A.R., Friedmann,T., Air,G.M., Barrell,B.G., Brown,N.L., Fiddes,J.C., Hutchison,C.A., Slocombe,P.M. and Smith,M. (1978) *J. Mol. Biol.,* **125**, 225.

8. Beck,F., Sommer,R., Auerswald,E.A., Kurz,Ch., Zink,B., Osterburg,G. and Schaller,H. (1978) *Nucleic Acids Res., 5,* 4495.
9. Grierson,D. (1982) in *Gel Electrophoresis of Nucleic Acids − A Practical Approach,* Rickwood,D. and Hames,B.D. (eds.), IRL Press Ltd., Oxford and Washington, DC, p. 1.
10. McConkey,E. and Hopkins,J. (1969) *J. Mol. Biol., 39,* 545.
11. Stanley,W.M. and Bock,R.M. (1965) *Biochemistry (Wash.), 4,* 1302.
12. Pearce,T.C., Rowe,A.J. and Turnock,G. (1975) *J. Mol. Biol., 97,* 193.
13. Mondal,H., Sutton,A., Chen,V.J. and Sarkar,S. (1974) *Biochem. Biophys. Res. Commun., 56,* 988.
14. Berns,A., Jansson,P. and Bloemendal,H. (1974) *Biochem. Biophys. Res. Commun., 59,* 1157.
15. Stravnezer,J., Huang,R.C.C., Stravnezer,E. and Bishop,J.M. (1974) *J. Mol. Biol., 88,* 43.
16. Morrison,M.R. and Lingrel,J.B. (1976) *Biochim. Biophys. Acta, 447,* 104.
17. Hamlyn,P.H. and Gould,H.J. (1973) *J. Mol. Biol., 94,* 101.
18. Grunstein,M. and Schedl,P. (1976) *J. Mol. Biol., 104,* 323.
19. Scazzochio,C., personal communication.
20. Brownlee,G., Sanger,F. and Barrell,B.G. (1968) *J. Mol. Biol., 34,* 379.
21. Loening,U.E. (1968) *J. Mol. Biol., 38,* 355.
22. Hames,B.D. (1981) in *Gel Electrophoresis of Proteins − A Practical Approach,* Hames,B.D. and Rickwood,D. (eds.), IRL Press Ltd., Oxford and Washington, DC, p. 1.

Suppliers of Specialist Items

Many of the larger companies have subsidiaries in other countries whilst most of the smaller companies market their products through agents. The name of a local supplier is most easily obtained by writing to the relevant address listed here, which is usually the head office.

Amersham International Ltd, White Lion Road, Amersham HP7 9LL, Bucks., UK.

BDH Chemicals Ltd., Poole BH12 4NN, Dorset, UK

Beckman Instruments Inc., Spinco Division, Palo Alto, CA 94304, USA

Bethesda Research Laboratories Inc., P.O. Box 577, Gaithersburg, MD 20760, USA

Bio-Rad Laboratories Inc., 2200 Wright Avenue, Richmond, CA 94804, USA

Boehringer Mannheim GmbH Biochemica, Postfach 31020, D-6800 Mannheim 31, FRG

Branson Sonic Power Inc., Eagle Road, Danbury, CT 06810, USA

BRL, see Bethesda Research Laboratories Inc.

Calbiochem-Behring, La Jolla, CA 92037, USA

Cambridge Biotechnology Laboratories Ltd., Uniscience Ltd., 12-14 St. Ann's Crescent, London SW18 2LS, UK

Collaborative Research Inc., 1365 Main Street, Waltham, MA 02154, USA

Difco Laboratories Ltd., P.O. Box 14B, Central Avenue, East Molesey, Surrey KT8 OSE, UK

Du Pont Instruments, Peck's Lane, Newtown, CT 06470, USA

Eastman-Kodak Co., 343 State Street, Rochester, NY 14650, USA

Eli Lilly Co., Indianapolis, IN, USA

Enzo-Biochem Inc., 325 Hudson Street, New York, NY 10013, USA

Ernst Leitz Wetzler GmbH, D-6330 Wetzler, FRG

Fisons Chemicals Ltd., Bishop Meadow Road, Loughborough LE11 ORG, UK

Flow Laboratories Inc., 7655 Old Springhouse Road, McLean, VA 22102, USA

Fluka AG, CH-9470 Buchs, Switzerland

Fuji Photofilm Co., 26-30 Nishiazabu 2-chome, Minato-ku, Tokyo 106, Japan

Gallenkamp Co. Ltd., Technico House, Christopher Street, London, EC2P 2ER, UK

Gibco-Biocult, The Dexter Corp., P.O. Box 200, Pine and Mogul Streets, Chagrin Falls, OH 44022, USA

Kodak, see Eastman-Kodak

Leitz, see Ernst Leitz Wetzler GmbH

May and Baker Ltd., Liverpool Road, Barton Moss, Eccles, Manchester M30 7RT, UK

Merck GmbH, D-6100 Darmstadt, FRG

Microinstruments (Oxford) Ltd., see Oxford Microinstruments Ltd.

Miles Laboratories Inc., Research Products Division, P.O. Box 2000, 1127 Myrtyle Street, Elkhart, IN 46515, USA

Minnesota Mining Co., 3M Center, St. Paul, MN 55101, USA

MSE Scientific Instruments, Manor Royal, Crawley RH10 2QQ, UK

Nasco Inc., 901 Janesville Avenue, Fort Atkinson, WI 53538, USA

New England Biolabs Inc., 32 Tozer Road, Beverley, MA 01915, USA

New England Nuclear Research Products, 549 Albany Street, Boston, MA 02118, USA

Olympus Optical Co., 43-2 Hatagaya 2-chome, Shibuya-ku, Tokyo, Japan

Oxford Microinstruments Ltd., 7 Little Clarendon Street, Oxford OX1 2HP, UK

Oxoid Ltd., Wade Road, Basingstoke RG24 0PW, UK

Pharmacia Fine Chemicals AB, P.O.Box 175, S-75104 Uppsala-1, Sweden

Pierce Chemical Co., P.O.Box 117, Rockford, IL 61105, USA

P-L Biochemicals Inc., see Pharmacia Fine Chemicals AB

P and S Biochemicals, 38 Queensland Street, Liverpool L7 3JG, UK

Research Instruments Ltd., Kernick Road, Penryn, Cornwall TR10 9DQ, UK

Schleicher and Schuell GmbH, Postfach 4, D-3354 Dassel, FRG

Searle Bioscience Ltd., Harbour Estate, Sheerness, Kent ME12 1RZ, UK

Serva Feinbiochemica GmbH, Postfach 105260, D-6900 Heidelberg 1, FRG

Sigma Chemical Co., 3500 Dekalb Street, St. Louis, MO 63118, USA

Singer Instrument Co., Treborough Lodge, Roadwater, Watchet, Somerset TA23 0QL, UK

Sorvall, see Du Pont Instruments

South African Snake Farm, P.O.Box 6, Fish Hoek, Cape Province, Republic of South Africa

Sterile Systems Inc., P.O. Box 387, Logan, UT 84321, USA

Wellcome Reagents Ltd., Beckenham, Kent, UK

Whatman Laboratory Products Ltd., Springfield Mill, Maidstone ME14 2LE, UK

Worthington Biochemical Corporation, P.O.Box 650, Halls Mill Road, Freehold, NJ 07728, USA

WP Instruments Inc., P.O.Box 3110, 60 Fitch Street, New Haven, CT 06515, USA

Xenopus Ltd., Homesdale Nursery, Mid Street, South Nuffield RH1 4JY, UK

Forthcoming

Plant cell culture
a practical approach
Edited by R A Dixon

This book provides the understanding required to rationalise the varying responses of plant cells to culturing and manipulation. At the same time it offers effective and practical protocols for establishing and using plant cell cultures in research.

1985; 250pp (approx); 0 947946 22 5 (softbound); £13.50/US$24.50

Nucleic acid hybridisation
a practical approach
Edited by B D Hames and S J Higgins

A practical laboratory-bench manual of techniques for identifying and analysing the structure of specific gene sequences. This book is unique in bringing together the techniques' major applications at both the theoretical and the practical levels.

1985; 250pp (approx); 0 947946 23 3 (softbound); price to be announced

Immobilised cells and enzymes
a practical approach
Edited by J Woodward

The first book to deal with the practical side of immobilising cells and enzymes. Easy-to-follow 'recipes' describe the major techniques involved. Individual chapters cover specific areas, eg the use of immobilised cells in hormone detection and quantification.

1985; 250pp (approx); 0 947946 21 7 (softbound); £12.00/US$21.50

Animal cell culture
a practical approach
Edited by R I Freshney

After an introductory chapter dealing with basic techniques, this book provides detailed protocols both for traditional areas like organ culture, characterisation and storage, and for those in developing fields. These include serum-free media, cell separation and *in situ* hybridisation.

1985; 250pp (approx); 0 947946 33 0 (softbound); price to be announced

⬡ IRL PRESS

IRL Press Ltd, PO Box 1, Eynsham, Oxford OX8 1JJ, UK
IRL Press Inc, Suite 907, 1911 Jefferson Davis Highway, Arlington, VA 22202, USA